The Regulation of Agricultural Biotechnology

The Regulation of Agricultural Biotechnology

Edited by

R.E. Evenson

Economic Growth Center
Yale University
New Haven, Connecticut
USA

and

V. Santaniello

Department of Economics and Institutions
University of Rome 'Tor Vergata'
Rome
Italy

CABI Publishing

CABI Publishing is a division of CAB International

CABI Publishing
CAB International
Wallingford
Oxfordshire OX10 8DE
UK

Tel: +44 (0) 1491 832111
Fax: +44 (0)1491 833508
E-mail: cabi@cabi.org
Web site: www.cabi-publishing.org

CABI Publishing
875 Massachusetts Avenue
7th Floor
Cambridge, MA 02139
USA

Tel: +1 617 395 4056
Fax: +1 617 354 6875
E-mail: cabi-nao@cabi.org

A catalogue record for this book is available from the British Library, London, UK.

Library of Congress Cataloging-in-Publication Data

The regulation of agricultural biotechnology / edited by R.E. Evenson and V. Santaniello.
p. cm.
Selected papers from a conference of the International Consortium on Agricultural Biotechnology Research (ICABR), held July 2002 in Ravello, Italy.
Includes bibliographical references.
ISBN 0-85199-742-2 (alk. paper)
1. Agricultural biotechnology--Law and legislation. 2. Agricultural biotechnology--Research. I. Evenson, Robert E. (Robert Eugene), 1934- II. Santaniello, V. III. International Consortium on Agricultural Biotechnology Research.
K3925.B56R44 2004
631.5'23--dc21

2003014503

ISBN 0 85199 742 2

Typeset by AMA DataSet Ltd, UK.
Printed and bound in the UK by Biddles Ltd, King's Lynn.

Contents

Contributors

Barkley, A.P., *Department of Agricultural Economics, Kansas State University, Manhattan, KS 66506-4011, USA.*

Binenbaum, E., *School of Economics, Adelaide University, SA 5005, Australia.*

Blakeney, M., *Queen Mary Intellectual Property Research Institute, CCLS, Queen Mary, University of London, Mile End Road, London E1 4NS, UK.*

Demont, M., *Department of Agricultural and Environmental Economics, Katholieke Universiteit Leuven, De Croylaan 42, 3001 Leuven, Belgium.*

Eaton, D., *Agricultural Economics Research Institute (LEI), Wageningen University and Research Centre, PO Box 29703, The Hague 2502 LS, The Netherlands.*

Esposti, R., *Dipartimento di Economia, Universitá Politecnica delle Marche, Ancona, Italy.*

Fulginiti, L., *Department of Agricultural Economics, 307C FYH, University of Nebraska, Lincoln, NE 68583-0922, USA.*

Gaisford, J.D., *Department of Economics, University of Calgary, 2500 University Drive NW, Calgary, Alberta T2N 1N4, Canada.*

Harmsen, H., *The MAPP Centre, Aarhus School of Business, Haslegaardsvej 10, DK-8210 Aarhus V., Denmark.*

Heumueller, D., *School of Law, Stanford University, Stanford, CA 94305, USA.*

Hobbs, J.E., *Department of Agricultural Economics, University of Saskatchewan, 51 Campus Drive, Saskatoon, Saskatchewan, S7W 5AB, Canada.*

Hopkins, R.F., *Department of Political Science, Swarthmore College, 500 College Avenue, Swarthmore, PA 19081, USA.*

Isaac, G., *Chair in Biotechnology Management, College of Commerce, University of Saskatchewan, 25 Campus Drive, Saskatoon, Saskatchewan, S7N 5A7, Canada.*

Janni, O., *Consiglio Nazionale delle Ricerche, Instituto Sperimentale di Studi Socio-Economici sull'Innovazione e le Politiche delle Ricerca, Milan, Italy.*

Jensen, B.B., The *MAPP Centre, Aarhus School of Business, Haslegaardsvej 10, DK–8210 Aarhus V., Denmark.*

Johnson, D.D., *Field Crops Branch, Market and Trade Economics Division, Economic Research Service, US Department of Agriculture, 1800 M Street, N.W., Rm 55206, Washington, DC 20036–5831, USA.*

Josling, T., *Institute for International Studies, Stanford University, Encina Hall Central, 616 Serra Street, Stanford, CA 94305, USA.*

Katz, P., *Crowell & Moring LLP, 1001 Pennsylvania Avenue, NW Washington, DC 20004–2595, USA.*

Kerr, W.A., *Van Vliet Professor, Department of Agricultural Economics, University of Saskatchewan, 51 Campus Drive, Saskatoon, Saskatchewan, S7W 5AB, Canada.*

King, J., *Economic Research Service, US Department of Agriculture, 1800 M Street NW, Washington, DC 20036–5831, USA.*

Klein, K.K., *Department of Economics, Universitry of Lethbridge, 4401 University Drive, Lethbridge, Alberta, T1K 3M4, Canada.*

Klotz-Ingram, C., *Economic Research Service, US Department of Agriculture, 1800 M Street NW, Washington, DC 20036–5831, USA.*

Knudsen, O., *The World Bank, 1818 H Street, NW Washington, DC 20433, USA.*

Krissoff, B., *Economic Research Service, US Department of Agriculture, 1800 M Street NW, Washington, DC 20036–5831, USA.*

Kuchler, F., *Economic Research Service, US Department of Agriculture, 1800 M Street NW, Washington, DC 20036–5831, USA.*

Ladewski, L., *Swarthmore College, 500 College Avenue, Swarthmore, PA 19081, USA.*

Lin, W., *Field Crops Branch, Market and Trade Economics Division, Economic Research Service, US Department of Agriculture, 1800 M Street, Rm 55206, NW Washington, DC 20036–5831, USA.*

Macdonald, P., C&M International, 1001 Pennsylvania Avenue, NW Washington, DC 20004, USA.

Mackenzie, G., *Crowell & Moring, Brussels Office, 71 Rue Royale, B-1000, Brussels, Belgium.*

Menrad, K., *Fraunhofer Institute for Systems and Innovation Research, Breslauer Str. 48, 76139 Karlsruhe, Germany.*

Moss, C.B., *Department of Food and Resource Economics, University of Florida, McCarthy Hall, PO Box 110240, Gainesville, FL 32611-0240, USA.*

Naseem, A., *Department of Agriculture, Food and Resource Economics, Rutgers, The State University of New Jersey, 55 Dudley Road, New Brunswick, NJ 08901, USA.*

Oehmke, J.F., *Department of Agricultural Economics, Michigan State University, East Lansing, MI 48824–1069, USA.*

Paarlberg, R.L., *Professor of Political Science at Wellesley College and Research Associate at the Weatherhead Center for International Affairs, Harvard University, Cambridge, Massachusetts, USA.*

Pardey, P.G., *Science and Technology Policy, Department of Applied Economics, University of Minnesota, 1994 Buford Avenue, 231 Classroom Office Building, St Paul, MN 55108, USA.*

Perrin, R., *Department of Agricultural Economics, 314A FYH, University of Nebraska, Lincoln, NE 68583-0922, USA.*

Phillips, P.W.B., *Department of Agricultural Economics, University of Saskatchewan, 51 Campus Drive, Saskatoon, Saskatchewan, S7N 5A8, Canada.*

Pray, C., *Department of Agriculture, Food and Resource Economics, Rutgers, The State University of New Jersey, 55 Dudley Road, New Brunswick, NJ 08901, USA.*

Price, G.K., *Economic Research Service, US Department of Agriculture, 1800 M Street NW, Washington, DC 20036–5831, USA.*

Raper, K.C., *Department of Agricultural Economics, Michigan State University, East Lansing, MI 48824, USA.*

Reiss, T., *Fraunhofer Institute for Systems and Innovation Research, Breslauer Str. 48, 76139 Karlsruhe, Germany.*

Scandizzo, P.L., Via dell'Alpinismo 24, 00194 Roma, Italy.

Schmitz, A., *Department of Food and Resource Economics, University of Florida, McCarthy Hall, PO Box 110240, Gainesville, FL 32611-0240, USA.*

Schmitz, T.G., *Department of Agribusiness, Arizona State University, Tempe, Arizona, USA.*

Schimmelpfennig, D., *Economic Research Service, US Department of Agriculture, 1800 M Street NW, Washington, DC 20036–5831, USA.*

Smyth, S., *Department of Agricultural Economics, University of Saskatchewan, 51 Campus Drive, Saskatoon, Saskatchewan, S7N 5A8, Canada.*

Sonne, A.-M., The *MAPP Centre, Aarhus School of Business, Haslegaardsvej 10, DK-8210 Aarhus V., Denmark.*

Sorrentino, A., *Dipartimento di Studi Aziendali, Tecnologici e Quantitativi, Universitá della Tuscia (Viterbo), Italy.*

Tollens, E., *Department of Agricultural and Environmental Economics, Katholieke Universiteit Leuven, De Croylaan 42, 3001 Leuven, Belgium.*

Tongeren, F. van, *Agricultural Economics Research Institute (LEI), Wageningen University and Research Centre, PO Box 29703, The Hague 2502 LS, The Netherlands.*

Weaver, R.D., *Department of Agricultural Economics, 207D Armsby Building, Pennsylvania State University, University Park, PA 16802, USA.*

Wolf, C.A., *Department of Agricultural Economics, Michigan State University, East Lansing, MI 48824, USA.*

Zepeda, L., *Department of Consumer Science, University of Wisconsin, 1300 Linden Drive, Madison, WI 53706–1524, USA.*

Acknowledgements

The chapters in this volume were originally presented at the Sixth International Conference of the International Consortium on Agricultural Biotechnology Research (ICABR), held at Ravello, Italy, in July 2002. They have since been edited and revised.

The editors acknowledge sponsorship by the following:

- CEIS – University of Rome 'Tor Vergata'
- Economic Growth Center, Yale University.

Editors' Introduction

The first successful agricultural biotechnology (agbiotech) products were introduced in 1996. These products were the glyphosate resistant products (e.g. Roundup Ready® soybeans) and the insect toxicity products (e.g. *Bt* cotton). The initial products achieved rapid farmer acceptance in selected countries (notably, the USA, Canada and Argentina) and were hailed as major advances in production technologies. Yet, today the agbiotech revolution is 'stalled'. Few new agbiotech products have been introduced to the market after 1998 (one of the few is Monsanto's new *Bt* rootworm maize product (*New York Times*, 26 February 2003)). The initial products have been subject to rising levels of criticism from consumer interest groups and from political interest groups because of food safety and environmental concerns and because these products have been associated with a small number of private multinational companies.

It is difficult to know why we have not observed more agbiotech introductions, including products with improved consumer quality attributes, from the private sector. It is also difficult to know why public sector agricultural research programmes have not introduced new agbiotech products. This includes the International Agricultural Research Centers (IARCs) who have not introduced new products with 'traits' enabling plant varieties to be utilized in high abiotic stress (drought, submergence, and high salt) environments for poor farmers.

These failures on the part of both the private and public sectors may be explained simply by the fact that the first successful products on the market were a matter of 'luck' and that the subsequent 'dry spell' in product introductions is simply due to the nature of the invention/innovation process. If so, the delay between the first generation products and future products could be quite long.

An alternative explanation, particularly for the public sector research centres (including the IARCs) is that they were caught 'sleeping at the switch' by the private sector and that they have not had time to get their research act together (this explanation has credence, in part, because many of the early innovations in both agbiotech and medical biotechnology originated in the molecular biology departments in Universities, not departments in the Colleges of Agriculture, and many of these departments were in private universities). The failure of the IARCs to provide aggressive leadership in using recombinant DNA (rDNA) techniques to develop host plant resistance to insect pests and plant diseases and for host plant tolerance to abiotic stress (drought, alkalinity, etc.) is at least partly explained by a lack of investment in developing these products.

The years since 1996, however, have clearly shown that consumer and political interest group sentiments have become increasingly hostile to the first generation products. Whether this hostility would have been stemmed by the continued introduction of new agbiotech products remains an open question. But the years since 1996 have also shown that the regulatory systems in place for food safety, environmental issues and consumer information (labelling) were inadequate to deal with agbiotech products.

This volume is addressed to the problems of regulatory inadequacy for agbiotech products and to directions for reforms achieving a regulatory system supported by consumers and other political groups.

These reforms are currently underway in many (but not all) countries. They are taking place in the context of International Trade Conventions and regulatory standards (particularly the World Trade Organization Trade Related Aspects of Intellectual Property Rights (TRIPS), Sanitary and Phyto-sanitary Measures (SPS) and Technical Barriers to Trade (TBT) agreements). They are not necessarily moving in the direction of a fully 'harmonized' regulatory system for agbiotech products, as many countries prefer different regulatory systems (including the banning of agbiotech product sales). But all countries will have to weigh international benefits against national interests in this regard.

The volume is organized as follows:

The first chapter in the volume (Paarlberg *et al.*) offers a broad political analytical perspective on regulatory problems and on international prospects for a more harmonized international regulatory system.

Part 2 of the volume includes chapters discussing evolving regulatory systems. These include regulations associated with trade law, with intellectual property, consumer information and commercial policy.

Part 3 addresses the effect of regulatory systems on the processes of invention and innovation. Regulatory systems can provide incentives that can facilitate inventions and innovation. Regulatory systems can also impede invention and innovation by raising 'transaction costs'. These are of real concern to the agbiotech industry and are the basis for the policy stance of the agbiotech firms associated with first generation products. The initial opposition to labelling regulations by US firms was probably a factor leading the strong L&T (labelling and traceability) regulations implemented in the European Union (EU) in 2001.

Part 4 includes chapters addressing the impact of regulations on innovation through effects on market structures.

Part 5 addresses the relationship between regulatory systems and market development. The focus of the chapters in this part is on the transaction costs associated with identity preservation, segregation and traceability of genetically modified (GM) and non-GM crops. These costs are difficult to assess because of limited experience to date with these regulations.

The final chapters in Part 6 address economic interests associated with international trade of agbiotech products. This includes inherent conflicts between WTO TRIPS, SPS and TBT agreements.

Chapter 1 (Paarlberg *et al.*) provides an overview of political and economic forces shaping the evolving regulatory system for GM crops. The authors note that regulatory issues have been a major factor inhibiting the market introduction of new agbiotech products. The EU Commission initiated an informal moratorium on new GM crop imports to Europe in 1998. In 2001 the EU Commission prepared new L&T regulations for GM crops and derived food products. These L&T regulations were designed to alleviate health safety concerns and expected to result in the lifting of the moratorium.

The authors of Chapter 1 do note that 'with so many conflicting interests at stake, it is not surprising that the international regulation of GM crops remains in contention'. The chapter discusses governance dynamics and two alternative routes (political and scientific) to resolution. They note that the political alternative essentially depends on the outcome of the US–EU contest.

The authors also suggest an alternative to political bargaining based on 'rules principally grounded in science'. This alternative depends on the 'maturity' of the sciences involved. Typically, most fields of science (or sub-fields) that emerge in response to societal demands for scientific findings go thorough a 'folk science' phase. During this phase, they are typically unable to resolve interest group differences through accepted scientific experimental evidence. This acceptance requires a growing body of evidence based on scientific experimentation. Ravetz (1971) notes that in the folk science phase, many scientific groups produce 'pseudo-science' for interests groups. (This is sometimes referred to as 'Jane Fonda Science' in recognition of her activist roles and attempts to 'capture' science for her interests.)

Chapter 1 discusses the politics and political strategies of this debate (e.g. the use of the fear of multinational firm dominance of the seed industry) in a way that is often absent in economic studies. The processes by which international agreements on regulatory harmonization are reached are discussed. One feature of this debate seems to be the concern over acceptability of exported products from developing countries, even though the actual level of exports of many developing countries is very low. But, it remains the case that the poorest developing countries also have very low levels of manufactured goods exports and hence have a potentially high stake in access to Organisation for Economic Co-operation and Development (OECD) country food markets. Recent development in agricultural protection in OECD countries markets (the US 2002 Farm Bill and the European Common Agricultural Policy (CAP) towards new EU members) effectively make it very difficult for developing countries to obtain access to markets for products where they have comparative advantage.

Interestingly, the debates over regulation of medical biotech products are not contentious and do not have the political dimensions characterizing agbiotech regulation.

The six chapters in Part 2 document the state of play in regulatory development. Chapter 2 (Katz *et al.*) makes a number of important points regarding the implications of pre-existing laws and regulations on agbiotech regulation. These include trade laws and food safety regulations. It is often assumed in discussions, for example, of labelling options that a company can simply decide to voluntarily label a food as 'GM free'. But, all labelling implies some form of legal liability and food sellers cannot ignore this liability.

Chapter 2 also provides a comparative discussion of US regulation of GM foods and EU regulation regimes. Readers will find these comparisons to be instructive regarding the different philosophies underlying regulation. The US Food and Drug Administration (FDA) uses the GRAS (generally recognized as safe) procedure to determine whether a food additive requires pre-market approval. This was also the basis for the FDA's decision that genetic modification of foods did not require labelling, ruling that GM foods do not differ in any material way from other foods.

The EU, by contrast, regulates GM foods under the 1997 Novel Foods Regulation. 'Novel' foods are subject to separate authorization and labelling rules. The EU system does apply the 'substantially equivalent' concept to existing foods to assess food safety. Labelling requirements, however, call for an analysis of foreseeable 'risks' of GM foods and other new foods.

In Chapter 3, Blakeney addresses Intellectual Property Rights (IPRs) and their role in regulations. The WTO-TRIPS agreement, the Convention on Biodiversity (CBD) and the International Treaty on Plant Genetic Resources (PGFRA) are discussed in the chapter.

The TRIPS agreement is based on the principle that IPRs 'should contribute to the promotion of technological innovation and to the transfer and dissemination of technology'. But developed and developing countries have long held a difference of view on patent IPRs. This is reflected in the fact that prior to the TRIPS agreement, 50 of the 94 developing countries with populations over one million did not have functioning patent systems. And many of the more advanced developing countries did not give priority to patent systems.

There appear to be several reasons for this striking difference in patent system priorities. IPRs impinge on a country's capacity to imitate or copy an invention/innovation. But in a well functioning IPR system this impingement is far from binding. Firms legally 'invent around' the IPRs of another firm and bring imitation products into the market regularly. Where inventing around is not possible, licensing arrangements are typically developed. When developing countries engage in imitation (and this requires a high level of competence) they may or may not be infringing on the IPRs of a developed country invention, but the developed country can freely charge them with infringement. In the absence of a court process, the validity of the charge is not tested.

The line between infringement and imitation rights is not entirely one of IPR language because IPRs apply to only part of what might be termed knowledge. Much knowledge is the 'common heritage of mankind' and is excluded from patent protection. Developing countries often feel that they are deprived of access to this knowledge by IPRs.

Will the TRIPS and other IPR systems provide better access to common heritage knowledge? Will developing countries design IPR systems to truly stimulate national inventions and innovations (instead of being driven by efforts to limit licensing payments to foreigners?). The TRIPS *sui generis* option should allow for creativity in IPR design for plants. But creativity in IPR design has unfortunately not been a hallmark of developing country policy.

Chapter 3 also discuss the conflicts between CBD based IPRs and TRIPs based IPRs. The Food and Agriculture Organization (FAO) PGRFA treaty was designed to achieve relatively free exchange of genetic resources in the context of both sets of IPRs. This full free exchange of breeding materials was a central part of the Green Revolution (Evenson and Gollin, 2003). As a practical matter, genetic resources held by private seed firms are typically not exchanged. They are held in secrecy and protected by trade secrecy law (another IPR). Developing countries have a vital stake in free exchange and legitimate concerns over privatizations of these resources. The PGRFA agreement is an example of an international agreement that does meet the TRIPS objective.

In Chapter 4, Zepeda addresses developments in labelling policies. This discussion turns on the policy polarization evident in the USA and the EU policy stance on labelling. The FDA in the USA now recommends voluntary labelling, but opposes mandatory labelling. Some observers might argue that by opposing even voluntary labelling for a time, the USA may have 'pushed' EU labelling regulators into mandatory labelling with stringent thresholds for adventitious or accidental inclusion of GM ingredients.

The polarization of policy on labelling of GM food products is somewhat surprising given trends in the food industry. These trends are toward niche markets for products with labelling and certification support. The organic food market is a case in point. Organic foods are demanded not because they are 'safer' than non-organic foods. They are demanded because consumers attach value to the perception that organic production methods are more harmonious with natural ecological conditions. Thus, consumers of organic food products do state a preference for production methods, but do not challenge FDA and other food safety systems. Nor do they insist that FDA regulations ban non-organic foods.

In food markets, the 'customer is always right' and the consumer is entitled to full information including labelling information.

In Chapter 5, Hobbs *et al.*, discuss international issues impinging on agbiotech regulation. In particular, the evolving WTO/General Agreement on Tariffs and Trade (GATT) rules have developed two mechanisms to address problems of 'avoidance' of international rules regarding tariffs – specifically, the imposition of quantitative trade restrictions under the guise of safety and risk issues.

The first of these is the Agreement on SPS. The SPS agreement was developed to provide guidelines for claims that safety issues justify quantitative trade restrictions. This agreement calls for the use of best available scientific information determining risk.

The second agreement is the Agreement on TBT. The TBT agreement deals with packaging and labelling regulations as barriers to trade.

Chapter 6 (Esposti and Sorrentino) utilizes the political economy model of Bagwell and Staiger applied to WTO negotiations. The chapter develops conditions for resolution of political conflicts. It is noted that there is a discrepancy between the 'principles and the practice' with WTO agreements and that both to SPS and TBT agreements could be seen as a structure within which conflicts between current regulatory rules and the WTO agreements can be realized.

In Chapter 7, Heumueller and Josling further addresses the conflict between EU regulations and the TBT agreement. The authors suggest principles by which TBT agreement measures can be made consistent with regulatory regimes. The chapter examines whether broad labelling requirements can be consistent with TBT principles.

GM foods raise issues with respect to both SPS and TBT. The WTO system has evolved over many years and has developed guiding principles regarding discriminatory treatment of different countries. These principles of trade liberalization can be used to achieve agreement on the development of regulatory systems.

Part 3 of the volume addresses the question of regulatory incentives and impediments to invention and innovation.

In Chapter 8 Knudsen and Scandizzo examine the effect of environmental liability on agbiotech R&D. They use a 'real options' approach to evaluate the merits of reduced uncertainty in the legal system regarding environmental liability. They find that ignoring environmental liability understates the consequences of liability during the development phase of innovation. Environmental liability thus deters development but has an ambiguous effect on research in the invention phase of the innovation process.

A welfare analysis of the consequences of reduced uncertainty regarding environmental liability is reported in the chapter. This analysis shows two different sets of results. The first is that reduced uncertainty enhances the Net Present Value of a research project already under way. The second is that reduced uncertainty increases the option value of research (invention) and lowers the option value to sue by the damaged party, but increases the likelihood of liability damages. This has the effect of delaying a project, not already started.

The US position has been that both the moratorium and the L&T regulations constitute violations of WTO, TBT and SPS regulations. This legal battle is further complicated by a scientific battle over food safety and environmental effects. And all of this is further complicated by a rising level of concern among developing countries that rich country non-governmental organizations are calling for regulatory systems that effectively pre-empt them from developing the capacity to employ rDNA methods and techniques.

Chapter 9 by Naseem and Oehmke addresses the issue of genomics R&D. As genome maps have been completed, systematic genomics and proteonomics R&D is now feasible. This research is very 'basic' in one sense, because its purpose is to discover specific gene functions. These 'pre-invention science' discoveries can then be converted into inventions and products.

Typically, private firms do not conduct very much basic R&D, concentrating instead on invention-focused research and innovation-focused development in their R&D strategies. Chapter 9 addresses the question of public and private conduct of genomics R&D. The issue is whether private firms will have incentives to conduct genomics research. The authors set up a model with first stage R&D (the racing stage) and second stage R&D (the innovation stage). The model suffers from the very unrealistic assumption of patent race models generally, i.e. that there is only one winner of the race. In practice, virtually all innovations are quickly imitated and patents are regularly 'invented around', so that no single patent provides 'race winner' rewards. Readers will find the analysis instructive, however, because it does suggest that private firms may have an incentive to initiate genomics R&D.

Chapters 10, 11 and 12 deal with intellectual property rights and their impacts on innovation. In Chapter 10, Eaton and van Tongeren address the issue of differentiated standards in IPRs by level of development of the country. This has long been a contentious issue because the World Intellectual Property Rights Organization (WIPO) has supported uniform (homogeneous) IPR standards for all countries.

Eaton and van Tongeren develop a model for a seed-breeding firm as a three-stage simultaneous move game. The model allows for North–South differentiated markets. Numerical simulations indicate that allowing different standards in the two markets increases net social welfare. The chapter thus supports the flexibility built into the WTO-TRIPS argument allowing for *sui generis* IPR system for plants and animals.

The chapter implicitly raises the question of creativity in IPR legislation and in the court systems administering IPR legislation. Prior to the TRIPS agreement, Paris Convention rules applied to patent IPRs. These rules allowed for differentiated national IPR standards, but required that 'national treatment' be provided to foreign member inventors. One might have expected developing countries to have derived many different and creative IPR systems. But they have been fixated on the obligations to foreign IPR holders to the neglect of incentives for domestic inventions. In short, they have shown precious little creativity in IPR system development.

In Chapter 11, Janni addresses the relationship between IPRs and biodiversity conservation. There is an underlying premise to this chapter and to the discussion of biodiversity in general, that agbiotech techniques have raised the implicit value of genetic resources outside the cultivated species (including closely related species where wide-crossing methods can be used in conventional breeding). The successful agbiotech products to date and the potentially successful products (Golden Rice) are based on genetic resources outside the conventional breeding range (some, however, may use rDNA techniques to enhance conventional breeding methods as in bacterial leaf blight resistance in rice).

The question is whether IPRs can provide incentives for bioprospecting and conservation programmes. And, also, whether these IPRs will be sufficient to achieve conservation objectives. Conventional plant breeding programmes rely on *ex situ* collections (gene banks) of landraces (farmer selected varieties) mutants, sports and related species. Curiously, the *ex situ* option for biodiversity conservation plays a small role in biodiversity policy discussions. This is probably because *in situ* conservation is also seen as habitat conservation. But, in practice, if one is serious about biodiversity conservation, *ex situ* collections have huge advantages over IPR based conservation. By maintaining replicated collections with long-term funding, conservation is not subject to political changes and disruption. (Vietnam, for example, lost its rice germplasm collections during the Vietnam War, but they were replaced from replicated collections at the International Rice Research Institute (IRRI) and elsewhere.)

In Chapter 11, Janni notes that two IPR systems the CBD system and the TRIPS system provide conflicting sets of incentives. The CBD provision of 'farmers' rights' could provide incentives for bioprospecting (and conservation) provided that these rights are associated with real prospectors. In many countries this association is not developed.

The TRIPS system could also provide patent rights to bioprospectors and this raises questions of 'foreign prospectors' stealing national resources. But in practice, the Costa Rican INBio agreements with Merck and Diversa are examples of effective institutional mechanisms to achieve biodiversity conservation.

In Chapter 12, Binenbaum and Pardey address the challenges (and opportunities) that IPRs present to the IARCs. The IARCs operated in situations where IPRs were of minor concern until the expansion of IPR scope associated with the development of agbiotech methods took place. Each major crop centre maintains *ex situ* collections of genetic resources (gene banks). These are mostly collections of landraces (farmers' varieties) mutants sports and related species (typically other non-cultivated species in the genus). For many years the IARCs responded to requests for gene bank accessions by simply sending these materials to requesting agents (including private firms). In addition, the IARCs devoted resources to 'pre-breeding' evaluation of these materials for potential breeding value.

Advanced or 'elite' breeding lines (many to be released as varieties) were also exchanged freely. IARCs have maintained International Nurseries where elite lines are effectively 'delivered' to National Agricultural Research Service (NARS) breeders enabling the NARS breeders to observe performance and select valuable parents (see Evenson and Gollin, 2003).

Chapter 12 addresses the implications of IPR expansion in the context of four underlying themes: partnerships, technology positioning, market segmentation and incomplete contracting.

Partnerships between IARCs and other research organizations and between IARCs have been of growing importance to the IARCs. Many of these IARC partnerships are formed to achieve R&D collaborating with advanced biotechnology centres. Some are designed to facilitate the transfer of IPR protected materials. But for many IARCs, partnerships come at a high price – the cost of managing a collaborative arrangement can be high – in some cases, eliminating the research that the IARCs were designed to conduct.

Chapter 12 also discusses options for IARCs associated with purchasing rights to IPR protected methods (process) and products on behalf of poorer developing countries. The IARCs have long been committed to extending technology to poor farmers in marginal areas. This was a challenging objective for conventional breeding methods. Agbiotech methods offer promise for furthering this objective, but the IARC system has yet to deliver on this promise.

Part 4 of this volume includes four chapters addressing the effects of regulations on innovation as mediated through industrial organization.

In Chapter 13, Weaver addresses the question of patent-based incentives for R&D and industrial structures. A distinction between 'local dominance' of a technology and 'universal dominance' of a technology is introduced. It is well known that crop varieties typically have a high degree of location specificity (to soil, climate and economic conditions) and have, at best, local dominance. However, many agbiotech products (such as *Bt* genes) can be inserted into a large number of locally dominant varieties and this may constitute a form of universal dominance.

Weaver derives market structure conditions where technologies that are not universally dominant can be incentivized by patents. This requires a 'restricted monopoly' or weak monopoly structure in industries. Alternative pricing strategies are evaluated and labelling regulations considered in this context.

In Chapter 14, Oehmke *et al.* analyse the mergers and acquisitions (M&A) feature of agbiotech firms. The agbiotech industry has only four of five firms generating meaningful revenues from agbiotech sales. Yet the industry is characterized by a high level of M&A. The chapter is empirical and designed to test four propositions derived from a related 'neo-Schumpeterian' model. These propositions are:

1. That the number of firms in the industry will move in a cyclical pattern – first increasing then decreasing (through M&A).
2. That the level of incentive actively moves co-cyclically.
3. R&D intensity (R&D/sales) moves counter-cyclically.
4. M&A activity moves counter-cyclically.

The model is tested using transgenic field trial data of USDA's Animal and Plant Health Inspection Service (APHIS) for three crops: soybeans, cotton and maize. General support for the first three propositions is reported.

Chapter 15 by Menrad and Reiss analyses the European agbiotech innovation system utilizing the National System of Innovation (NSI) conceptual framework. This framework provides a useful way to assess innovating capacity.

The chapter reports data from a study funded by the European Commission covering eight European research groups engaged in biotechnology research in three sectors, biopharmaceuticals, agro-food and equipment and supplies. The study reports assessments of the knowledge/skills network, the industry/supply network, the finance/industrial development network and the demand/social acceptability network.

Chapter 16 approaches the innovation question in a different empirical framework. In this chapter Klotz-Ingram *et al.* analyse data for 15 biotech firms (mostly based in the USA) on innovative output and firm characterization. Innovative output is measured by patents obtained, by field trials conducted and by Plant Variety Protection Certificates (PVPCs) obtained.

Innovation output measures are first statistically related to innovative activity measures (level and growth of R&D expenditures, R&D intensity), then output measures are related to financial characteristics; firm size (sales and number of employees), shareholders, equity and earnings per share and gross profit margins. Preliminary analysis and interpretations are reported.

Part 5 (Chapters 17–20) addresses the relationship between regulatory costs and farmer and consumer welfare. Most regulations are implemented to address market failure or market imperfection problems. Regulations are often seen as 'second best' solutions to these problems. To the extent they do provide solutions, they are generally regarded to be welfare improving. But not all regulation is welfare improving. The chapters in Part 5 show that under some conditions, regulations can actually reduce the welfare of both producers and consumers.

Chapter 17 (Perrin and Fulginiti) analyses pricing strategies for IPR protected traits of GM crops. The analysis treats traits as durable or semi-durable goods. The authors show that monopoly pricing for durable goods differs from standard monopoly pricing because buyers have incentives to wait to purchase a good if they expect the price to decline. The chapter concludes that the durable goods feature of traits reduces the value of patent rights incentives. The authors also note that weak IPRs such as breeders' rights provide even lower incentives for plant breeding research.

Chapter 18 (Smyth and Phillips) clarifies the types of regulatory requirements and costs associated with three elements of regulations: to achieve product differentiation. These are:

- Identity Preserved Production and Marketing (IPPM).
- Segregation.
- Traceability.

IPPM systems are not new. They have been used in many niche markets for specialized products including organic food, specialized rape (canola) varieties, and high oleic maize. These systems are necessary if market premiums are to maintain in markets. Most of these systems are voluntary.

Segregation differs from IPPM in that segregation systems are required to ensure that potentially hazardous crops (to humans) are prevented from entering supply chains where products are destined for human consumption. This requires 'that crops be kept separate to avoid commingling during planting, harvesting, loading, unloading, storage and transport' (Lin, 2002).

Traceability has also been used for many years in the food industry. Products with unacceptable levels of bacteria or pesticide or chemical residues need to be removed from markets quickly. Traceability systems allow retailers to identify the source of contamination.

Smyth and Phillips note that all these regulatory devices have been used in the food system, but that there are options. IPPM systems are essential for revenue management and premium pricing. In some markets (as noted in Chapter 19) sufficient supply is available relative to demand that premiums will not emerge. Mandatory IPPM systems in such a situation, will impose costs on both producers and consumers.

Segregation systems are essential for risk management. Imposing segregation regulation in low-risk market situations can impose costs that exceed benefits.

Traceability systems serve a similar purpose regarding food safety.

In Chapter 19, Schmitz *et al.* provide the market analysis required to determine whether premiums will exist for non-GM foods and whether mandatory IPPM, Segregation and Traceability costs (treated as identify preservation (IP) costs in the chapter) are associated with consumer benefits. The authors develop several models (using graphic analysis (Figs 19.1–19.4) where feasible).

Chapter 19 first shows that demand must be differentiated if a premium is to be possible. Imposing IP costs in a situation where demand is not differentiated places a burden on both producers and consumers.

The authors then show that with differentiated demand a premium for non-GM crops can emerge, in the presence of IP costs. The premium (or discount for GM crops) can exceed IP costs and could not be achieved without incurring IP costs. In this chapter the differentiated demand indicates that consumers could obtain benefits exceeding IP costs, but that producers will bear part of the IP costs.

The soybean markets in the US are used to illustrate these points.

Chapter 20 (Price *et al.*) presents a case study of the EU L&T regulations and their impact on the US soybean sector. (Chapter 18 also includes a trade model and Chapter 19 assesses the burden of IP costs for maize and soybeans.)

In Chapter 20, regulatory requirements are reviewed and three scenarios are analysed:

1. The USA produces and exports conventional and EU approved soybeans to Europe.
2. The USA produces conventional EU approved and EU unapproved soybeans and segregates exports to Europe.
3. The USA exports only non-biotech soybeans to Europe.

Scenario 1 is relatively low cost, although EU L&T rules must be met. Scenario 2 requires the addition of segregation costs. Scenario 3 requires strictness in segregation and is high cost.

These scenarios are then analysed using cost estimates. Implication for farmers and marketing agents are discussed.

Chapter 21 (Lin and Johnson) reports a case study of the costs of segregation of non-GM maize and soybeans and an analysis of the 'burden' of the costs. Evidence for price premiums paid in Japan is utilized in the analysis.

The study concludes that Japanese consumers are willing to pay relatively high premiums for non-GM crops. These premiums allow a producer premium to non-GM producers, but do not fully cover segregation costs. These are partly borne by the non-GM producer in the form of lower premiums for non-GM grains.

Part 6 of the volume includes three chapters dealing with economic impacts of regulation.

In Chapter 22, Harmsen *et al.* utilize 'scenario construction' methods to assess three alternative scenarios.

The first scenario was a naturalness scenario where consumer mistrust of current regulatory system remains high. The second was a scenario in which GM products with feasible health qualities are developed. The third scenario was based on reduced disposable real income. Future impacts of the scenarios are analysed.

The first two chapters in the volume utilize more formal computable general equilibrium (CGE) models to perform scenario analysis.

In Chapter 23, Demont and Tollens analyse *ex ante* welfare effects of introducing transgenic herbicide tolerant sugarbeets in Europe. The CGE model includes the USA, the EU and the Rest of the World (ROW). Benefits are calculated for each region. Impacts on sugarcane growers (negative) are also analysed.

In Chapter 24, Barkley utilizes a CGE model to analyse the economic impacts of the production of GM maize and soybeans in the USA. Simulations for US adoption without EU import bans, and with EU import bans are analysed. Simulations for ROW adoption of cost-saving technologies are also considered.

In summary, this volume provides readers with an overview of regulatory system inadequacies and of directions for regulatory reform for agbiotech products. Several chapters provide cost data for meeting regulatory requirements. And several chapters analyse the public benefits and losses associated with regulations. While regulatory regimes are still evolving, the chapters in this volume should inform this evolution.

References

Evenson, R.E. and Collin (2003) *Crop Variety Improvement and its Effect on Productivity: the Impact of International Agricultural Research*. CAB International, Wallingford, UK.

Lin, W. (2002) Estimating the costs of segregation for non-biotech maize and soybeans. In: Santaniello, W., Evenson, R.E. and Zilberman, D. (eds) *Market Development for Genetically Modified Foods*. CAB International, Wallingford, UK, pp. 261–270.

Ravetz, J. (1971) *Science and its Social Problems*. Oxford University Press, Oxford, UK.

1 Regulation of GM Crops: Shaping an International Regime

Robert L. Paarlberg,[1] Raymond F. Hopkins[2] and Lisa Ladewski[2]
[1]Wellesley College and Weatherhead Center for International Affairs, Harvard University, Cambridge, Massachusetts, USA; [2]Swarthmore College, Swarthmore, Pennsylvania, USA

Introduction

Three controversies envelop new genetically modified (GM) crop prospects, limiting their development and creating incentives for global regulatory standards. *First*, trade factors have pushed the USA and the European Union (EU) toward a confrontation in 2003 over regulation of GM crops. The EU Commission, which initiated an informal moratorium on new GM crop imports in 1998, proposed a new regulation in 2001 requiring labelling and traceability of GM and GM-derived food products. The Commission hoped that this move would alleviate health safety concerns surrounding GM products in member states, allowing lifting of the moratorium on approvals. However in May 2003, the USA challenged EU's moratorium in the World Trade Organization (WTO) claiming that the proposed EU regulatory solution was too stringent and stigmatized GM use globally and would be unduly costly to American farmers.

In May 2003 Robert Zoellick, US Trade Representative, asked the WTO to declare Europe's actions illegal. Earlier, Zoellick had warned his European counterpart, EU trade commissioner Pascal Lamy, that the Bush Administration would seriously consider challenging the EU's moratorium and its likely safety rules in the WTO. Lamy's response in 2002 was to say that '[L]itigation in this field would be immensely counterproductive; it would be seen as a challenge to consumer fears and perceptions' (*Inside US Trade*, 2002). In 2003, similar heated words were exchanged. By 2004, the debate over whether European standards for approvals, tracing, and labelling of GM foods constitutes acceptable precaution (based on scientific foundations) or a violation of rules forbidding trade restrictions *not* based on science, had triggered considerable passion.

A *second* controversy over GM regulation has welled up among scientists. Among both public and private sector scientists who do research on transgenetic innovations, competing conclusions on the risk and gains of GM products over issues of biosafety and food safety exist. Whether genetically modified organisms (GMOs) produced already, and ones contemplated, are safe or not divide scientists; as a result government officials, often themselves scientists who normally rely on the conclusions of research, are themselves enmeshed in controversies over GM crops and their proper regulation.

This split in the scientific community is evidenced in several publications and political documents in the 1990s. Initial research revealed few biosafety risks for various GMOs, and the scientific community had a rather permissive attitude about GMO use (*The Times*, 2000; NRC, 2002). Some subsequent studies, done by scientists at research universities principally, raised concerns about ecological damage and possible unexplored risks. While only a few studies have raised warnings, for instance ones popularized by publication in *Nature* magazine discussed below, division among scientists about safety has resulted.

Such safety concerns prompted a *third* controversy. Contending public interest groups, both intergovernmental agencies and various non-governmental organizations (NGOs), are now battling over the net benefits of GM crops and placing demands on various government agencies, including national regulatory bodies, the Food and Agriculture Organization (FAO) and the World Bank to advance or pull back from use of GM technology (EU, 2002; NRC, 2002). Greenpeace, for example, citing scientific concerns, has been active in lobbying worldwide for a moratorium. In the media, coalitions approaching a social movement do battle with private economic firms such as Monsanto over the interpretation of GM crop benefits. Consumers, firms, NGOs, and various developing country leaders have weighed in with competing claims. This public debate not only argues about risks to the environment, but also attacks the effort of multinationals to protect their intellectual property rights at the expense of fair use and practice by farmers. Agricultural interests in developing countries have weighed in to demand rich country NGOs not pre-empt their right to use new technology.[1]

One effect of the campaigns of anti-GMO NGOs has been to amplify the view of some scientists that GM technology 'poses irreversible threats of biological pollution and a host of other risks' (Greenpeace, 2002), while tying this claim to other concerns such as a fear of a multinational take-over of the seed industry.[2] The public and cultural controversy therefore is rather diffuse, mixing together food and environmental safety, corporate power and world hunger issues in the GMO dispute.

During the coming years these conflicting arguments and the groups that support them are likely to come to a head in the ongoing dispute. The core scientific question is whether GM varieties should be treated with the *same* or with *more stringent* regulation than that of traditional varieties. Advocates of more stringent regulation favour use of the precautionary principle adopted by the Convention on Biological Diversity (CBD) in its Cartagena Biosafety Protocol (CBD, 2000), which allows prevention of GMO imports even when evidence of risk is uncertain. Until these conflicts between precautionary and more permissive approaches are reconciled, no coherent international regime surrounding GMOs can be built.

Currently, a variety of national and regional policies have conflicting impacts on GM research, intellectual property rights, safety and trade. Wide variations among countries exist in the scope and degree of prohibition for GM efforts; curtailed production and trade are potentially detrimental results of this discordant environment. In principle, this need not be.[3] In other areas, such as monetary affairs or nuclear power, international controversies surrounding economically or scientifically disputed regulatory decisions have been harmonized, with a resulting international agreement and formation of a cohesive regulatory regime. This paper reviews how an international regime for governing GM materials might be constructed from the current discord.

Background

Food security and technology

Technology in the form of 'the Green Revolution' greatly aided world food security in the 20th century. High yielding varieties, mostly

[1] See recent op-ed articles by the Ugandan and Nigerian Ministers of Agriculture.

[2] NGOs have protested GM crops for a variety of reasons, from food safety to unwarranted control over crops thanks to intellectual property rights rulings. For example, Greenpeace published a press release in April 2002 with the altruistic headline 'Greenpeace Stops US Shipment of Maize to Mexico to Eliminate Source of Genetic Contamination' (Greenpeace, 2002). Within this article, the NGO cites an article in *Nature* (2002), evidently ignoring the controversy surrounding the study's validity and using it as proof toward the Greenpeace argument.

[3] Biosafety and food safety concerns are the core issues for finding common interest regulations with global harmonization. If these are subordinated to the US–EU trade conflict over technical barriers to free trade, regulatory harmonization of GM technology use becomes defined in rather implacable ways.

developed by international public research centres, gave farmers, especially in Asia, new crop varieties with substantially higher yields. In many countries production growth helped development and accelerated food supplies ahead of burgeoning populations. Unfortunately, this growth had levelled off by the mid-1980s, leaving many goals of increasing world food security unachieved (Nottingham, 1998). Green Revolution crops were traditionally bred to yield larger grains per plant volume. The asymptotic levelling-off trend seen in the 1980s reflected the physiological limits of seed size – seeds that are too large will break the stem on which they are growing (OECD, 2001). In addition, Green varieties had other drawbacks. They demanded costly inputs to provide the larger yields, some of the inputs were environmentally damaging, and the benefits of this 'revolution' did not reach the poor very well. Several major parts of the world, particularly Africa, gained little, due to the choice of crops that were developed and the high costs of inputs (Pinstrup-Andersen, 2001).

The development of recombinant DNA (rDNA) technology in the 1990s promised to overcome these limitations. GM crops have been developed commensurate with the move toward 'precision farming', where such new technology is able to reduce indiscriminant use of chemical and water inputs. In contrast to the Green Revolution crops, these crops contain specific gene sequences artificially inserted into their genome. Therefore, while traditional breeding techniques can take years to develop a crop with the desired characteristics – it took 25 years to create a European sugarbeet with three desired characteristics (Dossier, 1990) – GM technology can introduce individual genes with specifically desired traits in months (Paarlberg, 2001). Both precision of the technique and the speed with which it can be developed for actual use offer major improvements for agriculture and for total food production.

Indeed, GM technology offers a wide variety of possibilities for new crops. Currently, insecticide-inherent and herbicide-resistant crops are most widely produced. However, researchers have envisioned a vast number of GM varieties that could help further improve productivity or nutrition, such as addition of micronutrients into staple grains (ERS, 2001). The development of Golden Rice with enhanced vitamin A content may be a first step in this direction, although some have argued this rice does not contain enough vitamin A per serving to help deficient consumers significantly (Porphyry, 2001). As GM technologies advance, many look hopefully toward this new technology for help in increasing global food security in the future (Conway, 1998).

GMO crop history

When the US Supreme Court ruled in 1980 that living organisms were patentable material, the biotechnology industry was given the opportunity to profit from development of new transgenic organisms. This ruling was echoed in Europe in 1998, authorizing an international profit-seeking venture in living organisms. Because of this opportunity, private companies have rushed to develop and patent new agricultural seed types, making any public venture pale in comparison (Flint, 1998).

Several GM crops are already produced commercially in large quantities, although the total agricultural area planted with these crops is still relatively minor. As of the 2001 growing season, the total area planted to GM crops worldwide was just 52.6 million ha (130 million acres), only 1.3% of total global cropland area. In addition, the geographical distribution of GM crop planting is confined to a small number of countries. Of this GM crop acreage 99% was confined to just four countries: the USA (68%), Argentina (22%), Canada (6%), and China (3%). Small numbers of farmers in nine other countries (South Africa, Australia, Mexico, Bulgaria, Uruguay, Romania, Spain, Indonesia, and Germany) were legally growing some GM crops

[4] These data are from the International Service for the Acquisition of Agri-Biotech Applications (ISAAA). ISAAA, an organization dedicated to the spread of GM crop technologies, bravely interprets these data as evidence that GM crops are overcoming international resistance (James, 2001a,b). There have been noticeable illegal GM plantings in Brazil and India, but so small as not to affect these data – at least 0.75 million ha in Brazil of soybeans and 10,000 ha in India of cotton.

in 2001, but in most countries the planting of GM crops was either not yet legal or farmers were deciding voluntarily not to use the technology.[4]

The most common transgenic crops grown are soybean, cotton, rape (canola), and maize. Most notably, as of 2001, approximately half of the worldwide soybean crop was GM. Therefore, although the cumulative acreage of GM is low, the acreage for some crops (such as soybeans) is relatively high. In 1999, considerable amounts of transgenic potatoes, squash and papaya were produced in addition to the four major crops. All of these GMOs possess one of four traits: herbicide tolerance, *Bt* insect resistance, *Bt* plus herbicide tolerance, or virus resistance. As of 2002, a total of 12 GM plant species had been approved for commercial production in the USA (Colorado State Univerity, 2002). In Argentina, Canada, China and Brazil, somewhat fewer crop varieties have been approved.

GM crop-derived foods have been in the market for some time in the USA and Americans regularly consume them. In the USA, it is estimated that 60–70% of foods on the market have at least some GM component, with maize- and soybean-based ingredients accounting for the bulk of these products (Cornell University, 2002).

Though modest amounts of herbicide-tolerant and pesticide-inherent soybeans and maize are being produced in four countries, GM crop production has yet to achieve its most ambitious humanitarian goal – to help those in food insecure regions through development of nutrition-enhancing and higher-yielding traits in useful crops. However, the potential remains, given the possibilities of GM technology (Pardey, 2001).

Demands for regulatory harmonization

A major barrier to achieving the initial promise and rapid diffusion of GM crops is the erection of regulatory barriers that inhibit or prohibit use of this new technology. Controversies over the risks and benefits of GM technology have led to diverse and sluggish regulatory practices in many countries, and a demand for greater harmonization of rules for GM use through international coordination (CBD, 2001; Paarlberg, 2001). As GM production is no longer spreading rapidly from one country to the next, the research, development and trade of these new and potentially valuable plants has slowed, and resulted in growing controversy. Substantial conflict now exists over what policies should govern biotechnologically engineered products, threatening to stalemate GM technology altogether.

Three differences between Green Revolution varieties and GMOs, although seemingly minor, have resulted in much greater public criticism of the latter: (i) GMO risks are harder to determine since they are developed over a shorter timeframe; (ii) private American or European companies claim intellectual property rights to most GMOs; and (iii) engineered genetic modification has cultural and ethical implications that many find troubling (Nation, 2001; Fukayama, 2002). These criticisms have fed debates about biosafety, trade, international property rights (IPR), and cultural arguments against the further development of GMOs.

Initially, private ownership of seed patents raised ethical concerns about the future fairness of availability of these crops to the world's farmers (NRC, 2002, p. 241). However, in retrospect, international IPR claims have not been a major factor slowing the spread of GM crops. Private international companies, led by the Monsanto Company, have shown a willingness to extend GM crop technologies even into developing countries offering few IPR guarantees. In some cases, as with Monsanto's openly pollinated *Bt* cotton in China, firms have been willing to tolerate significant local IPR piracy as a price worth paying to gain access to a large and potentially lucrative commercial seed market. In other cases – Monsanto's hybrid *Bt* cotton in India, for example – the company protected itself by introducing a hybrid variety, which could not be locally replicated. In still other cases (e.g. virus resistant sweet potato in Kenya, and the patents used in the development of Golden Rice) Monsanto was willing to permit the use of its PRs by developing countries on a royalty free basis. In fact, all four of the private companies holding patents on the technologies used in Golden Rice have agreed to make them available to developing countries on a royalty free basis.

The limiting factor in the uptake of GMOs around the world has neither been IPR barriers nor high seed costs but lack of biosafety approvals. This lack of approvals is caused by a number of factors discussed next, of which trade concerns loom large. The biosafety argument rests on the fact that GM crops are often developed quickly, one of the more positive yet risky aspects of GM technology. The timeframe of GM development leads to concerns about possible unknown biosafety hazards, as crops may be released before proper testing has been carried out to determine if they will produce detrimental environmental side effects. Increasing, scientists worldwide have been acknowledging the need for more careful testing than has sometimes been done. In 2002 in the USA, the National Academy of Sciences noted the need to have more careful monitoring of GM crops to enable appropriate data collection so that actual risks could be identified (NAS, 2002). Biosafety procedures are often used to withhold new GM crop approvals; increasingly, however, trade concerns are actually the largest factor.

GMO trade concerns in the developing world and developed world

Screening GM crops case by case for biological safety risks is a routine national policy function in all wealthy countries, and this case-by-case screening practice has now been established in most important developing countries as well. Yet as of May 2002, it was not yet legal for farmers to grow GM crops in most of the developing world. National biosafety screening systems, even those in place and functioning, are producing very few approvals of GM crops for commercial planting.

In developing Asia, no national governments have approved production of any GM food or feed crops. The only significant biosafety approvals yet given in Asia are for an industrial crop, *Bt* cotton, which has been released previously to farmers in China, Indonesia, and in 2002 in India. In Africa and the Middle East, only the government of South Africa has yet approved the commercial growing of any GM crops (*Bt* cotton and *Bt* maize). In South America, where the government of Argentina was quick to go ahead with several important GM food and feed crop approvals in the mid-1990s (soybeans, maize and cotton), these same Argentine authorities after 1999 imposed an effective freeze on new approvals (so as to avoid losses in export sales to Europe), and in several other important agricultural states in the region – including Brazil and Chile – no official GM crop approvals have yet been granted.

What explains this surprising regulatory caution toward GM crops in the developing world? Scientifically demonstrated biosafety or food safety risks have not been the key factor in slowing down the official approval process. The slowdown has come instead from factors such as weak bureaucratic and technical capacity, donor-induced caution, legal and political opposition from domestic and international NGOs, and most of all fear of lost export sales. None of these is technically a biosafety concern, but the most convenient way for officials to address these concerns has been, quite often, to slow down the biosafety approval process. We can draw this conclusion from a comparative review of recent experience in several key developing countries, including Kenya, Brazil, India and China (Paarlberg, 2001). China, originally an enthusiast for GM cotton, has not approved any significant GM food or feed crops, in part out of fear of lost exports sales in Japan, Korea and Europe.

A number of both rich and poor nations have frozen GM crop approvals directly in response to international market concerns. Argentine officials, who were at first aggressive in their approval of planting GM soybeans and corn, more recently have held back from approving any GM food and feed varieties if not yet approved by regulators in the EU. Between July 1998 and April 2001 no additional GM crops of any kind were cleared for commercial production by Argentine authorities (Burachik and Traynor, 2002 unpublished report). In this case Argentine officials were primarily reacting to a 1998 moratorium on new approvals in the EU. For similar reasons the USA and Canada are now both going slow in approving GM wheat, which could cause lost export sales in Japan in particular if mixed into bulk shipments. The Chair of the Canadian Wheat Board recently estimated that the first major exporting country to begin planting GM wheat could immediately lose one third of its foreign customer base (Raine, 2002). Under pressure from frightened wheat growers in February 2002, Monsanto announced it was pushing back the commercialization of its new GM wheat varieties in the USA until 2005 at the earliest.

Fear of lost export sales is thus emerging as the single most important constraint on additional plantings of GM crops around the world, in rich as well as poor countries. Because of adverse consumer views toward GM products in two of the world's wealthiest and most prominent food importing regions – Europe, and Japan/South Korea – all nations with food and commodity export aspirations have begun to feel it might be safest to remain GM-free.

Trade: the principal dispute

As suggested above, trade creates the basic incentive structure around which issues of regulation at the national level are now forcing global conformity. This is because food and farm commodities are so heavily traded across borders that an economic imperative exists to harmonize minimum regulatory standards across those same borders. If heavy importing countries such as those in Europe create prohibitory national-level GMO regulations, they will inhibit production of these crops in export countries, an outcome that is favourable to the importers. Regardless of harmonization among other countries' national regulations, if a major import market, such as Europe, is closed to GM products because of labelling or cultural barriers, the result will be a common global practice of aversion to GM products, as producers and research firms respond to the calculations of strategies that offer a profitable opportunity. Thus EU biosafety and food safety regulations, with strong labelling and tracing requirements, affect the calculus of the majority of those who would adopt or promote GM seeds and crops. In the case of GM foods and commodities, the *principal* dispute in regulation has arisen since the late 1990s between the emerging EU standard of strict labelling and traceability (L&T) for any GM product, which requires costly market segregation, and the permissive path adopted in the USA, based on the more traditional standard requiring a scientific demonstration of risk, avoiding costly mandatory segregation for otherwise like commodities.

The critical issue holding up GM crop technology today is how national governments, such as the EU and Japan will be 'permitted' either to block imports of GM commodities and foods at their border, or require stigmatizing labels on imported GM commodities and foods. The EU answer is an approach to risk assessment based on a 'precautionary principle'. The EU Commission, acting under the authority of a February 2000 'Communication from the Commission' that was subsequently supported by both the Council and the European Parliament, has been working in as many settings as possible to advance this approach. While inserting this more sweeping approach into the EU's own newly established general food law, the Commission is also pushing for an incorporation of this principle more broadly: within a revised Codex risk analysis framework, within hoped-for revisions of the Sanitary and Phyto-sanitary (SPS) and Technical Barriers to Trade (TBT) Agreements of the WTO, within the work of the Organization for Economic Co-operation and Development (OECD), and (less successfully) in discussions with the USA within the Transatlantic Economic Partnership (Coleman, 2002).

The US approach, requiring scientific evidence of risk, became politically unacceptable in the EU following the 1996 bovine spongiform encephalopathy (BSE – 'mad cow disease') crisis. Having learned not to trust EU authorities on the safety of BSE meat, consumers began to mistrust authorities on GM products as well. The US mad cow scare in January 2004 suggests that the general risk adversion in the USA will also grow. Consumer confidence falls when a new technology comes under attack from a variety of anti-corporate and anti-globalization European NGO activist groups such as Greenpeace. In response to such scares, EU food consumers, in the last decade, began shopping around for non-GM sources of supply, giving an advantage to farmers and retailers who produced non-GM products. These concerns led separate EU governments, and then finally EU authorities in Brussels, to place an informal moratorium on the approval of any new GM crops for commercial production and consumption, pending the development of a more 'precautionary' regulatory environment.

This informal 1998 moratorium on new GM crop approvals in the EU has continued to the present day. There are now a dozen or more new GM crop varieties stuck in the approval pipeline; all have been successfully screened by the EU

Scientific Committee on Plants but they have not been approved for commercial release. Hoping that reluctant EU governments might consent to restart the approvals if tighter labelling rules were in effect, the EU Commission in July 2001 proposed ambitious new regulations concerning 'traceability and labeling of genetically modified organisms and traceability of food and feed products produced from GMOs' (Commission of the European Communities, 2001). The USA opposed this move because, while the moratorium has led to loss of some bulk shipments to Europe, the new regulatory proposals threaten to be less temporary, and to create heavy costs on US producers if they wish to sell to Europe.

The new L&T regulation is not based on demonstration of or belief in actual GMO risks. EU Commissioner for Health and Consumer Affairs David Byrne states repeatedly that the GM crop varieties currently approved by EU regulators for the market pose no greater food safety or environmental threat than the corresponding conventional varieties. Commissioner Byrne even describes EU consumers who continue to fear GM foods as 'irrational' (Byrne, 2001). In the autumn of 2001 the EU Commission released an official summary of the results of 81 separate scientific studies, all of which concluded there was no evidence of any new risks to human health or the environment from GM foods (Busquin, 2001). Thus EU Commission research admits that its new L&T regulation will not give EU consumers something they *need* to know, only something they *wish* to know. In December 2001, a Europe-wide survey revealed that 95% of citizens there want to know whether they are eating GM foods or not, so they can – if they wish – choose not to eat them (Biotech, 2001). The EU Commission has adopted L&T in the hope that if consumer choice is guaranteed by it, political space will open up to re-start the new GM crop approval process. In the meantime, the moratorium continues. The Council of Environmental Ministers said they would not consider lifting the moratorium until the new L&T rules were actually operating. Some European governments have even hinted that they will block GM approvals until strict environmental *liability* legislation is also in place. All this implies considerable further delay and a political process that could stretch out for several years (*Inside US Trade*, April 2002).

While new labelling rules may eventually lead GM critics in Europe to withdraw their demands for moratoriums on GMO approvals, labelling and traceability will create barriers to profitability that slow producers or processed food firms in countries exporting to Europe from adopting GM seeds or crops. Barriers set up by technical tracing and labelling requirements took effect after October 2002 for the European market and for products entering the European market. The US response – the WTO suit – came 6 months later. The costs to the US food and farm economy of complying with the EU L&T system is substantial – probably in excess of the annual value of exports now at risk. In countries such as the USA or Canada, where GM crops have until now been widely grown in an unsegregated manner, moving to this kind of product segregation could increase commodity costs by 15–50%, and final retail food prices by 9–10% (GMA, 2001). In view of such anticipated costs, the Government of Canada officially branded the new EU regulation as 'discriminatory, very costly, unworkable, and unenforceable' (Government of Canada, 2001).

Even if the USA were to win it's formal WTO challenge, and even if the EU were then to comply by relaxing the L&T directive (and lifting its moratorium) there is little guarantee that EU importers would go forward to make voluntary purchases of unlabelled GM maize or GM soybean products from the USA. The spectacle of a US diplomatic effort to use WTO rules to coerce unwilling EU consumers into taking foreign GM foods unlabelled might only harden consumer resistance to GM foods in the EU.

Conclusion

With so many conflicting interests at stake, it is not surprising that the international regulation of GM crops remains in contention. Questions about international governance of GMO affairs are spurred on by differences between governments and firms who promote GMO use, such as the USA, and those who oppose it, as do many in Europe. What agencies and rules will prevail in setting the effective standards that firms and farmers must meet to produce and sell GMO seeds, crops, and their products?

So far, policies adopted within and among states address only some of the issues raised by GMOs. These administrative or legislative rules, mostly developed by a single state or economic union have yet to lead to international codes meant to bind all signatory states. However, the increasingly complex and institutionalized state-based regulatory approaches adopted within different countries, in an attempt to satisfy varying consumer, business and government desires, are proving inadequate to guide the international market in food.

Currently, conflict exists even between international regulatory bodies. Existing international rules, such as those of the WTO governing free trade and those of the CBD and CBP protecting environmental safety, are inconsistent. Thus, before development of a cohesive international GMO regulation regime, fundamental differences regarding GM products will have to be reconciled. If scientific, economic and cultural viewpoints continue to clash, broad benefit from GM technologies will be impossible. Ideally, a solution would allow for a more unified global regime. From the standpoint of food insecure people, industrial country farmers, biotech firms and environmental activists a more predictable, but updateable set of rules regulating GM is desirable. Whether the debate over the degree of risk, and research needed to assess this, can be resolved is doubtful. As a result, while the complementary goals of safety, fair trade and improved food distribution can be served better by 'solution' with such a process-oriented quality, the political path to transform the current impasse into a legal international formula that is sustainable (as discussed in Abbott *et al.*, 2000) remains uncertain. None the less, however complex and contentious the current GM regulatory situation is, cooperative gains based on science and justifiable caution are conceivable, as a review of current international regime governance controversies demonstrates.

Governance Dynamics

In searching for a sustainable positive outcome for the global governance of GMOs, development of policy in three different areas is important – biological research into risks, *trade regulations* commensurate with those risks, and consideration of *cultural concerns* about genetic modification (Table 1.1). The current stalemate in GM crop approval worldwide can be attributed to discord within and between all of these realms. First, scientists lack (and may never be able to obtain) sufficient knowledge in order to assess GMO biosafety risks to the degree requested by sceptics. Second, international trade interests polarize political leaders, as GMO exporting countries would like to continue producing them without restriction and importers would like to restrict imports as well as production. Finally, many consumers are averse to the use of DNA recombinant technology in foods, seeing DNA as the basis for their own lives and scientific power over life's DNA make-up as scary (Nottingham, 1998). These three areas of controversy are discussed in more detail below.

First, policy makers have looked to the biological research and statistical assessments published by *scientists* to help discern the relative risk of using certain GM varieties, often in comparison with traditionally developed varieties. However, scientific research on GM crops has not proved to be an objective measure upon which to base policy decisions for two reasons. First, scientific researchers have formed no consensus on the issue of risks in spite of many studies that detect none. Because several studies have claimed to demonstrate significant biosafety risk from particular GMOs, scientists are more cautious regarding transgenic crops than a decade ago. Most GMO research has indicated GMO safety, but the private companies that produce GM seeds have paid for the bulk of this. Therefore, many are sceptical of the objectivity of

Table 1.1. Approaches used in analysing GM policy.

	Science	Social science	Popular exchange
Domain	Biology	Economics	Media, personal conversations
Theoretical tools	Statistics	Policy sciences	Value structures, cultural beliefs
Objective	Risk assessment	Risk management	Reduced awareness of risk

these studies. Critics wonder whether the companies have done all of the relevant studies and disclosed all of the important information.

A standard answer to this problem has been to request further research. Moreover, at best, such research provides an 'assessment' of the risk – probabilities of various bad effects associated with the benefits. This approach is common for scientific assessment of building codes, new medicines, or GMOs. However, at what point is there enough research to be sure the assessment is correct? And what level of uncertainty about the risk is tolerable? Adoption and use of any technology involves taking some risk, which brings one to the second problem in shaping biosafety-based policy for GMOs. Controlled scientific experimentation is good for demonstrating the presence of specifically hypothesized risks, but no amount of experimentation can demonstrate the absence of all risk. There will always be an *n*th hypothetical risk not yet tested for, or an *n*th year of hypothetically risky exposure. Proof of safety is logically impossible to establish, at least if the test is a precautionary principle that advocates protection from unexpected or 'unknown' risks (e.g. risks that have not yet been hypothesized for testing, because the technology is so new). The only way to protect completely against unknown risks is to never do anything for the first time. Therefore, scientific research can potentially provide one with some risk assessment but it cannot 'prove' the safety of GMO use. This has been the case for all previous technologies, however, and is being used in this case primarily as a scapegoat to prevent further GMO production.

Is it sensible to invoke such a precautionary principle? Is it used in other areas of new technology – computers or pharmaceuticals? No.[5] Hence use of this principle, leading to regulatory blocks in many countries, including many in Europe but also in the developing world, brings one to suspect that mercantile-based *economic* considerations are the main concern inhibiting further GMO approval and production. Although regulatory policies are often formed under the pretence of biosafety or health concern, the apparent risks do not seem to be commensurate with the precautionary stance. Perhaps by coincidence, heavy agricultural importers, such as Japan and many countries in Europe, have the most popular opposition to GM crops and governments that support more prohibitive GMO regulation such as required labelling. The precautionary biosafety approach is consistent with these countries' policies that keep high costs on imported goods, and subsidize domestic production.

From the consumer's point of view, both potential health risks, discussed above, and *cultural* concerns play a role in choice of food. GM foods, which are those made with GM crop ingredients, have been referred to as 'Frankenfoods' and the GM technology has come to have many of the same connotations as mythical monsters created by mad scientists manipulating the essence of life (Lotz, 2001). Many food companies have catered to the consumers' concerns, labelling their products with 'Contains no GMO' if they do not contain these ingredients.

Not all popular expressions of preference mediated by culture are negative toward GM crop production – many articles in the press have argued against distinguishing safety standards for GMOs from traditional varieties and some residents, in both developing and developed countries, and hail the technology as the next step toward global food security (Conway, 1998, 1999; NRC, 2002). However, because the majority of GM crops now being grown commercially were developed to decrease chemical application or increase yield, there is little consumer benefit to choosing a product containing GMOs. Therefore, given a choice, consumers are likely to buy the package stating 'Contains no GMO' rather than an unlabelled one (or one labelled 'Contains GMO', if Europe's new labelling and traceability regulations become the norm). Until GM products are actually viewed as better than others by consumers, people's fears about consuming GMO ingredients will undoubtedly preside.

The debate among and within these three orientations – scientific, economic, and cultural – creates a complex situation for policy making. A sustainable governance outcome will have

[5] In the case of nuclear power, some countries have opted for such a position, but not in the earliest stages when it was 'new'.

to take into account these differences, reshaping or advancing the interests they reflect. As GMO regulation has become a global problem a question before various participants is: how can ethical and 'enlightened' self-interested calculations reach a regulatory outcome that respective states can internalize into national regulation. Since the tension between gaining GMO benefits and protecting against its uncertain risks has no easy solution from science or any deductive processes, public agencies that negotiate bargains among competing claims and stakeholders are needed.

Given this debate what role can international agencies play in shaping sustainable worldwide regulatory regime norms? In answering this question we turn to existing organizations and procedures for policy coordination. Guidelines for creating a regime exist in a variety of locations: the WTO 1995 agreement, the 2000 CBP, and national decisions on regulatory policy in various countries whose influence on production and sales world wide is significant. National policies are crucial as anchors, barriers and examples. A mixture of diffusion and amendment of national policies must occur as part of a 'solution'. An effective regime rests on the capacity of national or regional entities to enforce policy; concurrently it builds on the greater sensitivity to the competing demands individual states experience (Hopkins, 1996; Cerny, 2001; Braman, 2002). A good case for examination is the efforts in Canada recently to work out a labelling standard. Because it included concerns of producers, distributors and consumers (Einsiedel, 2002), we forecast it has a greater prospect of sustainability than ones developed largely at the demand of consumers, as in Europe, or producers, as in the USA. Two competing solutions currently are promoted by national and international agencies: the first is the rather permissive solution in the USA, and the other is the more restrictive and precautionary alternative in the EU and several other countries. The challenge in creating a coherent regime is not only to satisfy strongly conflicting interests, but also to create it within the existing complex maze of multiple international institutions charged with harmonizing GMO policies already.

Multiple international agencies

The first components of a nascent regime addressing these diverse policy considerations can be identified. Three principal organizations affect international regulatory policy – the WTO, CBD, and Codex Alimentarius Commission. In addition, other agencies provide arenas for discussion or directly promote interests within the current debate. In Table 1.2 we sketch out these agencies, suggesting how each, with varying bases of authority, interacts with one other. Each set of agencies plays somewhat different roles in the policy process. For example, a large number of transnational and transgovernmental organizations play largely an advocacy role in policy, for instance, industrial firms and their trade associations, environmental NGOs, and international research and scientific organizations. Another important set of agencies serve largely to facilitate discussion and coordination of policy, such as the OECD, or play an informational and harmonization

Table 1.2. International agencies shaping GM policy.

Regulation	Harmonization	Advocacy
Convention on Biological Diversity (CBD)	Organization for Economic Co-operation and Development (OECD)	Consultative Group on International Agricultural Research (CGIAR)
World Trade Organization (WTO)	World Health Organization (WHO)	Scientific research/professional organizations
Codex Alimentarius Commission (Codex)	Food and Agriculture Organization (FAO)	Biotech industry organizations (e.g. BIO)
	Conferences (e.g. Rio '92)	Non-governmental organizations (e.g. Greenpeace, Rockefeller)

role in negotiations – such as the World Heath Organization (WHO) and the FAO. Basically all international agencies have weak authority; state level power remains crucial for international governance (Paarlberg, 2002). However, authority varies, among states and among international agencies. Hence, harmonization and advocacy agencies, while they have the least authority, may be helpful for conflict resolution among strong states because they pose little challenge to state-based power and may contain a helpful balance of scientific and economically trained staff to propose solutions sensible to both skills used in regulation formation.

Advocacy agencies: non-governmental intergovernmental organizations

Myriad groups, in many parts of the world, expressing views on the issue of GM crops and their regulation reflect the globalization era, and the quick path by which new technology activates political controversy. NGOs, ranging from social movements to foundations, business firms and intergovernmental groups all have varied but non-trivial access to government officials, both within states and intergovernmental agencies. Often there are official status agreements; more often than not personal ties operate. In any case, examples abound of promotion and response between advocacy groups and government agencies. In the European Commission, for example, a variety of European consumer groups, such as the European Consumer Organization (BEUC) and the European Organization of Consumer Cooperatives (EuroCoop), have been influential in successfully demanding tightened labelling and traceability as a precondition for accepting a restart of the blocked GM crop approval process.

At the international level, the number of influential NGOs that promote views on the safety and value of GMO crops is very large.[6] The concern over the power of GM to go beyond ethical uses has been a concern of advocacy organizations since at least 1963 (CHF, 1997).[7] EuropaBio, a trade group of European Bioindustries, promotes technical solutions to the EU via a technical advisory group, offering the 'current consensus' of industry to the Commission. Similar help from firms and associations occurs in the USA, and as input to bodies such as the OECD and the WTO. Food industry and farm advocates in the USA have played a strong role in advocating action against the EU moratorium and the new L&T regulation, citing the view that these are clear violations WTO trade rules. The European L&T regulation is thus challenged by the USA as violation of the SPS and TBT agreements. For example, a 20-page comment brief was submitted to the US International Trade Commission by a law firm representing US food industries argued that mandatory labels for foods derived from GMOs would violate WTO trade rules. Industry tends to see vital importance for GM development, and brings information confirming successful results quickly to the attention of relevant officials. Since the health and income of a powerful industry is in the balance, the progress made within this group is often the productive element driving new GM options and regulatory issues. Groups and governments advocating permissive adoption of GM technology even more clearly view the informal EU moratorium as a violation of the WTO's SPS Agreement, because it restricts imports while preventing the operation

[6] Among these are: Ag BioTech, Center for Food and Agricultural Research, the Consumer's Union, the Council for Responsible Genetics, Food First: Institute for Food and Development Policy, GE Food Alert, Greenpeace, Institute for Agriculture and Trade Policy, Pesticide Action Network of North America, Union of Concerned Scientists, AgBios: Agriculture and Biotech Strategies (Canada), AgCare (Canada), Canadian Food Inspection Authority, CropGen (UK), Consultative Group on International Agricultural Research, GeneWatch (UK).

[7] In *The New York Times*, 'Probing heredity's secrets', 12 September 1963, an editorial states: 'Geneticists are on the threshold of a historic breakthrough in their efforts of probe the secrets of heredity . . .. Is mankind ready for such powers? The moral, economic, and political implications of the possibilities are staggering, yet they received little organized public consideration. The danger exists that scientists will make at least some of these God-like powers available to us in the next few years, well before society – on present evidence – is likely to be even remotely prepared for the ethical and other dilemmas with which we shall be faced'.

of science-based risk assessment procedures, and because it violates the WTO guarantee against quantitative trade restrictions.[8]

Finally, various bodies are concerned with promoting public sector research and diffusion of GM crops as a way to promote food security and economic development. The Consultative Group on International Agricultural Research (CGIAR) system, prominent in the success of the 'Green Revolution' is a major advocate of moving ahead with GM, particularly allowing for crops to be developed with genetic engineering that would fit the needs of farming conditions in poor countries, where crop yields are low (Pinstrup-Andersen, 2001). Research science organizations also come forward to promote GM opportunities. Such group's positions, including the Rockefeller Foundation and other early backers of high-yielding varieties, counter the more fearful environmental protection NGOs. The public views advanced by these groups, through writings and conferences, create space for action by authoritative bodies. Since these advocacy groups are split, however, their effect is muted.

Inter-governmental organizations: harmonization

Organizations that contain specialists from eclectic fields could assist in harmonization (Fig. 1.1). Such organizations – thanks to the mix of professional capacity that they house – offer viable solutions easily acceptable to a wide range of people involved in GMO research, trade and regulation. Basically this group of intergovernmental organizations functions to aid harmonization; indeed some of the organizations exist principally to assist in policy coordination of member governments. The OECD is an example. Formed from an earlier body focused on European recovery after the Second World War, the OECD is a kind of intellectual clearing house and research arm of 30 industrialized state members. Collecting information, promoting shared policy goals on issues such as foreign assistance, trade, ageing and corruption, it holds conferences and intergovernmental exchanges to encourage policy harmony and cooperation among members. Food safety is one of its directorates, and in this sphere it has

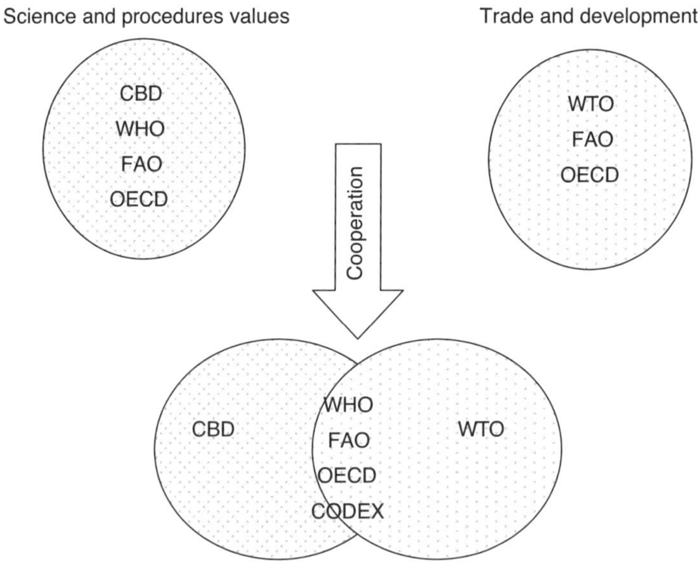

Fig. 1.1. How existing agencies surrounding GMO regulation overlap and provide opportunities for harmonization. CBD, Convention on Biodiversity; WHO, World Health Organization; FAO, Food and Agriculture Organization; OECD, Organization for Economic Co-operation and Development; WTO, World Trade Organization; Codex, Codex Alimentarius Commission.

[8] Contained in Article 11 of the General Agreement on Tariffs and Trade (GATT).

laboured to harmonize regulatory standards, such as its recommendation of a 'substantial equivalence' approach in 1993. Indeed, the OECD's Working Group on Harmonization of Regulatory Oversight in Biotechnology is currently addressing the problem of GM regulatory contention in this way (OECD, 2000). The group is primarily made up of members of ministries and agencies interested in environmental effects of GM crop use and has regularly included members from such agencies as the CBD. This particular group focuses on the international harmonization of safety studies. The working group states that if all countries use similar methods to assess risks and share their information freely, efficiency in safety analysis will be improved. In addition, cooperation could stimulate better understanding between research groups in different countries.

One of the substantial projects undertaken by the Working Group was the development of BioTrack Online, the OECD's main information exchange system, containing information on member country regulations as well as on GM crop field trials and product approvals. The group hopes that this service will be a first step towards an international harmonization of GMO risk regulation and testing.

In addition to the OECD, some UN agencies, such as the WHO and FAO, act as harmonizing international bodies, in spite of also having 'authority' via assessed budgets. Technical experts in member governments work closely with specialists in these bodies, and they often engage in promotion as well as policy coordination. Because food safety, research and biotechnology affects agriculture and medicine so prominently, both agencies have been active in holding meetings, preparing proposals and seeking common positions among members on GM issues.

The FAO has exhibited an interest in regulatory harmonization. Louise Fresco, the Assistant-Director General of the FAO Agricultural Department, notes that one cannot 'talk intelligently about GMOs if debate remains at the level of generalities' (Fresco, 2001). To avoid this, the FAO has been developing a worldwide inventory of GM crop applications, allowing a more focused study of particular products. Indeed, GM is merely a technology – it is the particular crops that must be tested for risks, not the process used to create them. Addressing each crop separately, as the FAO is proposing, is essential for risk assessment.

The FAO has also participated in multilateral programmes, such as the International Undertaking on Plant Genetic Resources, which is an attempt to make GM knowledge readily available, especially to developing countries. As a result, the position of the FAO strikes a delicate balance among several constituencies, debating GM largely on its merits. Its joint venture with the WHO for harmonizing food safety standards, Codex, has been active in this effort. Because Codex can propose language for national standards, and helps set regulations via consultations and issuance of a standard set of codes, it has become involved in the dispute over GM regulation, as we see shortly (Codex, 2002).

The WTO

The single most authoritative rule making body for trade, the WTO, does not yet have any rules in place that are specific to GM products. In part this is because both the SPS agreement the TBT agreement were negotiated within the WTO in 1986–1993, before any GM crops had been commercialized. However, although the SPS and TBT agreements do not specifically address GMOs, they are potentially applicable.

Within the WTO, the SPS Agreement permits governments to limit imports in some cases for the purpose of protecting human, animal or plant life or health, and the TBT agreement permits other actions – including packaging, marking, or labelling requirements – for the purpose of maintaining product quality or uniformity. Under the SPS agreement nations are permitted to use import restrictions to pursue any level of health or environmental protection they wish, but these import restrictions must be appropriate to that standard, they must be consistent with internal policy so as not to discriminate against imports, and they must be based on a sound scientific assessment of risk (Roberts, 1998). Article 2.2 of the SPS agreement states that measures taken by states must be 'based on scientific principles' and must not be maintained 'without sufficient scientific evidence'. However, in the absence of sufficient research, the SPS allows for temporary import restrictions, with the burden on Members to seek further research.

Import restrictions or provisions without scientific evidence of risk on an open-ended 'precautionary' basis are not allowed.

The European moratorium is clearly in contradiction with SPS regulations. Although the new L&T directive appears to be less restrictive, the USA has argued that it is against WTO rules as well. The L&T regulation seems to be a TBT agreement violation under both Article 2.1 ('like products') and Article 2.2 ('legitimate objective'). A US food industry lawyer submitted a brief to the US International Trade Commission, which argued that mandatory GM labelling on processed foods (where no traces of novel DNA and/or protein are detectable) violates the TBT's 'like products' rule (Keller and Heckman, 2000). The government of Australia agrees (Government of Australia, 2001), as do respected business and trade law experts in Canada (Isaac, 1999).

European critics of the WTO approach are now asserting that GM varieties are not in fact 'like products'. However, their argument rests on consumers' feelings that these are not like products and not on any qualitative difference between products. The TBT (Article 2.2) states that regulatory measures that disrupt trade must be necessary to fulfil a legitimate objective and a consumer desire for more information *per se* has not been viewed traditionally as a legitimate objective under this provision.

Although the USA argues that the L&T directive is in contest with WTO regulations, the WTO may be an unpromising venue within which to fight this battle. The EU Commission may not be able to sufficiently weaken the L&T at this point enough to satisfy the USA and WTO. In addition, the USA has little international diplomatic support in its opposition to L&T. Finally, the EU previously demonstrated that the WTO does not have substantial power to enforce a ruling. The WTO ruled, following a contest brought by the USA and Canada against the EU's ban on hormone-treated beef, that the EU's ban was against WTO regulations. However, the EU continued the ban and responded to the ruling by accepting trade retaliation instead.

Thus, although the WTO's SPS and TBT agreements can be applied to GMO regulation, the WTO does not have the power to ensure all parties' commitment to these principles. The EU, with the support of a more precautionary and less science-based set of regulations, the CBP, is now prepared to challenge the WTO approach.

The 2000 CBP

The authority of the WTO to rule on GM food and crop trade issues was challenged in 2000 when a new international agreement was reached on the trans-boundary movement of living GMOs (LMOs) within the CBD. This new agreement, the CBP, gives importing states more leeway than they would have under WTO rules. The CBP permits blocking of imports or requiring of labels on a 'precautionary' basis, without any firm scientific evidence of risk. The WTO is now under pressure from a number of European states to adjust its own rules to accommodate this more precautionary approach.

Drafted within the CBD, the 2000 CBP is nominally intended to protect biological diversity within GMO importing countries, yet it is an agreement that focuses almost entirely on trade (on the 'trans-boundary movement' of living GMOs, known as LMOs). Because the CBD originally emerged from negotiations launched by the United Nations Environment Programme (UNEP), the 2000 CBP was negotiated primarily by representatives of national environment ministries, rather than by trade ministries, science and technology ministries, or agricultural ministries. In part because of its emphasis on the natural environment, the CBP explicitly excludes governance of trade in pharmaceuticals for human use.

The assumption behind the CBP was that poor countries lacking biosafety capacity within their borders would need stronger means to stop potentially dangerous LMO movements into their countries at the border. The terms of the CBP were originally drafted to resemble the Basel Convention on trans-boundary movement of hazardous wastes (Gupta, 2000), and under the CBP – just as under the Basel Convention – importing countries are offered generous options for blocking or requiring labels on trans-boundary product movements. Under an Advance Informed Agreement (AIA) procedure in the CBP, governments that import LMOs intended for 'environmental release' for the first time (e.g. GM seeds or GM plant materials) are permitted to require prior notifications from exporters regarding biosafety. For LMO shipments intended for direct use as food or feed, or for processing, the AIA provision does not apply, but potential exporters are none the less obliged to provide timely information about such LMOs (through a newly created international Biosafety Clearing-house), and labels are required on shipments identifying them as possibly containing LMOs and as 'not intended for intentional introduction into the environment'. The Conference of Parties of the CBD has been instructed to produce more precise identification requirements for such LMO shipments within 2 years after the CBP comes into force.

The CBP endorses scientific risk assessment, but it also endorses 'the precautionary approach' under conditions of scientific uncertainty. In the body of the text it states repeatedly (in Articles 10 and 11) that 'lack of scientific certainty due to insufficient relevant scientific information and knowledge' should not prevent states from taking precautionary import actions against Limos (2000 CBP). The difference between this precautionary approach and the less restrictive approach required under the WTO's SPS agreement was obvious to all at the time the CBP was being negotiated. The preamble to the CBP sought to resolve these differences by stating that the CBP 'does not imply a change in the rights and obligations of a Party under any existing international agreement' (such as the SPS Agreement in WTO). But the preamble also asserts that the CBP is not to be considered 'subordinate' to those other agreements. Exporters of GM crops, led by the USA, fought hard at the time to insert a so-called 'savings clause' into the operational part of the CBP upholding the authority of existing WTO rules, but these exporters were blocked from doing so by the EU and by most developing countries (*Inside US Trade*, February 2000). The capacity of the USA to influence these CBP negotiations has been weakened by the non-participation of the USA in the CBD, owing to a Senate failure to act when the CBD was submitted for ratification in 1993.[9]

The Codex

Another agency, the Codex, though recognizing the more precautionary approach, is also intimately connected to the WTO and agreements reached there. The Codex is an intergovernmental body established in 1962, responsible for managing a joint FAO/WHO Food Standards Programme. So far the Codex has developed more than 200 food standards for commodities and more than 40 codes of hygiene and technological practice. The Codex has a nearly universal membership of 163 member states (Borgen and Veggeland, 2002). Prior to 1995 the Codex was neither a powerful nor a prominent instrument of global governance, since its standards had no force in international law. All this changed when the new SPS agreement entered into force on 1 January 1995. In an appendix to the SPS agreement (Annex A) the Codex is given responsibility for maintaining the international standards relevant to food safety that should be recognized by the WTO.[10] Henceforward, it would presumably be up to the Codex to determine whether an SPS barrier imposed by a state was sufficiently 'based on scientific principles' to be legal under WTO rules.

Since taking on this important new responsibility the character of the Codex as an institution has changed dramatically. Winning trade disputes in the WTO can now require that an argument about science first be won in the Codex. Accordingly, the activities of the Codex have now become more highly politicized along traditional trade interest lines. Whereas Codex meetings before 1995 were attended primarily by low level technocrats with backgrounds in food science and food safety, now these meetings are also attended by politically instructed delegates from ministries of trade and industry, finance, and foreign affairs (Borgen and Veggeland, 2002).

Under these more politicized circumstances the traditional Codex method of working by consensus has limitations. A Codex food

[9] The CBP has not yet entered into force. Entry into force will happen 90 days after 50 countries have ratified the treaty. As of March 2002, only 13 countries had notified their ratification, but EU member states were expected to do so by year's end, and the threshold of 50 could be met by the end of the year or early 2003.

[10] Annex A states: 'For food safety, the relevant standards, guidelines and recommendations are those established by the Codex Alimentarius Commission relating to food additives, veterinary drug and pesticide residues, contaminants, methods of analysis and sampling, and codes and guidelines of hygienic practice'.

standard can be adopted only after eight stages of consultation have been completed, and this process can generate paralysis in cases where the USA and the EU disagree. Codex procedures are also complicated by overlaps between different committees. GM food safety issues are currently being considered inside the Codex by a Committee on Food Labelling, a Committee on Food Import and Export Certification and Inspection Systems, by a Committee on General Principles, and also by a special Task Force on Foods Derived from Biotechnology.[11] Within the Food Labelling Committee, the US delegation has recently supported a Codex guideline for mandatory labelling only when the biotech-derived foods differ significantly from corresponding conventional foods in terms of composition, nutritional value, or intended use, while the EU supports draft language that would also permit mandatory labelling based on methods of production, including rDNA methods, even if there is no detectable presence of transformed DNA or protein in the end product (GAO, 2001). As of May 2002, the Food Labelling Committee review of GM food issues remained only at an early stage (at step three) in the eight-step Codex approval process. Disagreements persisted over issues of purpose, scope, definition of terms, labelling provisions, threshold levels, exemptions, and label declarations (*Food Traceability Report*, 13 May 2002, Vol. 2, No. 19).

Even as groups such as Codex struggle with the application of traditional health and safety risk concerns to GM, the EU has been moving to ask that some non-traditional risks be considered as well. Beyond human health or the environment, the EU now wants the WTO and Codex to consider 'risks to other legitimate factors' such as social values (Isaac, 1999). The EU believes mandatory labelling should be permitted to inform consumers regarding process and production methods (PPMs), such as use of rDNA to improve the genetics of plants. The WTO until now has not permitted this kind of process standard for mandatory labelling, but the EU hopes the WTO will change. In December 2001 the EU tabled a proposal to the WTO agricultural negotiations that would permit mandatory labelling for PPMs in instances where 'societal values or concerns' were at stake (*Inside US Trade*, December 2001). Under this 'consumer's right to know' approach, mandatory labels could be required even in instances where there was no scientifically established product safety or product quality concern.

Harmonization prospects

International governance seldom emerges from any one institutional source or process. Hunger policy, for example, like nuclear non-proliferation or ocean pollution policy, is embedded in a variety of instruments from treaties to ad hoc working groups (Cerny, 2001). The scope, adequacy, and effectiveness of any set of policies vary with circumstances and the level of authoritative commitment among the principal agents entrusted to carry out the policy. GM regulation has so far taken divergent paths, expressing different preferences among 'stakeholders' and the competencies of different international bodies for negotiations. Table 1.2 and Fig. 1.1 acknowledge these alternative paths through which the current controversy has evolved; they also sketch a way to see relations among them.

Given these multiple jurisdictional overlaps between existing global agreements and institutions, and given the unsettled and conflicted rules that are now in place within those institutions (e.g. the WTO's provisional approach versus the CBD/CP's precautionary approach), and given the weakness of the Codex for imposing any one approach over the objections of either the USA or the EU, what are the prospects for developing an internationally harmonized system to govern trade policy toward GMOs? One system may evolve from a test of strength between the USA and the EU. The EU policies toward GMOs – the informal moratorium on new GM crop approvals and the impending regulation on GMO labelling and traceability – provide the occasion for such a test within the WTO. Such an outcome, however, may not be

[11] France recently proposed to the Committee on General Principles that yet another working group be formed, tasked to develop a definition of traceability for use by Codex. European delegations supported this French proposal but it failed to gain support from most non-European delegations.

satisfactory: it may not lead to harmonization, and is likely to lack respect for new technology benefits. The principal alternative we see is allowing harmonization to arise from a more incremental approach, one in which different interests and perspectives are fully incorporated.

Two regulatory mindsets, one resting on economic and legal approaches and the other on science and procedural traditions assessments, inform the outlook of officials within these myriad of agencies described above, including the major three agencies with most authority. While both mindsets seek to weigh the advantages and risks attendant to GM crop introduction, their methodology and views lead to differing judgements. Moreover, in harmonizing agencies, and most critically the Codex, both viewpoints are represented. Indeed, the Codex includes many properties for developing standards that serve well a balancing of competing considerations. Thus, it offers a possible arena for less confrontational harmonization of GM crop regulation than the path of WTO dispute settlement. Its use to settle WTO challenges, a task authorized by the Uruguay Round agreement, however, exposes a danger. After 1995, in the role of a referring agency for WTO SPS challenges, Codex meetings have become 'politicized' – diplomatic officials appearing instead of scientists to discuss findings – and its results subject to similar shortcomings as WTO confrontations. Consequently, our suggestion of the Codex as a 'model' for harmonization rests on the Codex operating on its own, and not as a dispute referee.

In agencies that house specialists in both domains, collaboration between the groups could assist in harmonization within that agency, as suggested by Fig. 1.1. These agencies, furthermore, could provide groups to influence other international agencies.

Some global regulation will inevitably result. The central question is how soon, and with how much opportunities lost? Below we consider whether it will follow a WTO (US preferred) approach, the CBD (EU preferred) approach, or some compromise. At the moment a compromise seems unlikely given current rhetoric; and the US prevailing seems unlikely in the near term. Appraising alternatives, however, opens questions about desirable designs for regulation in terms of an international division of labour, incorporation of international agreement into national legislation, and responsiveness to profit and distributive needs of various groups who stand to gain from the successful use of GM in crops.

The Way Ahead: Alternatives

Two principal alternatives toward international harmonization of GMO policy exist. Both arise from the pressure of national or EU decisions whose effects have prompted efforts to coordinate and negotiate. Discussions of these occur within the WTO and other international fora, including transgovernmental networks (Hopkins, 1976; Slaughter, 2000). A harmonization of GM regulation is possible, perhaps analogous to ones found in other regulatory regimes that have been developed in the last few decades – nuclear power, pharmaceuticals, climate and electronics (Cerny, 2001; Biermann, 2002).

First alternative

The first alternative, driven by commercial and trade interests, centres on the outcome of a US–EU contest. Regime norm setting would result from unilateral measures of the two dominant players and WTO rulings. The underlying principles will be trade norms in which arguments over 'safety' concerns would be interpreted in terms of whether they constituted allowable or illegal barriers to trade. This is the most likely alternative, given recent trends. The worldwide hesitation to proceed with GM food or feed crops in countries as diverse as Brazil and China reflects a growing commercial anxiety about loss of access to big import markets such as the EU, Korea, and Japan.

A harmonization of GM trade rules around EU standards could effectively halt or even reverse the spread of GM crop production in any country aspiring to sell food or feed commodities internationally. The EU's proposed L&T Regulation makes clear that products entering the Community from countries that plant GM will have to be segregated and identity preserved according to the Regulation, and imports from non-GM countries will have to be sampled and tested to certify the absence of GM content (Commission of the European Communities, 2001). Wealthy exporting countries such as the USA that now plant GM may eventually decide not to plant GM if this kind of import regulation becomes the international norm. It will be economically cheaper for the USA to stop planting GM altogether, allowing domestic farm production costs to increase slightly, rather than ask the entire food and farm sector to pay the burdensome segregation costs and accept the legal liabilities that will be associated with an L&T system. This is particularly true since embracing an L&T system will be no guarantee that export sales of GM products will continue to be made. EU import inspectors will not stop GM foods from entering the Community so long as they are segregated, identity preserved, and labelled, but many private importers will probably opt not to buy them once a GM label is attached. So if the EU standard is embraced as a universal trade standard, the safe and prudent option for exporting countries such as the United States will be to stop planting GM crops – in effect, following the practice of farmers in the EU.

Are there some prospects for compromise or adjustment is such a dyadic struggle? Several options, suggested by some observers of WTO and global governance issues, outline possible paths for harmonization; yet each of these is flawed in the current circumstances. While all players agree that avoiding a major disruption of trans-Atlantic GMO food and feed trade will require some form of self-conscious regulatory cooperation between the USA and the EU, actually achieving this has proved illusive.

In the abstract, this needed cooperation might take one of three different forms: mutual recognition of regulatory systems, explicit regulatory harmonization, or informal mutual regulatory adjustment. The *mutual recognition* approach requires a formal international agreement in advance not to place restrictions on product trade despite differing internal product regulations. This highly decentralized, pro-trade approach to regulatory cooperation can work for well-established and familiar products, particularly those traded among similar countries enjoying high levels of political trust. For example, within the EU itself a mutual recognition agreement allows wine produced in any EU country to be sold in all 15, even where production standards differ between countries. For GMO trade across the Atlantic, however, this mutual recognition approach has little promise. Within Europe GMOs are not a familiar and trusted product, and the US approach to GMO regulation – because it does not embrace the precautionary principle – is politically unacceptable.

A second possible approach to regulatory cooperation would be *explicit regulatory harmonization*. This would require agreement on a single international standard for the approval of GMOs, and agreement on a single standard for the labelling of food and feed products containing or derived from GMOs. We have seen the limits of this approach as well. Agreement on a single international standard is unlikely on GMOs, given the entrenched differences not only between the USA and the EU, but also between the WTO and the CBP. And we have seen that the Codex, because of increasing internal politicization, has so far failed to bridge these differences.

A third possible approach, calling for formal or informal *mutual regulatory adjustments* (short of formal harmonization), falls somewhere in between the first two. In order to preserve a mutually convenient high volume of trans-Atlantic food and farm commodity (especially animal feed) trade, the USA and the EU might formally or informally agree to move their existing regulatory systems toward each other slightly. For example, the EU might agree to modify its L&T regulation to accept a threshold of GMO contamination above 1%, making the task of product segregation more affordable for the USA and other exporters. The USA could at the same time agree to require at least some form of mandatory GM labelling above those higher thresholds, for at least some foods and feeds. For example, the USA could adopt mandatory GM labelling for packaged foods and food commodities (and even feeds) at a 5% threshold, where GM content is detectable through physical testing (e.g. not in processed foods). Unfortunately, this third approach was never seriously tried by either the USA or the EU. In the summer of 2001, as the EU was moving toward a formal promulgation of its proposed L&T regulation, several US government officials did contact their EU counterparts to suggest the kinds of adjustments that might be needed to make the regulation compatible with undisrupted trans-Atlantic GM trade, but these suggested changes were nearly all rejected at the EU end.[12] And no offer of regulatory adjustment (e.g. movement toward some form of mandatory labelling) was ever made at the US end.

It may now be too late to secure the regulatory cooperation needed to avoid a major trade disruption. Indeed, a significant disruption has already taken place, in the form of bulk shipment exclusions from the EU market and also in the numerous decisions by exporters to hold back on the deployment of GM technologies not yet approved in the EU. The unresolved issue is whether a formal WTO challenge from the USA will only enflame these disruptions further. With or without such a challenge, we may now see a growing exclusion of GM-contaminated foods – and perhaps also feeds – from the important EU market, and probably also from the Japanese and Korean markets. This exclusion will be both formal (e.g. growing out of the continuing approval moratorium, or out of the new L&T regulation) and informal (e.g. growing out of private consumer preferences not to eat GM foods).

The high market segregation costs associated with the L&T approach will be even more discouraging to GM crop producers in the developing world. Building the costly parallel infrastructure needed to segregate GM from non-GM commodities will not be affordable in much of the developing world, and the sophisticated testing capacity needed to enforce segregation will not be available. The L&T regulation permits only a 1% contamination threshold, even for GMOs that are approved, and the Government of Canada has described this low tolerance level as 'costly and unworkable, particularly from a bulk commodity perspective. To determine the adventitious presence of GMOs, particularly at very low levels, such as 1%, will require time consuming and costly tests in modern state of the art labs, and may be particularly onerous for developing countries' (Government of Canada, 2001). The Government of Australia also describes the L&T Regulation as 'particularly burdensome for developing countries' (Government of Australia, 2001). The Grocery Manufacturers of America likewise warn that 'Companies in *developing* nations with potentially less infrastructure and resources would confront significantly more difficulties and costs in exporting to Europe as they develop biotech products in the future' (GMA, 2001). The obvious solution to this problem for developing countries will be to remain GM-free, never giving farmers permission to grow any GM crops in the first place.

If the past has been disappointing are there political or economic forces that might eventually move the world toward harmonization around this EU standard? EU officials can be expected to try to defend and advance their approach politically, within all the key intergovernmental venues discussed here – including WTO, CBP, Codex, OECD, and International Organization for Standardization (ISO). In each of these intergovernmental settings, however, US officials may retain sufficient political influence to mount blocking actions. In the Codex, as noted earlier, the USA has so far been able to neutralize EU efforts to push a strict L&T approach. Even within the CBP, where the US is not an official party, the fully precautionary EU approach has so far been held at bay.[13]

[12] As a first step in trying to get the EU regulations changed, on 1 June 2001, the US Undersecretary of State Alan Larson telephoned EU Commissioner for Health and Consumer Protection David Byrne to ask that the EU reconsider its proposed traceability regulations. But when a new version of the regulation was proposed by the Commission at the end of July, only a few adjustments had been made. The requirement that commodity handlers specify which GMO varieties were contained in shipments was relaxed to allow operators to specify only what varieties the shipments 'may contain'. This gives handlers an option to avoid strict market segregation among GM varieties by 'over-reporting' what their shipments might contain. Yet they will remain obliged to segregate GM from non-GM if they wish to avoid a GM label (*Inside US Trade*, 8/3/01, p. 10).

[13] In March 2002, the USA successfully blocked an EU effort within a technical group meeting to consider minimal shipping documentation needed to accompany trans-boundary movements of GMO commodities. The EU asked that shipping documents be required to list all specific GMO varieties contained rather than simply say 'may contain' GMOs but the USA (supported by Canada, Argentina, and others) was successful in putting off a decision on such a move (toward the EU L&T approach) for at least 2 years (*Inside US Trade*, 22 March 2002, pp. 28-29).

EU-based NGOs such as Greenpeace can also be expected to continue using what political muscle they have to advance national and international regulatory systems unfriendly to the planting of GM crops, but US-based biotech firms are now countering with expensive advertising and public relations offensives of their own. So in both the political and NGO/multinational corporation realms, this test of strength between the USA and the EU could be a stand-off.

Within the international marketplace, however, the EU holds a stronger hand. In any international commodity market, it will be the biggest importers that will have the greatest leverage (the option to buy from someone else), giving them the greatest influence over international product standards. In today's international agricultural markets, the EU is the biggest importer. In 2000, the EU 15 as a group imported from the rest of the world US$54.8 billion in agricultural products. The EU and Japan (another GM-averse country) together imported US$91.0 billion in agricultural products from the rest of the world in 2000. These two GM-averse importers thus offered to the rest of the world a commercial market more than twice as large as the commercial market offered by the GM-friendly USA – which imported a total of just US$44.9 billion worth of agricultural products in 2000 (FAOSTAT Database). Ultimately it could be this combined European and Japanese purchasing power that will pull international GMO trade standards in the direction of EU preferences. The *Wall Street Journal* recently acknowledged Europe's influence over global food safety standards, but mistook the reason for this influence because of a failure to appreciate the greater size of EU agricultural imports compared with US imports.[14]

The EU's proposed L&T directive demonstrates the precautionary approach that these countries take toward GMOs. If an outright contest between states over proper international GMO regulation occurs, further prohibitive measures are likely to become the global norm. In order to allow GM technology to develop but also to protect people and the environment from possible risk, we propose another path toward harmonization that may be a more promising, egalitarian one.

Second alternative

The second alternative would be for rules principally grounded in science and with a commitment to their regular updating. Policy findings and regulations will not be authored by science, but the formula for governance can be driven by science-based risk assessment, and reasonable standards for its management. To reduce uncertainty for producers and manufacturers harmonization of regulations would still be needed, and occur through the decentralized adoption by states of compatible rules. The broad norms for such a regime, and the actual rules reached, however, would be constrained not only by the L&T debate within WTO principles, but also the search for biosafety and other risk reductions by science. One governance prospect is the discussion and outcomes that occur for advising on regulation under the Codex. In this forum, as it has deliberated in the past 2 years representatives have tried to balance both WTO trade concerns and CBD bio and ecology safety protections.

This latter alternative is promising from lessons learned in rule making in other areas of global governance. For example, international institutions work best when several desiderata are taken into account collectively. The FAO has a broader base for calling on specialists, a wider range of countries prospectively represented; it has clear ties to the enforcement agencies for food and bio safety in countries, compared with the WTO. It also can meet and act with less cumbersome procedures, and take uncertainty into account more formally. While economic actors may prefer rigid and durable rules to enhance certainty of the results of their investments, consumers, producers and long-term food security interests may be best served by a governance

[14] In an April 2002 article, The *Wall Street Journal* pointed out accurately that 'Americans may not realize it, but the rules governing the food they eat . . . are increasingly set in Brussels'. What this article failed to mention is the connection between this rule setting power and the greater size of European food import markets.

process that regularly requires updating of rules themselves, based on changes in scientific evidence, hence building in flexibility for change (Koremenos *et al.*, 2001, pp. 775–779; 1032–1041).

Such changes would not constitute a new regime, as the EU tracing and labelling plan does. It could, however, allow for the high costs of a segregated use of GMO crops, should this become a practice, to be progressively reduced as experience and further research gave more scientific evidence that the GM products presented no evidence of causing harm. Further, their benefits as a crop could also be taken into account in the risk assessment of the early phase of scientific appraisals. The risk management tasks would still engage lawyers, and officials specialized in the codes and enforcement practicalities, but their efforts would be less subject to premature and hard-to-reverse qualities.

For this second alternative, one promising development was the outcome of the March, 2002 Meeting of the Codex Intergovernmental Task Force on 'Foods Derived From Biotechnology'. The principles it adopted for risk analysis and guidelines for conducting safety assessments of foods derived from biotechnology may allow compromise. The meeting included the concept of traceability in the standards without explicitly referring to it. The principles for risk analysis would require authorities to take into account uncertainties identified in the safety assessment and implement 'appropriate risk management' measures, treating pre-market safety assessment on a case-by-case basis. This compromise is to be submitted to the Codex for adoption at its meeting in July 2003.

This more process-based approach would require regular updating of regulations and a release, setting standards as well as approvals on an ongoing basis as further information becomes available. Risk assessment changes would be tied to rules about risk management, with issues about safety central, but not based on 'principles' that allowed for withholding new technology for commercial reasons that relied on uncertainty as a precautionary standard. The varying standards used over time with drugs – allowing quicker release and less caution for those with most promise – is one example of how a flexible, yet science-based approach might work. The alternative, a blocking of GM crops on principle seems to ignore – and will most certainly not benefit from – the potential global gains from further development of the technology. Halting GM technology development now will eliminate this tool from the global repertoire. In a conservative application of a risk-based but not 'principle'-based approach, such as current policies surrounding pharmaceuticals and nuclear power, the issue of harm to users would be seriously considered in the formation of regulation. This approach would ensure that both potential risk and benefit of different GM crops were considered, allowing appropriate use and development of this technology now and in the future.

Conclusion

Turning the clock back to 1995 might be an escape from the trade confrontation. Costs to US agriculture, although not the agbiotechnology industry, would be politically tolerable, especially in light of the new US Farm Bill. However, it could be a costly long-term outcome for developing country farmers and perhaps impossible anyhow, as GM crops have already been released to the environment. If GM technology is prohibitively restricted now, its potential to help food-insecure regions produce more food in the future will be eliminated. A less restrictive path toward regulatory harmonization could allow compromise between national economic interests and consumers' cultural and safety concerns without eliminating the possibility for further GMO development. Inclusion of a variety of players in the formation of regulation, perhaps especially including the consumers themselves in the process, will be essential in achieving a satisfactory outcome.

Whichever path is followed, international agreement will be reached by conventional bargaining over texts and intragovernmental procedures for regulation (Keohane *et al.*, 2000; Cerny, 2001). Negotiations will centre on common practices for government action, such as labelling and safety approval procedures. Outcomes will be resolved by narrowing differences, or where necessary by arbitration such as under WTO dispute rules. International agreements will need to be formed and used to harmonize current practices and reduce uncertainty that is currently inhibiting the development and production of GM crops.

Both those worried about incautious and callow introduction of 'Franken-foods' and those eager to expand the use of new GM technology, especially for use by poor-country farmers, want stability in a regulatory regime, and accept the value of careful drafting of codes and statements. At the WTO, and we propose at the Codex and CBD arenas, draft documents, with controversial items bracketed, should allow principal alternatives – including the two discussed above – to be explored. Clearly we favour a more participatory and scientifically based outcome, one that allows flexibility, along with the inevitable escape clauses, as the next outcome in the GM regulatory regime.

If we can move from the fragmented and contested situation of uncertainty and competing standards of risk management, principally between the USA and Europe, to an international regime with more sustainable harmony in both procedures and principles, the potential in GM can be rejuvenated. The alternative to the trade-driven, EU-dominated regulation standards is to empower harmonizing agencies of the international system – OECD, Codex, and Third World oriented IGOs, such as CGIAR – to take the central position in hammering out the standards and regulations for risk assessment and risk management, including safety precautions and labelling requirements. This would occur while paying more attention to the risks attendant to the new technology than occurred during the last decade. At the June 2002 Plus 5 World Food Summit, for instance, delegates recognized the value of biotechnology products for help to impoverished farmers, but also avoided endorsement of any regulatory principles. Beneficial outcomes of such a process would require adoption and subsequent ratification by the majority of international parties. Incorporating the agreed principles into national legislation and policy would follow.

For GM regulation, a single WTO venue, and single debate process is not likely to produce harmonization. The alternatives we envisage are either a model of national decisions controlling outcomes and forcing adaptations to the standard of the most powerful – Europe, in this context or, alternatively, an international forum, such as Codex, allowing key national representatives to meet under the auspices of a joint umbrella that is specialized, focused and sustainable. Unlike the WTO or 'Summit' meetings, regime building in this controversial realm requires an incremental and non-confrontational approach.

References

Abbott, K.W., Keohane, R.O., Moravcsik, A., Slaughter, A.M. and Snidal, D. (2000) The concept of legalization. *International Organization* 54(3), 401–419.

Biermann, F. (2002) Institutions for scientific advice. *Global Governance* 8(2), 195–220.

Biotech (2001) Europeans want right to choose on GMOs. See www.checkbiotech.org 19 December.

Braman, S. (2002) Informational Meta-Technologies, International relations, and genetic power: the case of biotechnologies. In: Rosenau, J.N. and Singh, J.P. (eds) *Information Technologies and Global Politics*. State University of New York, Albany, pp. 91–112.

Borgen, S.O. and Veggeland, F. (2002) Changing the Codex: the role of international institutions. Paper presented at the *10th National Conference in Political Science*, Solstrand Fjordhotel in Os, Norway, 7–9 January 2001.

Busquin, P. (2002) EC-Sponsored Research on the Safety of GMOs. At http://europa.eu.int/comm/research/quality-of-life/gmo/index.html

Byrne, D.A. (2001) A European Approach to Food Safety and GMOs. Speech of EU Commissioner for Health and Consumer Protection. *Europa* 12 October 2001.

Cerny, P. (2001) From 'Iron Triangles' to 'Golden Pentangles'? Globalizing the Policy Process. *Global Governance* 7(4), 397–410.

Codex Alimentarius Commission (2002) Joint FAO/WHO Food Standard Programme Report of the Intergovernmental Task Force on Foods Derived from Biotechnology. *Alinform* 03/34, Yokohama, 4–8 March.

Coleman, R.J. (2002) The U.S., Europe, and Precaution: a Comparative Case Study Analysis of the Management of Risk in a Complex World. Address by Mr Robert J. Coleman, Director General for Health and Consumer Protection, European Union Commission. Grand Hotel-Bruges, 11–12 January, 2002.

Colorado State University (2002) Information and data on 'Transgenic Crops: an introduction and

Resource Guide' At http://www.colostate.edu/programs/lifesciences/TransgenicCrops

Commission of the European Communities (2001) Proposal for a Regulation of the European Parliament and of the Council concerning traceability and labelling of genetically modified organisms and traceability of food and feed products produced from genetically modified organisms and amending Directive 2001/18/EC. COM(2001) 182 final, Brussels, 25.7.2001.

Conway, G. (1998) *The Doubly Green Revolution: Food for All in the Twenty-first Century*. Comstock Publishing Associates, Ithaca, New York.

Cornell University (2002) 'Genetically engineered organisms: public issues education project'. At www.comm.cornell.edu/gmo/crops/eating

Dossier, P. (1990) *Miracle or Menace? Biotechnology and the Third World*. The Panos Institute, Washington, DC.

Einsiedel, E.F. (2002) A proposed voluntary labelling standard for GM food. Ravello, International Consortium on Agricultural Research, July, 2002.

Flint, J. (1998) Agricultural Industry Giants Moving towards Genetic Monopolism', in 28.06.1998, in *Telopolis*. http://www.heise.de/tp/english/inhalt/co/2385/1.html 28 June.

Fresco, L. (2001) Genetically modified crops. *FAO Agriculture 21 Magazine*. February. http://www.fao.org/ag/magazine/011sp/htm

Fukuyama, F. (2002) Gene regime. *Foreign Policy* March/April, No. 129, 57-63.

[GAO] General Accounting Office (2001) *International Trade: Concerns over Biotechnology Challenge U.S. Agricultural Exports*. Report to the Ranking Minority Member, Committee on Finance, US Senate. GAO-01-727. United States General Accounting Office (GAO). June, 2001.

Government of Australia (2001) *Australian Response to the European Communities' TBT Notifications on GM labelling and traceability*. http://www.foodtraceabilityreport.com/document_km.asp?id=849

Government of Canada (2001) Comments on proposed EC Regulations concerning genetically modified food and feed. http://www.foodtraceabilityreport.com/document_km.asp?id=850

Greenpeace (2002) http://www.greenpeace.org. 'Greenpeace Stops US Shipment of Maize to Mexico to Eliminate Source of Genetic Contamination'. 6 April 2002. Which cites an article in *Nature* (2002) questioning a study of GM diffusion into maize varieties in Mexico. http://www.greenpeaceusa.org/ge/overviewtext.htm

[GMA] Grocery Manufacturers of America (2001) Re: Traceability & labelling provisions of proposed regulations on genetically modified food & feed and traceability & labelling. Letter to the EU Commission, 21 December, 2001.

Gupta, A. (2000) Creating a global biosafety regime. *International Journal of Biotechnology* 2(1–3), 205–230.

Hilos, T. (2002) World food summit endorses biotechnology to combat hunger. 12 June *Inter Press Service*.

Hopkins, R.F. (1976) The international role of 'domestic' bureaucracy. *International Organization* 30(3), 405–432.

Hopkins, R.F. (1996) Reform in the international food aid regime. In: Haas, P.M. (ed.) *Knowledge, Power and International Cooperation*. University of South Carolina Press, Columbia, pp. 225–264.

Inside U.S. Trade (2002) 1 February and 5 April.

Inside U.S. Trade (2001) EU paper calls for WTO approval for process-based labelling schemes. 14 December, p. 13.

Inside U.S. Trade (2000) Vol. 18, 4 February, p. 25.

Isaac, G.E. (1999) Food Safety and Trade Policy: Agricultural Biotechnology Issues. Prepared for Seminar on Food Safety and Trade Policy, Consumers' Association & International Trade Policy Unit, London School of Economics, 1 July 1999. Revised October 1999.

James, C. (2001a) Global Review of Commercialized Transgenic Crops: 2000. ISAAA Brief No. 23-2001. International Service for the Acquisition of Agri-biotech Applications (ISAAA), Ithaca, New York.

James, C. (2001b) Preview: Global Review of Commercialized Transgenic Crops: 2001. ISAAA Brief No. 24-2001. International Service for the Acquisition of Agri-biotech Applications (ISAAA), Ithaca, New York.

Keller and Heckman (2000) Comments of Keller and Heckmann LLP on Non-Tariff Trade Barriers for Processed Foods and Beverages: Mandatory Bioengineered Food Labelling Requirements. Brief Before the United States International Trade Commission, Investigation No. 332–421.

Keohane, R.O., Moravcsik, A. and Slaughter, A.M. (2000) Legalized dispute resolution: interstate and transnational. *International Politics* 54(3), 457–488.

Koremenos, B., Lipson, C. and Snidal, D. (2001) The rational design of international institutions. *International Organization* 55(4), 761–799.

Lotz, H. (2001) Free trade and frankenfood. *International Studies Convention*, Hong Kong, July.

[NRC] National Research Council (2002) *Environmental Effects of Transgenic Plants: the Scope and Adequacy of Regulation*. Committee on Environmental Impacts Associated with Commercialization of Transgenic Plants, Board on Agriculture and Natural Resources.

Nottingham, S. (1998) *Eat Your Genes: Genetically Modified Food is Entering our Diet*. Zen Books, New York.

OECD (2000) *Report of the Working Group on Harmonisation of Regulatory Oversight in Biotechnology*. Organization for Economic Cooperation and Development. 25 May. http://www1.oecd.org/subject/biotech/report_workgroup.pdf

OECD (2001) Modern biotechnology and agricultural markets: a discussion of selected issues. *OECD Working Papers* No 8.

Pardey, P.G. (ed.) (2001) *The Future of Food*. IFPRI, Washington, DC.

Paarlberg, R. (2001) *The Politics of Precaution: Genetically Modified Crops in Developing Countries*. Johns Hopkins University Press, Baltimore, Maryland.

Paarlberg, R.L. (2002) *Governance and Food Security in an Age of Globalization*. IFPRI, Discussion Paper No. 36, Washington, DC.

Pinstrup-Andersen, P. (2001) *Seeds of Contention: World Hunger and the Global Controversy over GM crops*. Johns Hopkins University Press, Baltimore, Maryland.

Porphyry, R. (2001) Golden Chance/Levels of Vitamin A Required To Prevent Deficiency. *AgBioView Post*, 29 July http://www.biotech-info.net/golden_chance.html

Raine, M. (2002) Seed growers see little good in GM wheat. *The Western Producer* 17 January.

Roberts, D. (1998) Preliminary assessment of the effects of the WTO Agreement on Sanitary and Phytosanitary Trade Regulations. *Journal of International Economic Law* 1, 377-405.

Secretariat of the Convention on Biological Diversity (2000) *Cartagena Protocol on Biosafety to the Convention on Biological Diversity*: text and annexes. Secretariat of the Convention on Biological Diversity. Montreal, Canada.

Slaughter, A.-M. (2000) Governing the global economy through government networks. In: Byers, M. (ed.) *The Role of Law in International Politics*. Oxford University Press, pp. 177-205.

The Times (2000) 12 July. See the website: http://www.jcue.unc.edu/CHEM%20006D/pro-GMOlinks.html

2 The Evolving GMO Food Trade Policy Debate: Towards a Global Regulatory Regime?[1]

Philip Katz,[1] Penny Macdonald[2] and Gordon Mackenzie[3]
[1]Crowell & Moring LLP, Washington, DC, USA; [2]C&M International, Washington, DC, USA; [3]Crowell & Moring, Brussels, Belgium

Introduction

The lack of a comprehensive and consistent trade policy regime governing foods derived from genetically modified organisms (GMOs) has hindered the growth of the international trade in such products, and has helped dampen consumer perceptions of those products. Key players remain unable, or perhaps unwilling, to agree on how to devise an effective set of international rules, let alone the substance of those rules. As a result, GMO foods are still a comparatively marginalized commodity, held hostage by political sensitivities and broader concerns relating to the thorny subject of agricultural trade.

Trade policy players addressing this issue are confronted by fundamental questions, including: what is the best forum for GMO trade rules; how to balance domestic political sensitivities against essential market access obligations; and how to allay consumer concerns without resorting to protectionism. Progress is made even more difficult by the active involvement of environmentalists and anti-GMO consumer groups that have staked out clear positions against the trade. In many instances, these activists have succeeded in making the question of importing GMO foods a domestic political issue, thus ensuring its integration into political and election discourse. Unfortunately, the focus on domestic politics has often generated more heat than light – contributing little to considered debate on the underlying policy issues. The GMO issue has been one that aptly demonstrates the fact that trading rules cannot solve a case where public suspicions and emotions take over.

Progress in reaching global agreement on trade in GMO food has also been slowed by the amount of resources and attention devoted to the Doha Development Agenda, the new trade Round currently evolving under the auspices of the World Trade Organization (WTO) in Geneva. Not surprisingly, opening up a significant portion of the entire multilateral trade policy umbrella for negotiation has taken precedence over sectoral negotiations. Although there may be residual hope that the new Round will address issues associated with GMO food trade, immediate positive multilateral developments in the GMO policy debate seem quite unlikely. If

[1] As this chapter is based on a roundtable discussion, it is fairly informal in presentation, and we have for the most part dispensed with footnoted citations to statutes, regulations, and the like. We are happy to provide such sources in response to a request for more information.

 The Regulation of Agricultural Biotechnology (eds R.E. Evenson and V. Santaniello)

anything is to happen, it will require sustained, concerted effort by those countries seeking to make progress. At the very least, there must be US commitment to an international policy process, and solid support by the US biotech industry. Without that, the GMO food issue will either never be addressed, or will be traded off against other, competing issues.

This chapter first examines the current US and European regulatory regimes, then examines how international agreement might best be achieved. In the end, we believe that development of a workable international trade policy for GMO foods may depend less on a successful international framework and more on one side establishing a policy regime that others can embrace as a starting point for real negotiation.

US Regulation of GMO Foods

Because US domestic policies regarding GMO foods help to frame the international debate, it is worthwhile to start with a brief summary of US law. Generally, federal regulation of food in the USA is conducted by the Food and Drug Administration (FDA) and the Department of Agriculture. With regard to GMO food products, as opposed to plants, FDA has the primary role.[2] In general, one of FDA's central missions is to ensure that foods sold to US consumers are safe and properly labelled. In that regard, FDA concerns itself with finished food products, as well as components and ingredients of such products. As a general rule, food does not require prior FDA approval before being marketed. Rather, FDA takes regulatory action when it determines that a food already on the market is 'misbranded' or 'adulterated', two statutorily defined terms. Food is considered misbranded if, among other things, it is not labelled in accordance with FDA regulations, or if its labelling is false or misleading. Labelling can be misleading not only by virtue of statements made, but also by way of material omissions. Adulterated food includes food with an unsafe food additive (see discussion below), and food containing a deleterious substance that may be injurious to health.

There is one important circumstance, however, in which food does require premarket approval. Food additives, which the statute defines as substances that either directly or indirectly become a component of food and affect its characteristics, typically include products added directly to foods (such as texturizers and flavouring agents), as well as substances used in packaging or food preparation that can migrate to the food. Generally, food additives require FDA approval before they can be used in food products. If an article meets the definition of 'food additive', it is considered unsafe unless FDA has issued a regulation establishing the conditions for its safe use.

There is one key exception to all this – an article that would otherwise be a food additive is *not* if the substance is 'generally recognized as having been shown to be safe under the conditions of its intended use'. (The shorthand for this status is 'GRAS'.) There is a procedure by which FDA can pass judgement on whether a food/component/ingredient is GRAS and therefore not requiring premarket approval, but it is not mandatory. Rather, a company may simply satisfy itself that a product is GRAS and go forward with the marketing of the food. As a middle ground, FDA has created an informal process by which a company can – but is not required to – notify the agency that the company has determined a product to be GRAS. This procedure gives FDA an opportunity to evaluate the company's determination.

With regard to bioengineered food, FDA in January 2001 issued proposals and guidance that deal with the premarket notification and labelling issues, which are the two means by which FDA would most likely regulate GMO foods. These actions are based on a policy statement issued in 1992 and still in force, which stated that: (i) bioengineered food would be regulated no differently than foods developed through traditional plant breeding; and (ii) decisions regarding bioengineered foods would be based on the characteristics of the food, not how it was developed. As a class, therefore,

[2] The roles of the Department of Agriculture and the Environmental Protection Agency, although significant, are beyond the scope of this chapter.

bioengineered foods were not subject to premarket approval, nor did they require special labelling. Essentially, under the 1992 policy statement, GMO foods require premarket approval only if the genetic modification creates a substance that falls within the definition of a food additive. And in that regard, FDA stated its expectation that most substances resulting from genetic modification will be the same or very similar to substances commonly found in food – such as proteins, fats, oils, and carbohydrates – and therefore GRAS, which means that they would not be considered food additives and would not require premarket approval.

The 1992 policy has been challenged in court and upheld, and although legislation has been introduced to change the statute, nothing has passed. For example, FDA's determination that GMO foods need not be labelled in any special way was upheld in a September 2002 decision, in which the court concluded that consumer demand or desire for labelling to identify the GMO status of a food did not make the fact of genetic modification a material fact.[3] None the less, the actions taken by FDA in 2001 (see discussion below) clearly reflect FDA's recognition that there is strong consumer sentiment regarding these issues. At the same time, they reflect the fact that: (i) the agency generally considers GMO foods to be safe; and (ii) even if the agency's view were different, it considers itself restrained in what it can do by the statute.

In January 2001, FDA issued a draft guidance to industry as to the labelling of GMO foods. To a large degree, the guidance reaffirms the 1992 policy, stating that, as a class, GMO foods do not require special labelling, because they neither differ in any material way from other foods nor pose safety risks greater than those from traditional plant breeding. At the same time, the guidance acknowledges that there are circumstances in which a bioengineered food might be mislabelled. By way of example, FDA notes that where a GMO food contains an allergen that consumers would not ordinarily expect to find in that food, the label should disclose the presence of that allergen, and failure to do so might constitute a material omission. Similarly, if a GMO food had a significantly different nutritional property from non-GMO food of the same type, that fact should be disclosed.

Although FDA has determined that GMO foods do not require labelling to highlight that fact, the agency also recognizes that some manufacturers might want to label food to tout the absence of genetic modification. In that regard, the agency expresses concern that such statements be truthful and not misleading. With that in mind, the guidance states FDA's view that claiming a food is 'GMO free' is unlikely to be true, in light of the fact that even traditional plant breeding constitutes genetic modification. For that reason, FDA believes that a phrase such as 'not bioengineered' may be more accurate. Even then, FDA warns that such a statement, while true, can be misleading if, for example, it implies that the food is superior to GMO food.

Along with the labelling guidance, FDA issued a proposed regulatory scheme mandating premarket notification. The proposal would require submissions of data and information 120 days before marketing any plant-derived bioengineered food for humans or animals. This premarket notification would build on the existing, informal consultations between industry and FDA that have been taking place for years. The premarket notification procedure would give FDA time to review information and determine, before the product was introduced to the market, whether the agency concurred with the company's conclusion that the product is GRAS. Although the regulations would not require premarket approval of bioengineered foods, it would give FDA the opportunity to take steps to stop market introduction if the agency concluded that the product was not GRAS, but was a food additive requiring premarket approval.

The proposal reiterates FDA's presumption that bioengineered foods are, in fact, GRAS, but also identifies possible issues derived from advancing technology. For example, because genetic material from a wider range of sources is now possible, there is a greater possibility that the resulting food contains something not historically found in food, or perhaps something now found at significantly higher levels. There also is

[3] The premise for the lawsuit was that GMO food was not labelled, therefore constituting a lack of a material piece of information, thus rendering the food misbranded and therefore not legally sold.

an increased risk of unintended changes to the food, which might make the food adulterated or misbranded. FDA also notes that most of the genetic modifications the agency previously reviewed involved agronomic traits, that is, characteristics of the plant, not the food from the plant. More recently, however, FDA has been seeing proposed modifications to the food itself, such as changing the quality of a protein or fibre in the food, or increasing the carotenoid content or the amount of fruit solids. Such genetic modifications affecting the food itself are more likely to raise regulatory issues, FDA believes.

Clearly, FDA's position continues to be generally pro-GMO. At the same time, the agency has identified emerging issues that may well lead to outcomes in which particular GMO foods require special labelling or premarket approval.

The EU Regime

European Union (EU) or 'Community' legislation is adopted through a system of interaction between the three main EU institutions: the European Parliament, the Council of the European Union (i.e. the representatives of the 15 EU Member States at ministerial level) and the European Commission. In most cases, the European Commission initiates legislative proposals that are decided jointly by the Council and the European Parliament and the most common Community legislative measures are Regulations (acts that are binding in their entirety and are immediately applicable throughout the EU) and Directives (acts that require the modification or establishment of national measures, generally for harmonization purposes).

Although the Council cannot adopt such measures without the intervention of the other EU institutions, a large enough group of Member States can block any action by the Council. Moreover, with regard to GMOs, the relevant Community laws also allow for the intervention of Member States in the approval of individual marketing authorizations. The decision by a group of Member States to block such approvals has created a certain amount of institutional paralysis. In order to restart the approval process, therefore, the EU is expected to adopt in 2003 controversial new laws on labelling and traceability (i.e. tracking GMOs and products produced from GMOs at all stages of the placing on the market).

Community legislation in the area of GMOs first focused on ending the disparity among Member States on their initial authorization. In 1990, Community legislation ('Directive 90/220') was adopted to harmonize the national rules on the deliberate release of GMOs for experimental purposes, and for their placing on the market. Directive 90/220 established an approval process for assessing the risks to human health and the environment before any GMO or product containing GMOs could be released into the environment or placed on the market. Directive 90/220 did not apply, however, to foodstuffs that are derived from GMOs, such as tomatoes or tomato ketchup derived from GMO seed.

GMO food became regulated at an EU level with the 1997 adoption of the Novel Foods Regulation. 'Novel foods' (i.e. foodstuffs consisting of GMOs or foodstuffs that are intentionally genetically altered) are subject to separate authorization and labelling rules. Where a food is derived from, but no longer contains, GMOs, the food may be subject to a simplified authorization procedure, if the food is 'substantially equivalent' to existing foods with respect to composition, nutritional value, metabolism, intended use and the level of undesirable substances. In such cases, the company need only notify the European Commission when placing the product on the market and provide either scientific justification that the product is substantially equivalent or an opinion to the same effect, delivered from the competent authorities of a Member State.

The labelling of GMO foods has been regulated by a variety of Community legislative instruments. Directive 97/35 first made labelling mandatory for the presence of GMOs as such or in a product. Additionally, the Novel Foods Regulation provides separate labelling requirements for food and food ingredients that contain or consist of GMOs, although it does not contain any guidance on how the labelling rules would be applied in practice. Additionally, specific rules have been introduced for the labelling of food products containing genetically modified soybeans and maize. These labelling obligations are triggered by the presence of DNA or protein resulting from genetic modification, and the rules have provided the model for generally applicable rules governing the labelling of all foods and food ingredients (including additives and flavouring) derived from GMO. Under the these rules, the adventitious presence of minute traces of approved GMOs in ingredients obtained from non-GMO sources does not require labelling provided that its presence is below a *de minimis* threshold of 1% and the operators can demonstrate that they have taken appropriate steps to avoid the presence of GMO material.

A total of 18 authorizations for commercial release (either for experimental purposes or for placing on the market) of GMOs have been granted since Directive 90/220 entered into force. There are another 13 applications pending. However, no authorizations have been issued since October 1998 because of a *de facto* moratorium imposed by a group of five EU Member States (Denmark, France, Greece, Italy and Luxembourg). In June 1999, these countries announced that they would effectively block all new GMO approvals until the European Commission proposed satisfactory legislation for traceability and labelling of GMOs and products derived therefrom.

In response to the paralysis in GMO approvals, the European Commission announced in its *White Paper on Food Safety*, published in January 2000, that it would introduce a number of additional legislative proposals covering GMOs and GMO food and feed. Insofar as GMO approvals are concerned, Directive 90/220/EEC was replaced in March 2001 by Directive 2001/18/EC.

According to Directive 2001/18/EC, when placing a GMO on the market, the manufacturer or importer to the EU must now include in the notification to the competent national authority a full environmental risk assessment in which an evaluation is made of the foreseeable 'risks' which a release to the market of the specific GMOs involved may cause to human health and the environment. The national authority is to conduct its own investigation and prepare an assessment report that serves as the basis for consideration by the Member States.

If the national authority decides that a GMO should not be placed on the market, the assessment report will state the reasons and the notification will be rejected. If the report is in favour of placing the GMO on the market, the competent authorities in the other Member States or the European Commission may present reasoned objections to the assessment report. Where there are a number of objections, the European Commission will arrange meetings with a view to reaching an agreement on the conditions for marketing of the GMO. Once authorized, a GMO obtains free circulation throughout the EU. Individual Member States cannot veto such approvals. They can nevertheless temporarily prohibit distribution within the country, if there is a basis for asserting that the product, although approved, 'constitutes a risk to human health or the environment'. The European Commission has 3 months within which to determine whether the prohibition can stand.

Directive 2001/18/EC does not provide a complete framework for traceability. For that reason, adoption of this new Directive did not signal the end of the moratorium. In fact, since adoption of the Directive, Austria, Germany and Belgium have joined the five countries blocking new authorizations. With the goal, therefore, of finally ending the moratorium, the European Commission in July 2001 proposed a new regulation that would cover traceability and labelling of GMOs and products derived from GMOs. In particular, the proposal provides: (i) a new authorization process for GMO food and feed; (ii) a harmonized system for tracing GMOs; and (iii) rules for the labelling of GMO food and GMO feed (e.g. genetically modified soybean meal and any compound that includes in its composition such soybean meal).[4] After a year and a half of negotiations, the Member States finally reached, in December 2002, political agreement on the proposals.

The European Commission's intention had been to establish a single, streamlined authorization procedure for all marketing applications of a GMO and its possible food and feed uses through a 'one door – one key' procedure. It

[4] Despite calls by certain consumer and environmental groups, these proposals do not include provisions addressing the question of liability for 'damage' caused by GMOs. The European Commission has instead issued a separate, more general proposal on environmental liability.

proposed to centralize filings at the European Food Authority (the recently created EU entity responsible for reviewing a wide range of food safety issues), with the European Commission preparing draft decisions. This proposal, however, met significant problems. The legislative procedure required to approve such a change towards a centralized procedure would have excluded the role of the European Parliament in the legislative process and would have required unanimity by all 15 EU Member States. As a result, a decentralized authorization procedure will remain in place, though subject to new strict time limits.

As regards traceability and labelling, the draft regulation is aimed at: (i) ensuring the traceability of GMOs from operator to operator;[5] and (ii) the creation of a comprehensive system of labelling for GMOs.

Under the current system, GMO labelling is based on the detectability of genetically modified DNA or protein in the final food product (i.e. highly processed foodstuffs produced from GMO material such as highly refined oils do not need to be labelled). Under the new rules, labelling will be extended to all genetically modified food or feed, *irrespective of detectability of the genetically modified DNA or protein in the final food product*. As a result, products such as sugar and highly refined oils produced from GMOs will have to be labelled. The new rules also require genetically modified feed, currently not covered by labelling rules, to be labelled according to the same principles. It does not cover, however, the labelling of meat or milk from an animal fed with GMO feed, nor would it cover non-food derivatives (e.g. tobacco).

According to the new rules, business operators must transmit and retain information about products that contain or are produced from GMOs at each stage of the placing on the market. In particular, the requirements are that:

- Operators shall have systems and procedures in place to identify to whom and from whom products are made available.
- For GMOs intended for deliberate release in the environment, operators must transmit specified information, including their unique code, on the identity of the individual GMO(s) incorporated in a product.
- For GMOs intended for food, feed or processing, business operators may either transmit the specified information indicated above or transmit a declaration that the product shall only be used as food or feed or for processing, together with the identity of the GMO(s) that the product may contain.
- For food and feed produced from GMO(s) operators shall inform the next operator in the chain that the product is produced from GMOs.
- Operators shall retain the information for a period of 5 years and make it available to competent authorities on demand.

The European Commission intends to develop technical guidance on sampling and testing prior to the application of the proposed regulation.[6]

The new rules also introduce revised thresholds below which products containing adventitious or technically unavoidable traces of GMOs do not need to be labelled. In particular, it proposes to lower the permitted threshold of adventitious but approved GMOs from 1.0% to 0.9%. The new legislation also introduces a tolerance threshold for GMO material in food that has not been authorized. Initially, the European Commission had proposed a threshold of 1% provided that the GMO material had received a favourable EU scientific risk assessment and that the operator can demonstrate that its presence was technically unavoidable. The Council has decided, however, to lower this threshold to 0.5% and to limit its application to 3 years. The European Commission will review the operation of this clause and make a proposal for its extension if appropriate. The European Commission has accepted that compliance with the new thresholds will require in some cases changes in farming practice by obliging farmers

[5] The European Commission takes the view that it is building on the general system of traceability found in general EC law governing food (e.g. for beef products) and feed.

[6] On 4 December 2002, the European Commission announced the launching of a European network of GMO laboratories to improve traceability in the food chain (IP/02/1795).

to set up monitoring systems as well as covering additional insurance needs.[7]

Although political agreement on the draft proposals has been reached, the EU legislative process has not ended. The text must return to the European Parliament and then back again to the Council for a 'second reading'. The draft can only be adopted once both the Council and the European Parliament reach common agreement (the legislative time limits allow the process to last as long as until end of 2003). Some further changes to the current text cannot be ruled out. However, in this case, the European Parliament is unlikely to insist on a major departure from the current informally agreed text. Accordingly, while the policy debate in the EU is not quite over, the recently agreed text is likely to represent the core of any European starting point in the multilateral context.

One final point worth noting about the EU regime is that while the policy is not helpful in a trade context, at least there is a policy on the table. This has added a little more certainty to the framework within which scientific information can be gathered and may, over time as such information accumulates, contribute to a decrease in the emotional opposition to GMO foods. The fact is that even a relatively hostile policy legitimizes the place of GMOs in the trade arena because it concedes that they integrated into the market access debate.

The GMO Food Debate in a Multilateral Context

At the heart of the GMO dispute lies distinctly different reasoning (championed by the EU and the USA, respectively) regarding the status of GMO foods. The 'precautionary principle' lies at the heart of the divide – due in part to the lack of a clear definition of the term, which makes it easy to hijack, and in part because the term is commonly used to capture a range of views that oppose practical, science-based assessments. The question of labelling and transparency is fraught with disagreement about what information is relevant to consumers, and how (or whether) to accommodate consumer choice that may or may not be based on scientific principles. These differences will only become more complicated as technology advances and the range of genes used in GMO foods increases, as modified products become more interwoven with other modified genes.

In general, an effective system of international trade in GMO foods has been stymied by a global impasse on key questions relating to market access for GMOs, labelling, and other specifications. In seeking to break the stalemate, much of the debate has focused on how existing multilateral processes and mechanisms can be utilized to move forward. These organizations are only as strong as the commitment of their members, however, and in the context of GMO food, that commitment has not been sufficient. Because trade-related GMO issues have been so contentious and politically volatile within domestic political debate, many countries – including those most influential in the debate – have been reluctant to cede any effective regulatory authority to consensus-based international forums.

The USA may be uniquely positioned to take the lead in moving toward an effective trade system for GMO foods. When the USA takes multilateral processes seriously, it tends to act as a catalyst for others. But to take the first step, the USA would require some degree of assurance that there would be constructive debate and compromise, not merely a protectionist feeding frenzy that would, if anything, make the situation worse.

The WTO, the Cartagena Biosafety Protocol (CBP), and the Codex Alimentarius Commission (Codex) all have the potential to provide a mechanism for establishing GMO policy. But any optimism as to the ability of such multilateral mechanisms to provide an answer must be tempered by recognition of how polarized the GMO trade debate has become, and the fact that the USA and the EU have been unable or unwilling to find much workable middle ground. For this reason, the multilateral process

[7] The European Commission's Joint Research Centre has examined various scenarios on the co-existence of genetically modified, conventional and organic crops in Europe. See: http://www.jrc.cec.eu.int/GECrops/ and http://www.jrc.es/welcome.html

needs to be gradually drawn in the direction of devising a binding trade regime. As the disastrous Seattle WTO meeting in 1999 so amply demonstrated, multilateralism is only as strong as the willingness of individual countries to accept some form of compromise that may have domestic political consequences.

At present, no country or group of countries has yet been willing seriously to relinquish control of the GMO debate to an effective international forum. This is not surprising, given the undeniably difficult economic, political and social issues that are interwoven into questions of GMO propagation, movement, and consumption. In addition, many of the central issues are new and complex, and the debate has all too often been based on lack of information and misinformation.

In that context, then, it is also not surprising that multilateral policy processes have failed to provide a solution. The Convention on Biodiversity and the CBP have not obtained the support or commitment necessary for them to craft policies addressing these issues of significant national sensitivity and interest, and the WTO already has enough politically fragile issues to deal with. Similarly, although the Codex has potential, it has been held hostage to completely divergent views. Nor have bilateral exchanges on trade-related GMO issues led to much progress. For the most part, they have taken the form of statements of differences rather than discussions on compromise.

Can the US Lead the Way?

So if existing multilateral forums have not done the trick, is there another way to move forward? One possible approach lies in one country developing a policy that other countries might be encouraged to adopt as their own. Given that the USA is the leading proponent of GMO trade, it is logical to explore first whether the USA might play that role. The difficult question, of course, is the extent to which the USA would need to compromise its current positions to develop a policy that would be acceptable to others. This is a question that begins with – but certainly extends beyond – US domestic politics.

The US government and its domestic GMO industry are far from reaching consensus on what US policy should be at present, let alone whether the goal should be to develop a model for others. Within the US bureaucracy there is debate between those who see some room for compromise and those who believe existing trade access rights should be used to force the issue of GMO exports. Similarly, while some GMO food companies favour a conciliatory, longer-term approach that seeks a comprehensive policy solution that will provide some degree of market certainty, others in industry are prepared to accept the risk of exacerbating tensions associated with market access, and have been urging the USA to force the issue.

Further, serious discussion about trade in GMO foods will require advocates to re-examine not only their substantive positions, but also the applicability or attractiveness of certain arguments for those positions. For example, touting increased agricultural yields is unlikely to be persuasive to many EU countries, where over-subsidized over-production is already a problem in some key commodity sectors. Some might suggest this is why the EU has adopted anti-GMO policies. Similarly, explaining to market-driven agricultural producers like Australia and New Zealand that GMO seeds can deliver better performing crops most likely would be effective only if there is an expectation of corresponding market demand for such products. And even where production-based arguments might be well-received, such as in developing countries, issues such as cost, sustainability, and marketability must also be addressed.

There has been an unhelpful but predictable lack of honesty in the GMO food debate. Many US industry representatives have noted that the EU is so far behind the USA in terms of scientific and technical ability with respect to GMO foods that their policy manoeuvres could well be a delaying tactic to give them time to catch up. That could be one reason why the USA and the EU are less polarized on the subject of biotech pharmaceuticals – their industries are more evenly on par and so economic interests more easily converge. Whatever the case, the key point is that in many instances the sensitivities used as a justification for policy are not necessarily or exclusively the ones really motivating the strategy underpinning the issue.

Given that the USA has championed GMO products and pushed hardest for their acceptance, development of a broadly acceptable policy will almost certainly take creativity on the part of the USA. One important component of the US approach will have to be continued vigilance on the part of officials to ensure that trade rights are not being unnecessarily or illegally constrained. In this regard, the WTO can be helpful, even though it is not well suited to serve as the forum for affirmatively establishing policy on trade in GMO foods. Through its binding mechanisms for addressing trade disputes, the WTO can be a means for helping to build policy through precedents and for creating *de facto* rules, not by specifying what is allowed, but by being very empirical about what is not permitted in the way of regulation. Accordingly, the best opportunity to move GMO trade issues forward may be for the USA to take all possible opportunities to rationalize the debate and cast the issue in general policy terms, not in terms specific to GMO foods.

In the nexus of the debate, then, sits the USA, the biggest producer and proponent of GMO products, and the most vociferous advocate of developing an export trade in these products. Warily eyeing the USA is the EU, which cites consumer concerns and agricultural sensitivities to explain the substance, magnitude, and nature of the debate, as well as Member States' resulting resistance to grant new approvals. On the sidelines is the WTO, limited in its ability to sort out policy debates when there is fundamental disagreement, but well equipped, through consideration of trade cases, to force issues at the heart of the policy debate, such as access to markets and reliance on scientific principles in excluding products or setting labelling requirements.

Another country which must be acknowledged as being a serious potential player is China. Its GMO industry is well established and its emphasis on export driven economic growth is increasing. As a new WTO member, preoccupied with implementing WTO commitments, China has played a comparatively low-key role in the GMO debate. But when progress starts to be made and China is in a better position to devote more attention to the issue, it will undoubtedly be a major influence on any policy outcome.

Over the next few years, we expect to see an increasing number of trade cases involving GMO foods. The US Congress and Administration have become increasingly frustrated at what they see as a lack of reciprocity in terms of openness on the part of major US trading partners. Although there are relatively isolated exceptions of highly protected sectors, such as textiles and steel, the US government has made some (mostly symbolic) unhelpful policy moves, such as passage of the the Farm Bill. All in all, the USA is comparatively a very open market. Moreover, in uncertain economic times, the pressure to expand foreign trade opportunities only grows. In that context, there can be little doubt that the GMO food industry will be seeking every opportunity to urge the US government to take aggressive steps. This does not necessarily mean that the USA sees a dispute as a panacea because the WTO is notoriously unable to compel countries to reflect its recommendations in a substantive way (the beef hormones case against the EU and the Foreign Sales Corporation case against the USA both being cases in point). But, while a case does not guarantee a practical solution, it does provide leverage in negotiations and can play an extremely important public relations role. Disputes are, therefore, a very important step in the process.

To be sure, the prospects of immediate progress are dim, if only because the USA and EU approaches are fundamentally and philosophically opposed. And the failure of both sides to address the real issues driving their respective agendas has not been constructive. No doubt this is immensely frustrating for private enterprises that must rely on international bureaucratic processes to facilitate and expand their business. None the less, there is reason for some optimism. The tone of the rhetoric on both sides has softened. Discussion within the EU and its Member States has focused less on whether GMO foods should ever be permitted and more on what standards are necessary to safely permit trade in GMO foods. Many in the USA consider this to be little more than a shift in rhetoric cloaking an intention to establish unreachable standards, but time will tell. At the same time, US policymakers have demonstrated an appreciation of the seriousness of the issues and a willingness to take serious enforcement steps against domestic violators of US standards. Additionally, to the extent trade cases at the WTO become the vehicle for addressing these issues, the shape of the debate may be presaged by the fact that the WTO has, in discussing non-GMO technical barriers to trade and phytosanitary access issues, repeatedly made reference to the necessity for relying on 'sound science'.

Clearly, there is a long way to go. The issues are complex and novel, the views of the participants are quite opposite and strongly held, and the domestic politics are daunting, which means international policy-setting forums are most probably not up to the task. None the less, there is a means to make some, albeit grudging, progress, using model policies and the mechanisms of deciding trade discrimination cases. Further, we believe the growth of GMO foods will only exacerbate the need for an international policy. It may be an ad hoc policy established via a series of trade cases, or the participants may conclude that their interests are better served by affirmatively negotiating an encompassing policy. However it happens, change is in the wind.

3 International Proposals to Regulate Intellectual Property Rights in Plant Genetic Resources

Michael Blakeney
Queen Mary Intellectual Property Research Institute, Queen Mary, University of London, London, UK

Abstract

The issue of securing intellectual property rights in genetic resources has been identified as a key issue for the round of World Trade Organization (WTO) trade negotiations launched in November 2001 by the WTO Doha Ministerial Meeting. The same subject was included in the Food and Agriculture Organization Treaty on Plant Genetic Resources, which was opened for signature in November 2001.

This chapter reviews the five key issues for developing countries in their negotiations on intellectual property rights and access to genetic resources. These are:

- the link between intellectual property protection and development;
- the mandatory obligation in the WTO Agreement on Trade Related Aspects of Intellectual Property Rights to introduce the protection of plant varieties;
- options for the *sui generis* protection of plant varieties;
- ethical issues relating to the patentability of life-forms; and
- relationship between the conservation and sustainable use of genetic material with the concepts of traditional knowledge and farmers' rights.

The Link Between Intellectual Property Protection and Development

A number of developing countries had noted the tension between the development and technology transfer objectives of the Trade Related Aspects of Intellectual Property Rights (TRIPS) Agreement and the way in which the Agreement made it possible for rights owners to impose unreasonable terms for technologies.

The objectives of the TRIPS Agreement are stated in Article 7 in the following terms:

> The protection and enforcement of intellectual property rights should contribute to the promotion of technological innovation and to the transfer and dissemination of technology to the mutual advantage of producers and users of technological knowledge and in a manner conducive to social and economic welfare, and to a balance of rights and obligations.

Given that technology transfer to facilitate economic development is stated as the objective of the TRIPS Agreement, World Trade Organization (WTO) Members are urged in a South Centre Report (Stillwell and Monagle, 2000) to 'examine as part of the Article 71.1 review the impact of implementing the TRIPS Agreement on the transfer and dissemination of technology and the related trade and development prospects of developing countries', with a view to 'operationalizing these provisions'. For example, in relation to Article 66.2, it is suggested that developed countries should 'provide more specific information on any existing schemes including the precise incentives, number of applying firms, and the effectiveness of these measures'. To the extent that intellectual property rules do not promote technology transfer, it is suggested that 'WTO Members should consider the establishment of additional mechanisms to facilitate access by developing and least-developed countries (LDCs) to technologies on a reasonable basis in order to fully implement the TRIPS Agreement, and to harmonize its operation with the broader objectives of the WTO Agreement'.

Submissions to the TRIPS Council by a number of developing countries, noting the difficulties faced by developing countries to obtain access to foreign technology, have argued that the TRIPS Agreement should be reviewed to consider ways and means to operationalize the objective and principles in respect of transfer and dissemination of technology to developing countries, particularly LDCs.

The relationship between the TRIPS Agreement and development has also been raised narrowly in the contexts of the implementation of the Agreement and more transcedentally in the context of the human rights to health and nutrition.

A number of developing countries have questioned what they consider to be unreasonable pressures by developed countries to effect compliance with the TRIPS Agreement, for example by the introduction of legislation and the establishment of an intellectual property infrastructure. This pressure is contrasted with the perceived failure of developed countries to provide incentives for the transfer of technology, as required by Article 66.2 and to provide technical assistance to developing countries, as required by Article 67.

A number of developing countries (e.g. Cuba, Dominican Republic, Egypt, Honduras) have indicated that the transitional implementation period of 5 years, granted under Article 65.2 has been insufficient to undertake the complex and costly administrative tasks required under the TRIPS Agreement, such as the modernization of their administrative infrastructure (intellectual property offices and institutions, the judicial and customs system), as well as the promulgation of new intellectual property laws. They have, therefore, sought an extension of the transition period for the developing countries.

Opposed to the desire of developing countries to delay the implementation of the TRIPS Agreement are pressures from developed countries to initiate the review of the implementation of the Agreement under Article 71.1. The European Union (EU) has reminded negotiators that the TRIPS Agreement establishes minimum intellectual property standards 'from which to seek further improvements in the protection of IPR [Intellectual Property Rights]. There should therefore be no question, in future negotiations, of lowering of standards or granting of further transitional periods' (WT/GC/W/193). Similarly Japan has declared that 'We should not discuss the TRIPS Agreement with a view to reducing the current level of protection of intellectual property rights. To the contrary, the TRIPS Agreement should be improved properly in line with new technological development and social needs' (WT/GC/W/242).

The development implications of the impact of the TRIPS Agreement on access to genetic resources is raised in the context, first of allegations of the extensive appropriation by corporations in developed countries of intellectual property rights in genes and plant varieties, as well as in enabling technologies. This has raised the concern of developing countries that their research in plant genetic resources will be stultified (Correa (2001) estimates that only 6% of the 25,000 biotechnological patents granted between 1990 and 1995 were obtained in developing countries) and that such research will be concentrated in the hands of a few multinational industrial seed suppliers (see Lesser, 1998b). Also a number of notorious instances in which IPRs have been obtained by applicants from the North in relation to genetic resources obtained from the South, or from germplasm collections maintained by the Consultative Group on International Agricultural Research (CGIAR) in trust for the international community, have raised concerns that the international IP regime is being maintained in a way which encourages so-called 'biopiracy', instead of benefit-sharing. (see Blakeney, 1997b, 1998; Mooney, 1998; Dutfield, 2000).

Capacity building is required in developing countries to enable them to deal with the impacts of IPRs upon biotechnological research. Jackson (2000 at 843) proposes the establishment, in Geneva, of a Genetic Resource and International Trade Institute 'to provide technical assistance training and research on genetic resources management and the rapidly changing policy environment to developing countries'.

The 'biopiracy' concerns of developing countries have been raised in the TRIPS Council, the Convention on Biodiversity (CBD), Food and Agriculture Organization (FAO) and are being addressed in the various international fora which are considering the introduction in patent and Plant Variety Protection (PVP) regimes of a mandatory obligation, in biotechnology patent applications to identify source countries.

Concerns about the extent of IPRs obtained over material acquired from germplasm collections maintained by the CGIAR, will be addressed, in the first instance, in the context of the Treaty on Plant Genetic Resources.

The Mandatory Obligation in the TRIPS Agreement to Protect Plant Varieties

Article 27.1 of the TRIPS Agreement provides that, subject to two categories of exception, 'patents shall be available for any inventions, whether products or processes, in all fields of technology, provided that they are new, involve an inventive step and are capable of industrial application'. Excluded from patentability, by Article 27.2 is the exploitation of inventions 'which is necessary to protect ordre public or morality, including to protect human or plant life or health or to avoid serious prejudice to the environment . . .'. Article 27.3 permits the exclusion from patentability of :

> (b) plants and animals, other than micro-organisms, and essentially biological processes for the production of plants and animals, other than non-biological and microbiological processes. However, Members shall provide for the protection of plant varieties either by patents or by an effective sui generis system or by any combination thereof.

In relation to genetic resources, the following technical issues are suggested by the terminology of these provisions: (i) what is a patentable invention for the purposes of Article 27.3(b) and how does it differ from a plant variety? (ii) what are microorganisms for the purposes of Article 27.2? (iii) what are plant varieties for the purposes of Article 27.3(b)? and (iv) should there be a research exception in relation to patents over plant material?

What is a patentable invention and how does it differ from a plant variety?

Intellectual property law attempts to draw a distinction between inventions and discoveries. The latter are not protectable. This distinction may be made in the relevant legislation. Additionally, IP courts distinguish between the discovery of non-patentable material which exists in nature and patentable inventions. The general approach which patent offices have taken, following the US Supreme Court decision in *Diamond* v. *Chakrabarty* (447 US 303 (1980)), which held that 'anything under the sun', apart from human beings could be patented, is that gene-sequences are inventions when they have been isolated and purified (see Doll, 1998).

A number of patent offices in developed countries have permitted the patenting also of partial DNA sequences and expressed sequence tags (ESTs), International Association of Plant Breeders for the Protection of Plant Varieties (ASSINSEL) has submitted that where these are not associated with an expressed characteristic, they should not be patented and that crop traits, as such, should not be patented (ASSINSEL, 1999).

Of course, it is open to a court or a legislature to rule or provide that genetic material is not patentable, even in its isolated or purified form, because it is a mere discovery. Indeed, nothing in the TRIPS Agreement obliges countries to deem the isolation of genetic materials to be inventions. A number of developing countries exclude the patentability of genetic materials (Mexico), or of materials existing in nature (Argentina, Brazil and the Andean Group Decision 486).

In Europe the Directive on the Legal Protection of Biotechnological Inventions (98/44/EC of

6 July, 1998, [1998] OJ L213/13), specifically provides in Article 3.2 that 'Biological material which is isolated from its natural environment or produced by means of a technical process may be the subject of an invention even if it previously occurred in nature'.

Article 53(b) of the European Patent Convention (EPC) excludes plant varieties, as well as 'essentially biological processes' from the scope of patentable subject matter. This raises, in the first instance, the definitional distinction between plants and plant varieties. The International Union for the Protection of New Varieties of Plants (UPOV) Convention defines plant variety in terms of a plant grouping within a single biological taxon of the lowest known rank, which grouping can be:

- defined by the expression of characteristics (such as shape, height, colour and habit) resulting from a given genotype or combination of genotypes;
- distinguished from any other plant grouping by the expression of at least one of the said characteristics; and
- considered as a unit with regard to its suitability from being propagated unchanged.

The first consideration of the distinction between plant and plant variety by the Technical Board of Appeal of the European Patent Office (EPO) occurred in 1984 in the *Ciba/Geigy* determination (Case T 49/83 [1984] OJ EPO 112). This concerned a plant which had been treated with a chemical compound to confer on the plant a degree of protection from the toxic side-effects of certain herbicides. The Examination Division had refused the patent application on the basis of Article 53(c). This was reversed by the Technical Board of Appeal, which, applying the definition of plant variety in the UPOV Convention, stated that Article 53(c), 'prohibits only the patenting of plants or their propagating material in the genetically fixed form of the plant variety . . . Plant varieties in this sense are all cultivated varieties, clones, lines, strains and hybrids' [at 114–115]. In this case the claims covered merely the application of a chemical treatment and not plant varieties as such.

This approach was applied by the Technical Board of Appeal in the *Lubrizol (Hybrid Plants)* case (Case T320/87 [1990] OJ EPO 71), where the Board held that 'the term "plant varieties" means a multiplicity of plants which are largely the same in their characteristics (i.e. homogeneity) and remain the same within specific tolerances after every propagation or every propagation cycle (i.e. "stability")' [at 79]. The Board then ruled that as the hybrids in issue were not stable, they did not fall within the excluded category of plant varieties.

The European Biotechnology Directive permits the patentability of inventions concerning plants, where 'the technical feasibility is not confined to a particular plant . . . variety' (*Directive on the Legal Protection of Biotechnological Inventions*, Article 4(1) para.2, 98/44/EC [1998] OJ L213/130). Patent claims can therefore be made in respect of plant groupings, or as stated in Recital 31 to the Directive,

> Whereas a plant grouping which is characterized by a particular gene (and not its whole genome) is not covered by the protection of new varieties and is not excluded from patentability even if it comprises new varieties of plants.

This qualification was addressed by the Technical Board of Appeal in *Novartis/Transgenic Plant* ([2000] OJ EPO 511). The application concerned a patent containing claims to transgenic plants comprising in their genomes specific foreign genes, the expression of which resulted in the production of antipathologically active substances, and to methods of preparing such plants. The EPO had denied registration, supported by the Technical Board of Appeal, on the ground that Article 53(b) denied the patentability of an invention which could embrace plant varieties.

In its decision of 20 December 1999, the Enlarged Board of Appeal indicated that it would favour the application because, in substance, it did not involve an application for a plant variety. This determination contains some useful guidance on the legal definition of plant varieties. The Enlarged Board of Appeal noted that the definitions of plant variety in the UPOV Convention and the EC Regulation on Community Plant Variety Rights refer to 'the entire constitution of a plant or a set of genetic information', whereas a plant defined by a single recombinant DNA sequence 'is not an individual plant grouping to which an entire constitution can be attributed'. It observed that the claimed transgenic plants in the application before it were defined by certain characteristics which allowed the plants to inhibit the growth of plant pathogens. No claim was made for anything resembling a plant variety. The tribunal noted that in the case of Plant Variety Rights (PVR) an applicant had to develop a plant group, fulfilling in particular the requirements of homogeneity and stability, whereas in the case of a typical genetic engineering invention, a tool was provided whereby a desired property could be bestowed on plants by inserting a gene into the genome of a specific plant. It observed that the development of specific varieties was not necessarily the objective of inventors involved in genetic engineering (see Llewelyn, 1999, 2000b).

An interesting question, raised by this case is the continuing role of PVR protection in the modern world of genetic engineering. To what extent will a *sui generis* system for the protection of plant varieties secure the rights of plant breeders in the face of innovations in patent law? To what extent ought PVRs be harmonized with patent rights? Given the developments in modern biotechnology, Farmers' Rights must be seen both in the context of patents as well as PVR protection.

The USA has never excluded biological material, including plant varieties from the scope of patentable subject matter. Plant varieties can be protected in the USA under a system of plant patents, or under a system of utility patents or under the Plant Variety Protection Act (PVPA). The Plant Patent Act (35 USC §§ 161–164 (1994)) makes available patent protection to new varieties of asexually reproduced plants. Under this scheme a plant variety must be novel and distinct and the invention, discovery or reproduction of the plant variety must not be obvious. One of the disadvantages of the scheme is that only one claim, covering the plant variety, is permitted in each application. In practice, this scheme has been in decline since the *Hibberd* decision of the Patent Office Board of Appeals and Interferences, opened up the normal patent system to applications which covered plant varieties (227 USPQ 443 (1985)).

Following the opening of the patent system by the *Hibberd* decision, it is possible for an applicant for a patent to protect plant varieties under the US Patents Act (35 U.S.C. §§ 100–103 (1994)) and breeding methods, provided they can be demonstrated to be the fruits of human ingenuity. This also has the advantage to the applicant that there is no farmer's privilege and only a limited experimental use exception. (See *Roche Products, Inc.* v. *Bolar Pharmaceuticals Co.* 733 F 2d 858 at 862–863 (1984).)

The PVPA deals with plants that are produced sexually, by means of seeds. Varieties are protectable if they are new, distinct, uniform and stable (7 USC §§ 2321–2583 (1994). In its 1994 amended form, this statute is in line with the 1991 version of UPOV.

What are microorganisms for the purposes of Article 27.3(b)?

Article 27.3(b) permits WTO Members to exclude from patent protection, plants and animals and essentially biological processes for the production of plants and animals. Members are specifically not permitted to exclude from patent protection microorganisms and non-biological and microbiological processes. The language used in Article 27.3(b) implies that a clear distinction can be made between plants and animals on the one hand and microorganisms on the other. However, there is no commonly accepted definition of 'microorganism' either in science or in patent office practice. The lack of any definition permits great variations between members in restricting this exclusion from patentability. For example, from the patent activity taking place in the USA, Europe and Japan, it has been observed that a very flexible interpretation is given to the concept of patentable subject matter, where the emphasis is on inclusion not exclusion (Adcock and Llewelyn, 2000).

A South Centre report on *Technical Issues on Protecting Plant Varieties by Effective* Sui Generis *Systems* (Mangeni, 2000, para. 24), urges the adoption, even of a narrow scientific definition of microorganism 'viruses, algae, bacteria and protozoa' to give some certainty to the exception.

The practice of patent granting offices in developed countries suggests that there is no perceived need for a definition. The key issue for protection being whether or not the invention meets the patent granting criteria and not its subject matter. One of the reasons for reluctance to use a definition, provided by the EPO was that 'it does not seem expedient to introduce such a definition as the rapid evolution in the field of microbiology would necessitate its frequent updating' (quoted in Adcock and Llewelyn, 2000).

In scientific practice the term 'microorganism' is ill-defined, because the scientific classification is continually evolving. In patent office practice, the only debate of any kind has taken place in Europe in the context of what constitutes a plant, which is patentable, and a plant variety, which is not. The result of this has been agreement within patent granting offices that excluded material is that which is protectable by a plant variety right as determined by the UPOV Convention, anything falling outside the scope of UPOV being protectable by a patent (Llewelyn, 2000a). The result of this is that patent protection is available for groupings of plants which encompass more that one variety provided the patentee has not claimed a plant variety as such. Anything, therefore, which does not take the form of a patent variety as such is patentable. It has not been regarded as necessary to provide any further definitions. The patent laws of developed countries are predicated on a presumption of patentability and the granting criteria are given a broad interpretation. Any exclusions are, therefore, given a restriction application. Thus where a country has adopted specific categories of excluded material, these exclusions are likely to be the subject of rigorous scrutiny particularly where the categories could be said to go beyond that which is permitted under the TRIPS Agreement (Adcock and Llewelyn, 2000).

Given the difficulties inherent in attempting a definition of microorganism, it may be more advisable for member states to introduce a higher threshold for patent protection in respect of living material. For example, Adcock and Llewelyn (2000) suggest that an invention involving biological material may not be regarded as novel: (i) if the information is already in the public domain; and/or (ii) the invention merely replicates biological material, or the function of biological material, which already occurs naturally. Mangeni (2000) suggests the exclusion from novelty of any information available to the public, whether in writing 'or though use including use by local and indigenous communities and through the deposit of information with deposit institutions, such as genebanks'.

An invention involving biological material will be regarded as lacking an inventive step if it: (i) merely identifies the biological material; and/or (ii) merely identifies the natural function of the biological material. An invention will demonstrate an inventive step if it takes the form of a significant technical application of an identified function of the biological material. This technical application must go beyond a mere simple replication of the natural function of the biological material, and the technical application must represent a significant technical advance on the prior art.

An invention involving biological material will be regarded as being capable of industrial application if it can be shown that it is capable of being used in a manner which provides a demonstrable public benefit. Public benefit means that the invention must be capable of being used in a manner conducive to public health and to social, environmental and economic welfare.

The current low thresholds for protection applied by the US and the European patent offices means that the courts are becoming the arbiters of patentability, as the revocation of the neem and turmeric patents demonstrate. The argument for raising the threshold for protection can be justified on the basis that it will result in greater predictability and certainty for the bioscience industry, ensuring that those inventions which deserve protection are protected and that this protection is less likely to be subsequently challenged in court (Adcock and Llewelyn, 2000). The re-opening of the neem and turmeric patents are cited as examples of courts being forced to reconsider the liberality of patent offices (see Das, 2000; Prakash, 2000). On the other hand, they may be considered to be examples of the necessity for patent offices to have access to data on traditional knowledge as part of the state of the art.

What is a plant variety for the purposes of Article 27.3(b)?

As noted above, a crucial issue in the establishment of a *sui generis* regime would be the definition of the protected subject matter. Article 27.3b of the TRIPS Agreement requires the protection of 'plant varieties', but does not provide (as in the case of inventions) a definition thereof. Therefore, national laws have ample room to determine what is to be deemed a plant 'variety' for the purposes of protection. There have been lengthy discussions on the concept of 'plant variety', particularly in the framework of UPOV. The scientific notion does not necessarily coincide with the legal concept. The law may require certain characteristics for a protected variety that may not be essential for a scientific definition.

Patent protection was not originally considered to be a particularly effective system for the protection of plant varieties. Prior to the development of modern biotechnology, the breeding of a new variety could not be said to involve an inventive step and such innovations as were made, could be considered to be obvious rather than inventive. However with the extension of patent protection to recombinant methods for producing transgenic plants and the resulting products, patents have begun to assume an increasing significance in plant variety protection. The broader ambit of patent rights is a particular advantage of this form of intellectual property protection, covering, as it does, plants, seeds and enabling technologies. PVRs are highly specific to the variety and their scope is limited by reference to the physical (propagating) material itself, combined with the description of the variety given in the documentary grant of the rights.

Research exception

Given the possibility of the application of patents to plant varieties, it would appear to be significant to secure within patent laws the same research exception that exists under PVR laws. Under European and US patent case law, the courts have permitted experimentation for the purposes of invention outside the scope of the patent. Although it should be noted that the scope of the exception has been interpreted differently in each jurisdiction. In Japan, the exception is created by statute (Patent Law, s.69(1)).

Correa (2001) refers to the recommendation by 'experts in the United States of an explicit research exemption for patents on plants and cites Barton (1993 at 9) for the suggestion that all biological materials required for IPR protection become "part of the national germplasm system"'.

Technical Issues Relating to the *Sui Generis* Protection of Plant Varieties

The principal technical issues that have been raised on the implementation of effective *sui generis* protection of plant varieties are: (i) what is meant by 'effective'? and (ii) what *sui generis* options are open to Member states?

'Effective' *sui generis* system

Article 27.3(b) provides no guidance on what is meant by 'effective', the debate in the TRIPS Council having focused upon which *sui generis* systems satisfy the obligation, One interpretation is that effective refers to the enforceability of the PVP rights granted by the relevant legislation (Blakeney, 1996a). Mangeni (2000) asserts that if the debate shifted to the meaning and implication of 'effective', the following would be among the conclusions reached:

- the *sui generis* system should be effective to *protect plant varieties* as such (including varieties developed by local communities and national/public research institutes);
- the rights of plant breeders should be protected as an international obligation as and when assumed by Members;
- the rights to be protected should be those set out in the obligations Members have assumed (for instance under the Agreement on TRIPS and equally the CBD);
- the rights should be protected in accordance with national objectives referred to in Article 7 and the principles in Article 8 (Agreement on TRIPS) as apply in the country should be protected within the overall framework of the CBD and the first recital of the preamble to the WTO Agreement – on sustainable development; and

Table 3.1. Developing countries' proposals for the review of Art. 27.3(b) of the TRIPS Agreement.

Countries/organizations	Patenting (life forms and biological processes)	*Sui generis* (plant varieties)
Kenya (WT/GC/W/23 of 5 July 1999)	Need 5-year extension of transition period. Harmonize TRIPS with CBD	Need 5-year extension of transition period. Increase scope of 27.3(b) to include protection of indigenous knowledge and farmers' rights. Harmonize TRIPS with CBD
Venezuela (WT/GC/W/282 of 6 August 1999)	In 2000, introduce mandatory system of IPR protection for traditional knowledge of indigenous and local communities, based on the need to recognize collective rights	
African Group (WT/GC/W/302 of 6 August 1999)	Review should be extended + additional 5 year transition hereafter. Review should clarify that plants, animals, microorganisms, their parts and natural processes cannot be patented	Review should be extended + additional 5 year transition after that *sui generis* laws should allow for protection of community rights, continuation of farmers' practices and prevention of anti-competitive practices which threaten food sovereignty. Harmonize TRIPS with CBD and International Understanding on Plant Genetic Resources for Food and Agriculture (IU) of FAO
LDC Group (WT/GC/W/251 of 13 July 1999)	There should be a formal clarification that naturally occurring plants and animals, as well as their parts (gene sequences), plus essentially biological processes, are not patentable. Incorporate provision that patents must not be granted without prior informed consent of country of origin. Patents inconsistent with CBD Art 15 (access) should not be granted. Need for extended transition period	*Sui generis* provisions must be flexible enough to suit each country's seed supply system. Need for extended transition period
Jamaica, Sri Lanka, Tanzania, Uganda, Zambia (www.foe.org/international/wto/gov 2 September 1999)	No patenting of plants without prior informed consent of government and communities in country of origin	
SAARC (South Asia Association for Regional Cooperation, WT/L/326, 22 October 1999)	There is a need to prevent piracy of traditional knowledge built around biodiversity and to seek the harmonization of the TRIPS Agreement with the UN Convention on Biological Diversity so as to ensure appropriate returns to traditional communities	
SADC (Southern Africa Development Cooperation WT/L/317, 1 October 1999)	Transitional period for implementation of 27.3(b) should be extended; and the 2000 review delayed. The review of 27.3(b) should harmonize TRIPS with CBD. Exclusion of essentially biological processes from patentability should extend to microbiological processes	The transitional period for implementation of 27.3(b) should be extended and the 2000 review should be delayed. The review of 27.3(b) should retain the *sui generis* option

Table 3.1. *Continued.*

Countries/organizations	Patenting (life forms and biological processes)	*Sui generis* (plant varieties)
Group of 77 (WT/MIN(99)/3, 2 November 1999)	Future negotiations must make operational the provisions relating to the transfer of technology, to the mutual advantage of producers and users of technological knowledge and seek mechanisms for a balanced protection of biological resources and disciplines to protect traditional knowledge	
Bolivia, Colombia, Ecuador, Nicaragua and Peru (WT/GC/W/362, 12 October 1999)	The Seattle Ministerial Conference should adopt a mandate to: (i) carry out studies in order to make recommendations on the most appropriate means of recognizing and protecting traditional knowledge (TK) as the subject matter of IPR; (ii) initiate negotiations with a view to establishing a multilateral legal framework that will grant effective protection to the expressions and manifestations of TK; (iii) complete the legal framework envisaged in paragraph (ii) above in time for it to be included as part of the results of the new round of trade negotiations	

- the protection should be consistent with international obligations that Members have assumed, for instance under the CBD.

Sui generis options

A *sui generis* option in the intellectual property context is usually taken to refer to a specially coined IP right outside the traditional categories of IP protection.

UPOV has advanced its system as the principal workable example of a *sui generis* plant variety protection system. It is interesting to note that the drafters of the TRIPS Agreement, who felt free to import into the agreement, provisions from other named international instruments, such as the Paris, Berne and Rome Conventions and the Washington Treaty on Integrated Circuits, in the area of plant varieties desisted from specifically importing the UPOV Convention.

A number of developing countries have made suggestions to the TRIPS Council on options for national plant variety legislation. These are conveniently summarized by GRAIN 2000, as shown in Table 3.1.

At 6 August 2001, some 49 states had acceded to the UPOV Convention (see Annex II). Of these, 29 states had acceded to UPOV 1978, 19 to UPOV 1991 and two states to UPOV 1961/1972. Mangeni (2000) states that the preference of developing countries is for UPOV 78 because of its reference to the right of farmers to save, replant and share seeds and because of the breeder's exemption to research, experiment and breed around the protected variety without undue claims from the breeder of the protected variety. He also asserts that it provides 'better protection for biodiversity, which developing countries consider beneficial for social justice in catering to local communities or the rural population and farmers, and for being supportive of domestic policies like promoting innovation and attaining food security'.

Ghijsen (1998) suggests that different PVP issues arise in relation to three distinct categories of plant: (i) open pollinated food crops; (ii) inbred lines and horticultural crops; and (iii) medicinal plants.

In relation to open pollinated food crops such as cereals and tubers, seed saving is important for farmers in developing countries.

Landraces are excluded from protection by the requirement that a new variety is distinct from 'varieties of common knowledge'. Similarly, material in germplasm collections, might be preserved from private exploitation through the publication of information about deposited materials, thereby placing them in the public domain. More importantly, the distribution of collected materials may be protected by means of material transfer agreements (MTAs) which prevent the seeking of IPRs in relation to those materials (or from essentially derived varieties).

For inbred lines and horticultural crops, such as ornamentals, fruits, vegetables and plantation crops, seed saving is not generally an attractive option for farmers.

In relation to medicinal plants, Ghijsen (1998) suggests the requirement of a certificate of novelty as a precondition for PVP. This would assure that important plants remain in the public domain. Another option, suggested by the PVP law of Ecuador, is that PVP cannot be obtained for wild species which have not been planted or improved by human intervention.

Leskien and Flitner (1997) have suggested a PVP seal, which allows the holder of the right to use the seal on seed packages of the protected variety. After the seed has been purchased the PVP right will be exhausted and any further transactions with the seed will be permissible. This suggestion has been criticised as an encouragement to piracy (Louwaars, 1998), although it should be noted that the exhaustion principle applies to trademarks.

The WWF and CIEL (2001) encourage the Secretariat of the CBD to compile further case studies and empirical evidence on the relationship between IPRs and TRIPS, particularly focusing on the relationship between IPRs and access and benefit sharing and the impact of IPRs on technology transfer. They urge that 'IPRs will need to be evaluated to ensure that they do not "run counter" to the objectives of the CBD'.

The South Centre report (Mangeni, 2000) recommendations include:

1. The existing provision in Article 27.3(b) for the protection of plant varieties by *sui generis* systems should be kept.
2. The provision should be construed and amended to be in conformity with other international obligations on the matter of protecting plant varieties, including those under the CBD.
3. The rights of breeders include research and experimentation, and innovation for purposes of seeking protection of consequent varieties, without compensatory claims.
4. Local farmers in developing countries have the right to save, share, sell, and replant seed, without compensatory claims from plant breeders.
5. IPRs that contravene the CBD and Article 27.3(b) as modified (to provide that plant varieties are to be protected by *sui generis* systems and that any intellectual property rights must be subject to the CBD and the FAO International Undertaking on Plant Genetic Resources) are not to be granted and are to be cancelled by the offices on their own motion or on petition.
6. Novelty is to mean universal novelty, in particular material for which an intellectual property right is sought is not novel if based on or it contains information available to the public through writing or use, including use by communities, or through deposit in deposit institutions.

The failure of the drafters of the TRIPS to define what was meant by *sui generis*, leaves considerable scope for nations in the range of legislation which they may implement in compliance with this provision.

Correa (2000, pp. 30–32) has suggested that a *sui generis* regime may provide for a dual system of protection which includes both 'modern' as well as farmers' varieties. Modern varieties would be protected under a UPOV-like legislation, requiring novelty, distinctness, uniformity and stability. For other cases (farmers' varieties) the requirements may be less stringent and be limited, for instance, to sufficient identification and distinctness. This distinction would take account of the more variable nature of farmers' varieties. Since the creation of the latter variety is generally a collective endeavour, the rights would be granted to the community that has developed and used the variety. In the case of farmers' varieties, national legislation may recognize a 'remuneration right', that is, an entitlement to receive compensation in all cases of use of a protected variety for propagating purposes outside the respective farming community or communities. This formulation would amount, in practice, to an open licensing system under which any interested party may utilize the protected variety for planting or multiplication, against a payment in favour of the titleholders.

Another possibility for protecting farmers' varieties, suggested by Correa (2000, pp. 32–33) would be through a regime that aims to prevent the misappropriation of such varieties. National legislation would provide that IPRs could not be obtained with respect to farmers' varieties (or derivatives therefrom). In the case of infringement of this rule, the conferred title should be declared void. Given the collective nature of these rights and the lack of legal personality for farmers communities, Correa (2000, p. 32) suggests the establishment of an 'ombudsman' empowered with the right to act on behalf of the communities so as to enforce their rights.

The features of the misappropriation regime, suggested by Correa (2000, pp. 32–33) are:

- it would recognize the informal, collective and cumulative systems of innovation of local and indigenous communities and farmers;
- no novelty, inventiveness or secrecy would be required;
- there would be no arbitrary time limit for protection;
- the conferred rights would be 'non-monopolistic' and would not hinder the non-commercial use and exchange of germplasm within and among communities;
- no registration, and therefore, administrative machinery, would be necessary;
- it would not oblige farmers or communities' members to keep secrecy or change their traditional practices;
- since no monopolies would be recognized, possession of the same knowledge by different communities would be perfectly legitimate;
- the rights against infringers would arise when a variety has been acquired in a manner contrary to certain rules, such as national access legislation or other accepted practices on the collection of germplasm.

Ethical Issues Relating to the Patentability of Life-forms

There is a substantial literature on the ethical implications of permitting the propertization of the 'building blocks of life' (e.g. see Suzuki and Knudson, 1989). A number of religions consider human intervention in relation to living material to 'violate' the divine creation, or at least to 'reduce the value of life and nature to the merely economic' (Bruce and Bruce, 1998, p. 231). Furthermore, the decisions in relation to these important matters are taken by entities, such as industrial property offices, which are outside democratic control. Similarly the proponents on either side of the patenting life debate, whether non-governmental organizations (NGOs) or life sciences companies are also outside the democratic process. This 'democratic deficit' counsels at least education of politicians, the public and the media in all aspects of the commodification of life (Bruce and Bruce, 1998, pp. 270–271, cited by Tansey, 1999, p. 19).

Researchers express the concern that biomedical and agricultural research are too important to be sterilized by the intervention of private IPRs. The decision of the National Institute of Health to file patent applications for gene sequences was described as 'sheer lunacy' by James Watson, the first Director of the Human Genome Programme, and cited as one of the reasons for his resignation (see McKeough, 1997).

A related concern is that the propertization of genetic resources has resulted in the concentration of proprietary biotechnologies in a few corporations (e.g. see Wells, 1994; Lesser, 1998b). In addition to the possible adverse impacts this market concentration might have upon the vigour of competition, the market dominance of these private corporations also has an important influence upon the sort of biotechnological research that is undertaken. For example, to what extent will the dominance of private corporations in biomedical and agricultural research direct that research towards Northern concerns and away from Southern health problems (see Wattal, 2000) and Southern food priorities (see Alston *et al.*, 1998). Will the owners of IPRs in key enabling technologies make them available to public research institutions on affordable terms? (see Leisinger, 1999.)

The concentration of proprietary technologies in the hands of a relatively small group of Northern life-sciences companies, has been exacerbated by the grant, by patent offices of over-broad patent claims, resulting in what Heller and Eisenberg (1998) have described as the 'biomedical anticommons tragedy'. This problem, as is mentioned above, can only to a limited extent be dealt with by policy directions to patent offices, as ultimately the interpretation of patent claims is a matter for the courts.

Article 27.2 of the TRIPS Agreement, it will be recalled, permits Members to disallow the exploitation of inventions 'which is necessary to protect ordre public or morality, including to protect human or plant life or health or to avoid serious prejudice to the environment . . .'. Member States would have to show that the commercial exploitation of the specific invention, would be contrary to *ordre public* or morality. In light of the interpretation and application of the equivalent provision within the European Patent Convention, and recently reinforced in the EU Directive on the Legal Protection of Biotechnological Inventions, it is unlikely that this exception would permit a general exclusion of living material from patentability. It is also questionable whether patent offices are the proper bodies to adjudicate the application of moral and ethical issues to the patent system (see Ford, 1997).

In any event, the patent offices have abstained from exercising moral judgements in this area. Thus for example in *Greenpeace* v. *Plant Genetic Systems NV* ([1995] 8 OJ EPO 545), in an opposition to an application for a patent directed to transgenic plants engineered to be resistant to the herbicide Basta, Greenpeace argued that it was immoral and therefore in breach of Article 53(a) of the European Patent Convention, to 'own' plants which were the common heritage of humankind. The Appeal Board of the EPO, sustained the Examination Division's view that it was not the proper forum for discussing the advantages and disadvantages of genetic engineering. Similarly in *Novartis/ Transgenic Plants* (Decision G0001 of 20 December 1999) the Extended Board of Appeal of the EPO, considered the debate over genetic engineering to be too controversial for it to sustain Greenpeace's opposition to the patent. The Extended Board of Appeal noted that the European Patent Directive on Biotechnology was an indication that the European Parliament considered there to be some benefit in genetic engineering.

On the other hand, the European Biotechnology Directive is an example of a legislature giving directions to its patent office on matters to be considered in evaluating the morality of an invention (see Leskien, 1998). The Directive excludes from patentability:

- the human body in its various stages;
- processes for cloning human beings;
- processes for modifying the genetic identity of human beings in the germ line;
- the use of human embryos for industrial or commercial purposes;
- processes for modifying the genetic identity of animals which are likely to cause them suffering without any substantial benefit to humans or animals, and also animals resulting from such processes.

The Treaty on Plant Genetic Resources

Background

The TRIPS Agreement is primarily concerned with Member nations' obligations to establish a legal infrastructure to permit the privatization of intellectual property rights. Article 27.3(b) permits the extension of those private rights to genetic resources. The TRIPS Agreement is seen by some to be in tension with the CBD, which is concerned with conserving the 'global genetic commons' for humankind (e.g. Buck, 1998; Lawson and Pickering, 2001). The Conference of the Parties (COP) to the CBD has reported that the value and benefit of genetic materials in the global biosphere may be realized and shared through IPRs (CBD, 1996a). The COP Secretariat has also noted the importance in the relationship between trade and biological diversity of the interrelationship between the TRIPS Agreement and the CBD (CBD, 1996b). The COP Panel of Experts on Access and Benefit Sharing concluded that IPRs were a significant influence upon benefit-sharing (CBD, 1999), but consensus has not been achieved within the COP on whether this influence has been positive or negative.

A significant recent development, which will have an important impact upon the negotiation of access to plant genetic resources, was the promulgation in November 2001 of the Treaty on Plant Genetic Resources. This Treaty was the enactment in a binding form of the International Undertaking on Plant Genetic Resources, which had been adopted by the 1983 the Conference of the FAO. The International Undertaking had originally been predicated on the principle that plant genetic resources for food and agriculture (PGRFA) should be freely exchanged as a 'heritage of mankind' and should be preserved through international conservation efforts. In subsequent years the principle of free exchange was gradually narrowed by the impact of intellectual property rights upon agriculture. In November 1989 the 25th Session of the FAO Conference adopted two resolutions providing an 'agreed interpretation' that plant breeders' rights were not incompatible with the Undertaking. The acknowledgement of plant variety rights obviously benefited industrialized countries, which were active in seed production. In exchange for this concession, developing countries won endorsement of the concept of 'farmers' rights'.

Subsequent negotiations on the text of the International Undertaking sought to reconcile the proposition that PGFRA were the common heritage of mankind with the sovereignty of states over their plant genetic resources and the Treaty on Plant Genetic Resources represents an attempt to reconcile this tension in a formal instrument.

Treaty provisions relating to access

The objectives of the Treaty, are stated in Article 1 to be 'the conservation and sustainable use of plant genetic resources for food and agriculture and the fair and equitable sharing of the benefits arising out of their use, in harmony with the Convention on Biological Diversity, for sustainable agriculture and food security'.

Article 4 of the Treaty requires signatories 'where appropriate' to 'promote an integrated approach to the exploration, conservation and sustainable use of plant genetic resources for food and agriculture'. Article 10.2 contains the agreement of the Contracting Parties to 'establish a multilateral system, which is efficient, effective and transparent, both to facilitate access to [PGFRA] and to share, in a fair and equitable way, the benefits arising from the utilisation of these resources, on a complementary and mutually reinforcing basis'. Facilitated access to PGFRA is to be provided in accordance with the conditions prescribed in Article 12.3. Paragraph (d) of this provision provides that the recipients 'shall not claim any intellectual property or other rights that limit the facilitated access' to PGFRA, or their 'genetic parts or components', in the form received from the Multilateral System. This, of course, does not prevent intellectual property rights being claimed in relation to germplasm which is modified by the recipient.

Article 13.1 recognizes that benefits accruing from facilitated access to PGFRA shall be shared fairly and equitably under this Article. Article 13.2 envisages that this sharing of benefits include the exchange of technical information, access to technology, capacity building and the sharing of monetary benefits from commercialization.

Article 28 provides that the Treaty enters into force, 90 days after accession by 40 countries. Until that date, the International Undertaking will remain operative.

Farmers' rights

Article 9 of the International Treaty on Plant Genetic Resources for Food and Agriculture implements the proposal which was developed under the International Undertaking for the recognition of farmers' rights. The policy behind this recognition is stated in Article 9.1, namely that

> The Contracting Parties recognize the enormous contribution that the local and indigenous communities and farmers of all regions of the world, particularly those in the centres of origin and crop diversity, have made and will continue to make for the conservation and development of plant genetic resources which constitute the basis of food and agriculture production throughout the world.

Article 9.2 envisages that 'the responsibility for realizing Farmers' Rights, as they relate to Plant Genetic Resources for Food and Agriculture, rests with national governments' and that national legislation should include measures relating to:

> (a) protection of traditional knowledge relevant to plant genetic resources for food and agriculture;
> (b) the right to equitably participate in sharing benefits arising from the utilization of plant genetic resources for food and agriculture;
> (c) the right to participate in making decisions, at the national level, on matters

related to the conservation and sustainable use of plant genetic resources for food and agriculture.

Finally, Article 9.3 provides that the Article shall not be interpreted 'to limit any rights that farmers have to save, use, exchange and sell farm-saved seed/propagating material'.

An assumption of Article 9.1 is that the landraces used by traditional farmers are a dynamic genetic reservoir for the development of new varieties and for the transmission of desirable genetic traits. The traditional knowledge of local and indigenous communities is similarly perceived. As a means of remunerating these groups for their past contributions to the development of plant genetic resources for food and agriculture production, there can be little argument, except about the quantum and distribution of this remuneration.

Inevitably, any calculation of the equitable share, which traditional farmers and indigenous communities might enjoy under a Farmers' Rights, or Traditional Knowledge regime will be arbitrary. However the intellectual property system is no stranger to arbitrary calculations, thus the 20-year length of a patent term is intended to provide an opportunity for the compensation of all inventors, whatever the area of technology. Similarly the 25 years of exclusivity which the UPOV Convention provides for new varieties of trees and vines, takes no account of variations in R&D costs between the different varieties.

The principal ways in which plant genetic resources are translated into food and agriculture production is through plant breeding and plant patenting. Standing at the heart of a Farmers' Rights regime is the concept of the equitable benefit sharing of benefits with farmers for their contribution to innovations in plant breeding and plant patenting.

It is estimated that about 6.5% of all genetic research undertaken in agriculture is focused upon germplasm derived from wild species and land races (see McNeely, 2001). As with plant breeding, the germplasm collections of the CGIAR Centres have constituted a useful source of genetic material.

Article 9.2 obliges the Contracting Parties to the Plant Genetic Resources Treaty 'to take measures', subject to their national legislation to protect and promote Farmers' Rights. The content of these rights is defined in the balance of that provision and embraces the protection of traditional knowledge, equitable benefit sharing and the right to participate in decision making. The Treaty leaves open the legal context within which Farmers' Rights are to be enacted.

National legislation on Farmers' Rights tends to combine one of the versions of UPOV with some of the access principles of the CBD. The African Model legislation for the Protection of the Rights of Local communities, Farmers and Breeders, and for the Regulation of Access to Biological Resources, which was adopted by the Organization of African Unity, Heads of States Summit at Ouagadougou in June 1998, adopts a *sui generis* regime based on UPOV 1991 (see Kongolo, 2000). However, most national statutes prefer access legislation combined with UPOV 1978 (e.g. Andean Community's Common System on Access to Genetic Resources, 1996; Costa Rica – Biodiversity Law 1998; India – Community Intellectual Property Rights Act 1999; Kenya – Seeds and Plant Varieties Act 1975).

Access to PGRFA and the Doha Negotiating Agenda

In the same month that the International Treaty on Plant Genetic Resources for Food and Agriculture was concluded, the WTO Ministerial Meeting in Doha issued a Ministerial Declaration which set out the negotiating agenda for the next trade round of the WTO. Article 19 of the Declaration instructed the Council for TRIPS, in pursuing its work programme, 'to examine, *inter alia*, the relationship between the TRIPS Agreement and the CBD, the protection of traditional knowledge and folklore, and other relevant new developments' raised by Members pursuant to a general review of the TRIPS Agreement.

Pursuant to this direction, the TRIPS Council has commissioned a number of studies on access to PGFRA in the context of the development objectives of the TRIPS Agreement. At the same time, the World Intellectual Property Organization has established an Intergovernmental Committee to consider the access issue in the light of case studies which it has undertaken on the utilization of traditional knowledge in the exploitation of intellectual property rights.

The primary significance of bringing the access issue within the purview of the TRIPS Agreement, is that this Agreement has been implemented by most countries, including the USA. Any access rules that are developed under TRIPS will thus be globally enforceable.

References

Adcock, M. and Llewelyn, M. (2000) *Micro-organisms, Definitions and Options under TRIPS*. Quaker United Nations Office Programme, Occasional Paper 2.

ASSINSEL (1999) Development of new plant varieties and protection of intellectual property. ASSINSEL Statement approved by the CSTA Board of Directors as a CSTA Position Document, 21 July, 1999. cdnseed.org/press/A.99.47IP.htm

Barton, J. (1993) Introduction: intellectual property rights workshop. In: Crop Science Society of America, American Society of Agronomy, Soil Science Society of America. *Intellectual Property Rights: Protection of Plant Materials*. CSSA Special Publication No.21, CSSA, Madison.

Blakeney, M. (1996a) *Trade Related Aspects of Intellectual Property Rights. A Concise Guide to the TRIPs Agreement*. Sweet & Maxwell, London.

Blakeney, M. (1997b) Access to genetic resources: the view from the south. *Bioscience Law Review* 3, 97–103.

Blakeney, M. (1998) Intellectual property rights in the genetic resources of international agricultural research institutes – some recent problems. *Bioscience Law Review* 1, 3–11.

Bruce, D. and Bruce, A. (1998) *Engineering Genesis – the Ethics of Genetic Engineering in Non-human Species*. London, Earthscan.

Buck, S. (1998) *The Global Commons*. Earthscan, London.

[CBD] Convention on Biological Diversity (1996a) *The Impact of Intellectual Property Rights Systems on the Conservation and Sustainable Use of Biological Diversity and on the Equitable Sharing of Benefits from its Use*. ENEP/CBD/COP/3/22, 22 September.

[CBD] Convention on Biological Diversity (1996b) *The Convention on Biological Diversity and the Agreement on Trade Related Intellectual Property Rights: Relationships and Synergies*. UNEP/CBD/COP/3/23, 5 October.

[CBD] Convention on Biological Diversity (1999) *Report of Panel of Experts on Access and Benefit Sharing*. UNEP/CBD/COP/5/8, 2 November.

Correa, C.M. (2000) *Options for the Implementation of Farmers' Rights at the National Level*. Trade-Related Agenda, Development and Equity (TRADE) Working Paper 8. South Centre, Geneva.

Correa, C.M. (2001) Access to plant genetic resources and intellectual property rights. In: Blakeney, M. and Drahos, P. (eds) *Biodiversity and Agriculture: Regulating the Biosphere*. Sweet & Maxwell, London.

Das, O. (2000) Patenting and the ownership of genes and life forms. The Indian experience. *The Journal of World Intellectual Property* 3, 577–588.

Doll, J. (1998) The patenting of DNA. *Science* 280, 689–690.

Dutfield, G. (2000) *Intellectual Property Rights, Trade and Biodiversity: Seeds and Plant Varieties*. Earthscan and IUCN, London.

Ford, R. (1997) The morality of biotech patents: differing legal obligations in Europe? *European Intellectual Property Review* 6, 315–319.

Ghijsen, H.C.H. (1998) Plant variety protection in a developing and demanding world. *Biotechnology and Development Monitor* 36, 2–6.

Heller, M.A. and Eisenburg, R.S. (1998) Can patents deter innovation? The anticommons in biomedical research. *Science* 280, 698–701.

Jackson, L. (2000) Agricultural biotechnology and the privatization of genetic information. Implications for innovation and equity. *The Journal of World Intellectual Property* 3, 825–848.

Kongolo, T. (2000) New options for African countries regarding protection for new varieties of plants. *The Journal of World Intellectual Property* 4, 349–371.

Leskien, D. and Flitner, M. (1997) Intellectual property rights and plant genetic resources: options for a *sui generis* system. *Issues in Genetic Resources* No. 6. IPGRI, Rome.

Lesser, W. (1997) Assessing the implications of intellectual property rights on plant and animal agriculture. *American Journal of Agricultural Economics* 79(5), 1584–1591.

Llewelyn, M. (1999) European plant variety protection: a reactionary time. *Bio-Science Law Review* 3, 211.

Llewelyn, M. (2000b) The legal protection of biological material in the new millennium: the dawn of a new era or 21st century blues. *Bio-Science Law Review* 4, 123–130.

Louwaars, N. (1998) *Sui generis* rights from opposing to complementary approaches. *Biotechnology and Development Monitor* 36, 13–19.

Mangeni, F. (2000) *Technical Issues on Protecting Plant Varieties by Effective Sui Generis Systems*. South Centre, Geneva.

McKeough, J. (1997) Patent abuse: rights in genetic and environmental resources. *Polemic* 8(1), 31.

McNeely, R. (2001) Biodiversity and agricultural development: the crucial institutional issues. In: Lee, D.R. and Barrett, C.B. (eds) *Tradeoffs or Synergies? Agricultural Intensification, Economic Development and the Environment*. CAB International, Wallingford, UK, pp. 399–408.

Mooney, P.R. (1998) *The Parts of LIfe. Agricultural Biodiversity, Indigenous Knowledge, and the Role of the Third System*. Development Dialogue. Special Issue, Uppsala.

Myers, G. (2000) The UN Biodiversity Convention, Biotechnology and Intellectual Property Rights. *Bio-science Law Review* 4(3), 131–147.

Prakash, S. (2000) WTO rules. Do they conserve or threaten biodiversity. *The Journal of World Intellectual Property* 3, 155–165.

Stilwell, M. and Monagle, C. (2000) *Review of TRIPS Agreement Under Article 71.1*. South Centre, Geneva.

Suzuki, D. and Knudson, P. (1989) *Genethics, the Ethics of Engineering Life*. Stoddart, Toronto.

Tansey, G. (1999) *Trade, Intellectual Property, Food and Biodiversity: Key Issues and Options for the 1999 Review of Article 27.3(b) of the TRIPS Agreement*. Quaker Peace and Service, London.

van Overwalle, G. (1996) *Patentability of Plant Biotechnological Inventions*. Bruylant, Brussels.

Watal, J. (2000) Pharmaceutical patents, prices and welfare losses: policy options for India under the WTO TRIPS Agreement. *The World Economy* 23, 733–752.

Wells, A. (1994) Patenting new life forms: an ecological perspective. *European Intellectual Property Review* 3, 111–118.

[WWF] World Wildlife Foundation and [CIEL] Centre for International Environmental Law (2001) *Biodiversity & Intellectual Property Rights: Reviewing Intellectual Property Rights in Light of the Objectives of the Convention on Biological Diversity*. Joint Discussion Paper. WWF, Gland, Switzerland.

Further Reading

Alston, J.P., Pardey, G. and Rosenboom, J. (1998) Financing agricultural research: international investment patterns and policy perspectives. *World Development* 26, 1045.

Blakeney, M. (1996b) The Impact of the TRIPs Agreement in the Asia Pacific Region. *European Intellectual Property Review* 18, 544–554.

Blakeney, M. (1997a) Protection of traditional medical knowledge of indigenous peoples. *European Intellectual Property Review* 19, 298–302.

Blakeney, M. (1999) Biotechnology, TRIPs and the Convention on Biological Diversity' *Bio-Science Law Review* 4, 144–150.

Blakeney, M. (2000a) Intellectual property in agriculture. In: O'Brien, L. and Henry, R.J. (eds) *Transgenic Cereals*. St Paul, Minnesota, American Association of Cereal Chemists, pp. 277–301.

Blakeney, M. (2000b) Plant variety protection, international agricultural research and exchange of germplasm: legal aspects of the new *sui generis* regimes. In: Padolina, W. (ed.) *Plant Variety Protection of Rice in Developing Countries. Impacts on Research and Development*. IRRI, Los Banos, Philippines, pp. 9–22.

Blakeney, M. (2000c) Protecting plant varieties: patents or sui generis protection? In: Padolina, W. (ed.) *Plant Variety Protection of Rice in Developing Countries. Impacts on Research and Development*. IRRI, Los Banos, Philippines, pp. 119–126.

Blakeney, M. (2000d) Protection of traditional knowledge under intellectual property law. *European Intellectual Property Review* 22, 251–261.

Blakeney, M. (2001) Intellectual property aspects of traditional agricultural knowledge. In: Blakeney, M. and Drahos, P. (eds) *Biodiversity and Agriculture: Regulating the Biosphere*. Sweet & Maxwell, London.

Blakeney, M. (2002) Intellectual property rights and global food security. *Bio-Science Law Review* 5, 1–13.

Buttel, F.H. and Belsky, J. (1987) Biotechnology, plant breeding, and intellectual property: social and ethical dimensions. *Science, Technology, & Human Values* 12(1), 31–49.

[CIDSE] International Cooperation for Development and Solidarity (2000) Biopatenting and the threat to food security – a Christian and development perspective. CIDSE, Brussels. http://www.cidse.org/pubs/tg1ppcon.htm

[CBD] Convention on Biological Diversity (2000) *Access to Genetic Resources*. UNEP/CBD/COP/5/21, 1 March.

Duessing, J.H. (1996) The role of intellectual property rights in the exploitation of plant genetic resources and for technology transfer under the Convention on Biological Diversity. In: Brush, S.B. and Stabinsky, D. (eds) *Valuing Local Knowledge: Indigenous Peoples and Intellectual*

Property Rights. Island Press, Covelo, California, pp. 132–147.
Evenson, R.E. (1996) Economic valuation of biodiversity for agriculture. In: Pan American Health Organization (ed.) *Biodiversity, Biotechnology, and Sustainable Development in Health and Agriculture: Emerging Connections*. PAHO, Washington, DC, pp. 153–166.
FAO (1995) *Trade Restrictions Affecting International Trade in Non-wood Forest Products*. FAO, Rome.
FAO (2001) *Report of the Commission on Genetic Resources for Food and Agriculture, 6th Extraordinary Session*. FAO Doc. CGFRA-Ex 6/01/REP, June.
Fowler, C. and Mooney, P. (1990) *The Threatened Gene: Food, Politics and the Loss of Genetic Diversity*. Lutterworth Press, Cambridge, UK.
Johnston, S. and Yamin, F. (1997) Intellectual property rights and access to genetic resources. In: Mugabe, J. Barber, C.V., Henne, G., Glowka, L. and La Viña, A. (eds) *Access to Genetic Resources: Strategies for Sharing Benefits*. ACTS Press, Nairobi, pp. 245–269.
Ten Kate, K. and Laird, S. (1999) *The Commercial Use of Biodiversity. Access to Genetic Resources: People, Plants and Patents Revisited*. Earthscan, London.
Lawson, C. and Pickering, C. (2001) The conflict for patented genetic material under the Convention of Biological Diversity and the Agreement on Trade Related Aspects of Intellectual Property Rights. *Australian Intellectual Property Journal* 12, 104–115.
Leisinger, K.M. (1999) Ethical challenges of agricultural biotechnology for developing countries. In: Persley, G.J. and Lantin, M.M. (eds) *Agricultural Biotechnology and the Poor*. Proceedings of an international conference, Washington DC, 22–22 October 1999. CGIAR, Washington, DC.
Leskien, D. (1998) The European Patent Directive on biotechnology. *Biotechnology and Development Monitor* No. 36, 16–19
Lesser, W. (1998a) *Sustainable Use of Genetic Resources Under the Convention on Biological Diversity: Exploring Access and Benefit Sharing Issues*. CAB International, Wallingford, UK.
Lesser, W. (1998b) Intellectual property rights and concentration in agricultural biotechnology. *AgBioForum* 1(2), 56–61.
Llewelyn, M. (2000a) The patentability of biological material: continuing contradiction and confusion. *European Intellectual Property Review* 22(5), 191–197.
Nijar, G.S. (1996) *In Defence of Local Community Knowledge and Biodiversity*. Third Network Paper No.1. Penang, Third World Network (TWN).
Petit, M., Fowler, C., Collins, W., Correa, C. and Thornström, C. (2000) *Why Governments Can't Make Policy. The Case of Plant Genetic Resources in the International Arena*. CGIAR, Washington, DC.
Price, S.C. (1999) Public and private plant breeding. *Nature Biotechnology* 17, 938.
Principe, P.P. (1989) The economic significance of plants and their constituents as drugs. In: Wagner, H., Hikino, H. and Farnsworth, N.R. (eds) *Economic and Medicinal Plant Research* Vol. 3. Academic Press, London and San Diego, pp. 1–17.
Union of Industrial and Employers' Confederations of Europe UNICE (2001) Farmers' rights and *sui generis* protection of plant varieties: UNICE Preliminary Comments. *UNICE News* May.
UN Commission on Human Rights (2000) Sub-Commission on the Promotion and Protection of Human Rights. *Fifty-second Session, Agenda Item 4, The Realization of Economic, Social And Cultural Rights, Intellectual Property Rights and Human Rights*. Doc. E/CN.4/Sub.2/2000/7, 17 August.
United Nations Commission on Human Rights (2001) *Economic, Social and Cultural Rights. The Impact of the Agreement on Trade-Related Aspects of Intellectual Property Rights on Human Rights*. Report of the High Commissioner. E/CN.4/Sub.2/2001/13, 27 June.
UNCTAD (1996) *The TRIPS Agreement and Developing Countries*. United Nations, New York and Geneva.
UNDP (2000) *Human Development Report 2000*. UNDP, Washington, DC.
van Wijk, J., Cohen, J. and Komen, J. (1993) *Intellectual Property Rights for Agricultural Biotechnology: Options and Implications for Developing Countries*. ISNAR Research Report 3. ISNAR, The Hague, The Netherlands.
Wells, A.J. (1994) Patenting life forms: an ecological perspective. *European Intellectual Property Review* 3, 111–118.
Winter, G. (1992) Patent law policy in biotechnology. *Journal of Environmental Law* 4(2), 167–187.
World Trade Organization – Committee on Trade and Environment (1996) *Environment and TRIPS*. WT/CTE/W/8. WTO, Geneva http://www.wto.org
World Trade Organization – Committee on Trade and Environment (1999) *The Relationship between*

the Convention on Biological Diversity (CBD) and the Agreement on the Trade-related Aspects of Intellectual Property Rights (TRIPS); with a Focus on Article 27.3 (b). WT/CTE/W/125. WTO, Geneva.

Wright, B. (1997) Crop genetic resource policy: the role of *ex situ* genebanks. *The Australian Journal of Agricultural and Resource Economics* 41(1), 81–115.

Wright, P. (1998) Intellectual property and farmers' rights. In: Evenson, R., Gollin, D. and Santaniello, V. (eds) *Agricultural Values of Plant Genetic Resources*. CAB International, Wallingford, UK.

4 Genetically Engineered Food Labelling: Global Policy Polarization

Lydia Zepeda
Department of Consumer Science, University of Wisconsin, Madison, Wisconsin, USA

Abstract

Current scientific evidence indicates that there are few potential risks of genetically engineered (GE) food. However, the policy environment regarding GE food regulation and labelling falls into two camps. This chapter explores international differences in GE food policy. Economic, social and political forces motivating these differences are examined. An argument for consistent international standards to facilitate trade will be discussed.

Current Policies and Proposals[1]

US and European Union (EU) policies represent different approaches to regulation of genetically engineered (GE) food and labelling. In the USA, a 1992 Food and Drug Administration (FDA) policy document outlined a policy that did not require premarket approval for GE crops[2] (Department of Health and Human Services (HHS), 1992). While the FDA document identified potental risks of GE foods to public health,[3] it placed the responsibility for investigating and reporting these problems on the companies developing GE foods. The FDA recommends voluntary labelling of foods with identified problems but opposed mandatory labelling for GE food in general. A subsequent proposal by the FDA in 2001 modified US policy, requiring premarket review of all new GE foods on the basis that while GE foods are developed with techniques that are generally recognized as safe (GRAS), the products created may not be GRAS and may have unintended affects (Department of HHS, 2001). The FDA reiterated its opposition to mandatory labelling of GE foods, however, it provided recommended language for voluntary labels (Center for Food Safety and Applied Nutrition, 2001).

In contrast, the EU requires mandatory labelling of GE foods (*Official Journal of the European Communities*, 2000, L006/13–17). The legislation includes guidelines for wording and format of labels as well as establishes a

[1] This chapter does not reflect the views of the University of Wisconsin. Any errors are entirely the responsibility of Dr Zepeda.

[2] In addition, the US Environmental Protection Agency (EPA) regulates GE *Bt* crops as pesticides (Anderson and Milewski, 1999).

[3] Some of the risks to public health from GE crops identified in the 1992 FDA policy document include: transfer of genes from allergens; transfer of genes from known toxicants; use of antibiotic resistance selectable markers in GE foods; and GE foods that make up a large proportion of livestock feed, especially field maize.

1% *de minimis* threshold for adventitious contamination. In 2001 the EU lifted a moratorium on environmental release of GE crops (*Official Journal of the European Communities*, 2001a, L106/1–38).

Other countries that have proposed moratoria on environmental release of GE crops or mandatory labelling legislation include Brazil, Thailand, Japan, South Korea, Taiwan, Australia and New Zealand (Bett, 1999; Reuters News Service, 1999). Mandatory labelling legislation has also been proposed both in the US Congress and within the state legislatures of California, Minnesota, Nebraska, Vermont and Wisconsin (Schmickle, 2000b; Tradewell, 2001). This proliferation of policies among large food producers and large food importers has highlighted the need for international guidelines on GE food labelling. Codex Alimentarius Commission (Codex) standards are the internationally recognized 'benchmarks against which national food measures and regulations are evaluated within legal parameters of the Uruguay Round Agreements' (Understanding the Codex Alimentarius, www.fao.org/docrep/W9114E/W9114E00.htm).

The proposed Codex draft recommendations for labelling of GE foods highlight international differences regarding labelling (Codex Committee on Food Labelling, 2001). The drafting group for the proposal includes representatives from Australia, Brazil, Canada, Germany, European Commission, India, Japan, South Africa, Thailand and the USA. The proposal outlines several provisions for labelling GE foods, from the transfer of known allergens or toxicants to a comprehensive labelling provision. One provision also includes ethical reasons as a basis for labelling. The proposal recommends the establishment of an unspecified *de minimis* threshold level for adventitious or accidental inclusion of GE ingredients. It also proposes establishment of unspecified exemptions for certain highly processed foods (presumably such things as oils that contain insufficient genetic material to verify whether they are GE). The proposal outlines appropriate wording and format of labels for GE food. However, while negative labels were discussed they were not considered, nor included in the proposal.

Among the full working group, India and Norway have taken the lead in supporting comprehensive labelling, with India advocating no exceptions, no threshold levels and the inclusion of ethical grounds as a basis for labelling. The US position is that it does not support labelling in order to facilitate consumer choice nor on the basis of ethical reasons. The US comments also call for the precise definition of threshold levels and exemptions, and for guidelines on negative labels (USDA, 2001).

Different Approaches to Labelling Policy

The divergence in policies does not reflect a divergence in consumer attitudes. The attitudes of US consumers towards GE foods are remarkably similar to consumers elsewhere. Most surveys indicate that a high proportion (82–93%) of US consumers want GE food labelled (Consumers Union, 1999).[4] Outside the USA, support for labelling is high as well, from 74% in the EC, 80% in Australia, 92% in the UK, to 98% in Canada (Consumers Union, 1999).

The approaches towards GE policy are quite different, however. Both the US and EU policies rely on scientific information to inform the decision-making process regarding GE policy. Differences in GE food policy stem from different approaches to food policy in general. Three principles of EU food policy distinguish it from the USA. First, the EU is explicit in how it uses scientific information to inform the policy process. Their approach is to separate scientific risk assessment from risk management (Carsin, 2000). The rationale is twofold: to facilitate full transparency and accountability, and to acknowledge that there are complex social, economic and political dimensions in managing risk. In the USA, policies are developed by

[4] In a very long question regarding FDA policy, a 1999 International Food Information Council survey found that 58% of those surveyed favoured the FDA labelling policy. The question is somewhat confusing since it seems to imply that the FDA does not support labelling under any circumstances, which contradicts the FDA's policy document.

appointed officials within regulatory agencies (FDA, EPA and US Department of Agriculture (USDA)) who are advised by staff scientists. While US policy is science based, decisions have been made against the recommendations of its own staff scientists (Reynolds, 2000; Eichewald *et al.*, 2001).

Second, for all food policy, the EU assessment of scientific information is based on the precautionary principle (*Official Journal of the European Communities*, 2001b, C96 E254). Under this principle, uncertainty about risk to health requires further scientific investigation and risk assessment. This approach explicitly recognizes and provides a mechanism for dealing with contradictory scientific information. In contrast the US approach can be characterized as innocent until proven guilty.

Finally, there is a different focus of food policy in the EU versus the USA. The primary purpose of the EU food policy is the 'protection of the interest of consumers and shall provide a basis for consumers to make informed choices in relation to the foods they consume' (*Official Journal of the European Communities*, 2001b, C96 E/254). The focus of the FDA is not consumers or their choices in the USA but the food itself, 'the FDA is responsible for ensuring all foods in the American food supply conform to the applicable provisions in the law . . . [it] has broad authority to regulate the safety and wholesomeness of food' (Department of HHS, 2001, p. 4708).

Differences in Economic Incentives

These differences in approaches and intent offer some explanation for the differences in policies. The applications and the location of GE food production offer further economic incentives that may influence GE policy. Applications of GE foods have focused on developed countries' foods and markets. The overwhelming majority of acreage of GE crops are maize and soybeans, used for livestock feed and processed foods and thus eaten by (in global terms) high-income consumers. It is estimated that 74% of the acreage of GE crops is in the USA, 15% in Argentina, 10% in Canada and 1% in the rest of the world (*Biodemocracy News*, 2000).

Clearly the USA has a large economic stake in GE food production and sales, while the benefits to the rest of the world are not as straightforward given the current applications. However, even within the USA, not everyone gains. So far, GE crops have not been developed with consumer benefits in mind. The majority of GE crops are herbicide tolerant or *Bt* crops. The cost savings of pesticide applications due to using *Bt* maize are estimated to be between US$2.80 and US$14.50 per acre (US$6.9–35.8 per ha) (Carlson *et al.*, 1997). However, given the acres planted to *Bt* maize in 1999 this was more than offset by the estimated loss to farmers of US$300 million in overseas sales attributed to unwanted GE maize (Schmickle, 2000a). The loss of export markets for the USA threatens to worsen; countries enacting bans or mandatory labelling of GE foods represent about 43% of the 1998 US agricultural export market (US Foreign Agricultural Service, 1999).

Thus companies that sell GE seed and those who work for them are clear gainers. Under a policy of mandatory labelling, they can pass the cost of labelling on to consumers. Liability costs on the other hand generally affect a single company, making it difficult for them to pass the costs on to consumers without becoming uncompetitive. Fuelling concerns about liability are insurance underwriters either seeking compensation for underwriting the risk of GE food or shifting liability (IATP/TWN, 1998; Madeley, 1999).

Public Perceptions of Risk

The essential components of any labelling policy are that it has clear standards, testing, certification and enforcement (Golan *et al.*, 2000). Labelling is particularly effective where consumer preferences differ, i.e. where there is not clear agreement about a product's characteristics (Magat and Viscusi, 1992). This permits consumers to align their preferences with their purchases. Consumer risk perceptions are also reduced where they perceive clear benefits from a product (Slovic, 1987). That most consumers would use labels to make purchase decisions, whether verifiable or not,[5] is probably unlikely. This does not mean that labels would not have an impact. Apart from making it possible to trace any potential problems, labels by themselves serve to reduce the perception of risks associated with GE food. Consumers can choose to incorporate the label information in their buying decision, or not. More importantly it permits informed consent, that is, it transforms risk perceptions from being 'involuntary' to 'voluntary' (Thompson, 1996). Theoretically and empirically, this reduces the perception of risk. A recent study demonstrated that availability of labels reduces risk perceptions towards GE food (Zepeda *et al.*, 2001), irrespective of whether people act on the information.

The mandatory labelling policy of the EU is aimed at addressing consumer concerns and allowing consumers to exercise choice. The use by the EU of the precautionary principle in risk assessment and management is to provide convincing evidence of the safety of foods in general and thereby enhance public confidence. The objective of the Codex is to establish international standards to facilitate consumer choice and international trade. In contrast, the position of the US government with regards to GE foods is that it does not support consumer choice as a basis for labelling policy (USDA, 2001).

In the USA, the definition of organic excludes GE ingredients. Organic sales have climbed, driven in large part by the demand for GE-free food. US food manufacturers are using voluntary GE-free labels for conventional foods as well to increase sales or prevent loss of sales due to consumer concerns about GE foods. Individual companies (Nestlé, Gerber, Heinz, FritoLay, MacDonalds and Iams) have banned all GE ingredients in some food lines, particularly those consumed by babies, children and pets (Bett, 1999; Schmickle, 2000b).

Conclusions

The USA and the EU represent two different political approaches to regulation of GE foods. In the EU food policy has been driven by the precautionary principle, the focus on consumer choice and the principle that risk management and scientific risk assessment ought to be separated. The resulting legislation requires strict evaluation of GE crops before approval and mandatory labelling of all GE foods.

In the USA GE food policy has been one in which companies are entrusted to determine if there are problems and to inform the FDA and the public. There is no system of standards, testing, certification or enforcement regarding problematic GE foods, the elements required for an effective labelling system (Golan, 2000). Further, US policy opposes mandatory labelling of GE foods. This has resulted in attempts to introduce labelling legislation to assuage consumers and protect agricultural export markets placed in jeopardy by the proliferation of labelling legislation among trading partners.

In both environments, policy makers are making decisions regarding regulation of GE food using scientific information. What greatly differs between the policy environments is the level at which decisions are made. In the USA, low-level political appointees make the policy decisions. In the European Community, elected officials make the decisions. In both cases, scientists provide advice and council, but clearly the incentives and accountability of policy makers are different.

Along with political differences, there are clearly economic incentives that influence the policy scenario. Not only does the USA have the majority of biotechnology firms it also has the vast majority of the world's area of GE crops. Clearly biotech firms have a strong incentive to oppose any kind of labelling in the litigious USA to minimize their liability exposure.

While there are clearly political and economic differences that have influenced GE food labelling policy, this chapter has not touched on the social or cultural differences that have been raised in the Codex Alimentarius working group on GE labelling. If GE food is ever to become a factor in the developing world it will be important to understand under what, if any, conditions GE food conforms with religious restrictions on food

[5] While for some GE foods labels might be difficult to verify, cheap tests (US$5.75) are available for some foods (Bett, 1999). The demand for developing such tests has spurred a growing industry.

and food preparation. Under what conditions is GE food Halal, vegan, vegatarian, Jain, kosher, etc.?

Consistent international standards are necessary to decrease consumers' perception of risk and thereby decrease loss of sales to farmers and to facilitate international trade of agricultural products and food, as well as to facilitate the development of agricultural biotechnology applications. Given that most GE food applications involve foods that are traded and indeed are the major trade commodities, differences in GE policies can cost farmers markets. Indeed, it is estimated that US farmers lost US$300 million in overseas sales in 1999 to GE maize alone (Schmickle, 2000a). Clearly, there are strong public welfare and efficiency arguments for having consistent labelling and clear market information.

References

Anderson, J.L. and Milewski, E. (1999) Regulation of plant-pesticides: current status. March. http://www.epa.gov/oppbppd1/biopesticides/otherdocs/ncipm_speech.htm

Bett, K. (1999) Mounting evidence of genetic pollution from GE crops: Growing evidence of widespread contamination. *Environmental Science and Technology* 1 December, www.purefood.org/ge/gepollution.cfm

Biodemocracy News (2000) 30, November, p. 2.

Carlson, G.A., Marra, M.C. and Hubbell, B. (1997) *The Economics of First Crop Biotechnologies*. North Carolina State University, Raleigh, USA.

Carsin, B. (2000) The role of science in EU regulatory policy. *AgBioForum*. www.agbioforum.org/vol3no23/vol3no23ar10carsin.htm

Center for Food Safety and Applied Nutrition, US Food and Drug Administration (2001) Guidance for industry: Voluntary labeling indicating whether foods have or have not been developed using bioengineering, draft guidance. January. www.cfsan.fda.gov/~dms/biolabgu.html

Codex Committee on Food Labelling (2001) Proposed draft recommendations for the labelling of food and food ingredients obtained through certain techniques of genetic modifcation/genetic engineering (proposed draft amendment to the *General standard for the labelling of prepackaged food*). Twenty-ninth session, Ottawa, Canada, 1–4 May, 2001. CX/FL 01/7. February.

Consumers Union (1999) Summary of public opinion surveys related to labeling of genetically engineered foods. www.consumersunion.org/food/summpollny699.htm

Department of Health and Human Services, Food and Drug Administration (1992) Statement of policy: foods derived from new plant varieties. *Federal Register* 57(104), 29 May, pp. 22984–23001.

Department of Health and Human Services, Food and Drug Administration (2001) Premarket Notice Concerning Bioengineered Food. *Federal Register* 66(12), 18 January 18, pp. 4706–4738. www.cfsan.fda.gov/~lrd/fr010118.html

Eichewald, K., Kolata, G. and Peterson, M. (2001) Biotechnology food: from lab to a debacle. *New York Times*, 25 January.

Golan, E., Kuchler, F. and Mitchell, L. (2000) *Economics of Food Labeling*. Economic Research Service, US Department of Agriculture. Agricultural Economic Report No. 793. December.

Institute for Agricultural and Trade Policy/Third World Network (IATP/TWN) (1998) Insurers wary of GMOs and GE products. 9 December. www.purefood.org/ge/insurege.cfm

Madeley, J. (1999) Insurance industry refuses to insure 'unpredictable risks' of genetic engineering. 27 January. www.purefood.org/ge/insurege.cfm

Magat, W. and Viscusi, W.K. (1992) *Informational Approaches to Regulation*. MIT Press, Cambridge, Massachusetts.

Official Journal of the European Communities (2000) Commission Regulation (EC) No. 49/2000 of 10 January 2000 amending Council Regulation (EC) No 1139/98 concerning the compulsory indication on the labelling of certain foodstuffs produced from GMOs of particulars other than those provide for in Directive 79/112/EED. L006, 1 November, 13–17

Official Journal of the European Communities (2001a) Directive 2001/18/ED of the European Parliament and of the Council of 12 March 2001 on the deliberate release into the environment of GMOs and repealing Council Directive 90/220/EED. L106, 4 April, 1–38.

Official Journal of the European Communities (2001b) Proposal for a Regulation of the European Parliament and of the Council laying down the general principles and requirements of food law establishing the European Food Authority. 27 March, pp. C96 E/247–268.

Reuters News Service (1999) GMO food regulations in Asia. www.gene.ch/genpost/1999/Jul-Dec/msg00033.html

Reynolds, R. (2000) FDA ignored own scientists in 1992 policy statement: genetically engineered foods were dubbed safe in spite of concerns. *Mother Jones*. 4 January.

www.motherjones.com/mother_jones/JF00/pandora.html
Schmickle, S. (2000a) *Star Tribune* 30 April.
Schmickle, S. (2000b) *Star Tribune* 1 May.
Slovic, P. (1987) Perceptions of risk. *Science* 236(17), 280–285.
Thompson, P.B. (1996) Food labels and the ethics of consent. *Choices* First Quarter, 11–13.
Tradewell, R. (2001) 25 April. Draft legislation 97.15, State of Wisconsin, USA.
'Understanding Codex Alimentarius' www.fao.org/docrep/w9114e/W9114eo1.htm
US Foreign Agricultural Service (1999) US Agricultural, Fish and Wood Products Exports, 1998. July. www.fas.usda.gov:80/info/factsheets/exptfy98.html
[USDA] US Department of Agriculture (2001) US government comments on agenda item 5(b) draft documents coming before the CCFL small working group on the labelling of foods derived from biotechnology. 1–4 May, Ottawa, Canada. www.fsis.usda.gov/OA/codex/CCFL_BIO.htm
Zepeda, L., Douthitt, R. and You, S.Y. (2001) *Consumer acceptance of agricultural biotechnology: the role of labeling and risk perceptions on food demand*. University of Wisconsin-Madison.

5 Conflict and Consensus-building: International Commercial Policy and Agricultural Biotechnology

Jill E. Hobbs,[1] William A. Kerr,[1] J.D. Gaisford,[2] Grant Isaac[3] and Kurt K. Klein[4]

[1]Department of Agricultural Economics, University of Saskatchewan, Saskatoon, Saskatchewan, Canada; [2]Department of Economics, University of Calgary, Calgary, Alberta, Canada; [3]College of Commerce, University of Saskatchewan, Saskatoon, Saskatchewan, Canada; [4]Department of Economics, University of Lethbridge, Lethbridge, Alberta, Canada

Abstract

The rules of international trade have been slow to adjust to the challenges presented by the commercialization of agricultural biotechnology. The controversy has centred on differing domestic perceptions and management of the uncertainty regarding issues of human health and environmental sustainability. The agreement to launch a new round of World Trade Organization (WTO) trade negotiations at Doha in November 2001 provides an opportunity to address international commercial policy conflicts pertaining to biotechnology. This chapter outlines the major international commercial policy issues pertaining to biotechnology. It explains the sometimes complementary and sometimes conflicting roles of the WTO and other international institutions in dealing with different ways of regulating trade in the products of biotechnology.

Introduction

The regulatory regime(s) governing international trade has been slow in adapting its rules to regulate the international movement of these products. This is a complaint both of those who are making investments in commercializing the products of biotechnology, and those who wish to impose a strict version of the 'precautionary principle' on the licensing of the products of biotechnology. The last round of multilateral World Trade Organization (WTO) negotiations had its agenda established in Uruguay in 1986 before the commercialization of biotechnology. A new round was not agreed until the WTO Ministerial Meeting in Doha, Qatar in the autumn of 2001. The intervening 15 years has seen both rapid advances in the ability to commercialize the products of biotechnology and the formation of opinions regarding the desirability of having those products available in the marketplace. These opinions differ both within countries and between countries. As a result, nations have been independently developing and implementing domestic regulatory regimes to govern their research, development, commercialization and marketing. Some of the regulatory regimes conflict with existing WTO commitments making it difficult to achieve a consensus regarding a revised international trade regime. The new 'Development Round' of WTO negotiations, however, provides an opportunity to devise a new international regulatory regime.

A Divergence in Policy Approaches

National policies towards genetically modified (GM) food appear to diverge on three key fronts, all of which have important implications for international trade: regulation of 'product' versus 'process'; risk assessment; and labelling.

A product vs. process regulatory approach

The USA, Canada and several other exporting countries have similar domestic regulatory approaches. These countries take a *product-based* approach, focusing on the safety of the product, rather than on the process by which it was produced. GM food products are subject to the same set of health and safety and approval regulations as conventional foods. The principle of 'substantial equivalence' is applied in determining whether a product should be approved for sale in the marketplace. This means that GM products are judged alongside similar non-GM products if it can be shown that the two are substantially equivalent, i.e. if the risk from consuming a GM product is no different than the risk from consuming a conventional version of the same food. If substantial equivalence cannot be demonstrated, however, the new characteristics are the subject of further safety assessments.

In contrast, in the European Union (EU), a separate regulatory framework was established to deal with GM food approvals. The 1997 Novel Foods Regulation introduced a pre-market approval system for novel foods (including GM food). Additional environmental legislation applies to organisms that will be released into the environment. A manufacturer or importer must show that commercialization of the GM food does not pose a risk to human health or to the environment. A key difference in the approach adopted by the EU is that the *process* by which the food is produced becomes important. If the food is genetically modified an entirely separate set of regulations applies. This has important implications in the international trade arena.

Risk assessment

A cornerstone of the EU approach to regulating GM goods is the *precautionary principle* and applies to situations in which uncertainty exists. Where risk assessments have been carried out but there are clear limitations to the science underlying the risk assessment, the precautionary principle is applied. This represents a fundamental difference between the regulatory approaches of the EU and that of Canada and the USA. The onus is on EU regulatory authorities *not* to approve GM products until it can be proved conclusively that they are safe (Perdikis, 2000). Proving absolute safety from a scientific standpoint is very difficult, if not impossible. This has led industry groups in exporting countries to accuse the EU of diverging from science as a basis for product approval and market access, pandering instead to protectionist political pressure.

Labelling

The EU has adopted a regulation requiring the mandatory labelling of all foodstuffs containing GM ingredients above a 1% threshold. Japan has adopted a similar policy, although with a 5% threshold, Australia and New Zealand are also introducing mandatory labelling. Officially, Canada and the USA remain opposed to mandatory labelling, arguing that it constitutes an unnecessary technical barrier to trade given the absence of scientific evidence showing a difference in the safety of GM and non-GM foods.

Diverging national policy approaches between trading nations create international trade tensions. The nature of the information problem facing consumers further complicates the situation.

The Consumer Information Problem

It is clear that there are consumer segments with different attitudes and that objections to GM products come from a number of different standpoints, including food safety, environmental and ethical concerns (Gaisford *et al.*, 2001). For those with food safety and environmental objections to GM foods, the problem is one of asymmetric information and uncertainty. For example, the food safety concern of many consumers is not 'Will I eat this food and be ill by lunch-time?', rather it is 'If I eat this food over a long period of time, will it have an unforeseen, deleterious effect on my health?'. The same can probably be said for environmental concerns, which generally relate to perceptions over the potential long-term negative impacts of GM crops as a result of out-crossing, etc.

A fear of the unknown presents a particular problem for policy makers. One cannot attach statistical probabilities to the likelihood of something completely unknown and totally unforeseen becoming a problem in the future – a state of *uncertainty* rather than risk (Knight, 1921). With insufficient information, designing an appropriate policy is extremely difficult.[1]

The second aspect of the information problem is the *information asymmetry* faced by consumers and firms downstream in the product's supply chain (i.e. retailers, further processors, etc.). Information asymmetry means that one party to the transaction (often the seller) has more information about the true quality characteristics of a good than the other party (i.e. the buyer). This is true for GM foods because the genetically modified organism (GMO) is a *credence* characteristic, meaning that consumers are not able to detect its presence either before purchase or after consumption of the food (Hobbs and Plunkett, 1999). The presence (or absence) of credence characteristics is usually signalled through product warranties, brand name reputation, labelling and certification, etc.

The conundrum facing policy makers is multifaceted. There exist different consumer preferences both across and within countries with respect to GM foods. Diverging regulatory approaches at the national level have led to different degrees of product approval and market access for GM foods. The inability of consumers or downstream firms to detect the presence of GMOs in complex food products leads to an information problem that is compounded by uncertainty.

A number of national policy responses regarding the production and import of GM food have emerged or are under consideration. These range from not allowing any form of product labelling about GMOs, to voluntary labelling (e.g. 'GM-free' or 'contains/may contain GMOs'), to mandatory labelling of food containing GMOs, to an outright ban on the production and importation of GM food. Diverging product labelling and market access policies for GM foods have become contentious trade issues. The WTO plays a pivotal role in guiding international trade relations between countries.

The WTO

The WTO was created as an outcome of the Uruguay Round of General Agreement on Tariffs and Trade (GATT) negotiations (1986–1993). The GATT was established in 1947 with the central objective of limiting the ability of domestic vested interests to obtain protection against imports. Initially its primary focus was the reduction of tariffs and it has met with considerable success in this regard. Progress on non-tariff barriers, however, was limited. At the Uruguay Round GATT negotiations important changes were made to the multilateral trading environment to facilitate the future reduction of non-tariff barriers to trade. The new Agreement on Sanitary and Phytosanitary (SPS) measures and the Agreement on Technical Barriers to Trade (TBT) are key for the international regulation of GM products.

It is important to remember that the WTO/GATT is not an international legal system. Rather it is a set of political compromises between sovereign nation states. The WTO operates on a set of core principles including non-discrimination, transparency, compensation and accepted retaliation.

[1] In situations of incomplete information or uncertainty, objective probabilities of various health and environmental outcomes are not yet known. In this setting, individuals may assign vastly different subjective probabilities to the various outcomes. Widely different individual opinions both within and across countries present significant challenges for domestic and international policy. Gradually, as new information becomes available, these problems may ease as subjective probabilities are updated (e.g. by *Bayes' Rule*) and converge on the underlying objective probabilities. In the short term, however, this is little consolation.

The agreement on SPS measures

The SPS Agreement was established in response to a concern that regulations with a legitimate domestic purpose of protecting human, animal or plant health and safety could also be misused to protect domestic producer interests by restricting imports. For this reason, an attempt was made to move the debate over SPS measures out of the political realm and into the realm of science. A core element of the SPS Agreement is that any trade restricting measure be judged on scientific criteria.

If a country feels that there is insufficient evidence to make a definitive judgement on the safety of a food, animal or plant product, there are provisions within the SPS Agreement which allow it to put in place temporary domestic regulations and trade restricting measures. In the meantime, the country is expected to seek out information to clarify the issue. The emphasis here is on 'temporary', meaning immediate food safety, animal or plant health concerns. The provision is not supposed to be used in situations where extensive information gathering over a long period of time is required to clarify a safety issue, such as might be the case with regulations concerning GM foods (Roberts, 1998).

The SPS Agreement recognizes that perceptions of safety are relative and may vary across countries; therefore, individual countries are allowed to specify their own acceptable levels of risk tolerance in their regulations. However, the risk tolerance levels allowed for one category of imports must be the same as those applied to imports of other products.

The framers of the SPS agreed to international cooperation in the design of food safety, sanitary and phyto-sanitary regulations. To this end, a major role was envisaged for international organizations outside the WTO process in developing internationally agreed scientific principles, standards and guidelines. These include the Codex Alimentarius Commission (Codex) for food safety, the International Office of Epizootics for animal health and the Secretariat of the International Plant Protection Convention for plant health. These organizations develop common sets of standards and recommendations based on relatively protracted, consensus-building negotiations (Kerr, 1999).

The agreement on TBT

As with the SPS Agreement, the intention behind the TBT Agreement is to prevent the nefarious use of domestic regulations by politicians to restrict imports, thereby protecting domestic industries. Unlike the SPS, however, the TBT Agreement is not rooted in scientific principles as an arbiter of international trade disputes. It deals with technical measures such as packaging and labelling regulations or product standards rather than the safety of the product. Standard WTO principles still apply, such as non-discrimination, transparency, etc. The TBT allows importers to require labelling if a product is sufficiently different from existing products to be considered novel. However, the novelty must be based on the characteristics of the product and not the production or processing method used to produce it (Isaac *et al.*, 2002). As biotechnology is a process, this distinction has become a contentious issue.

GM Foods within the WTO Framework

Under the current WTO structure, questions over market access and labelling of GM foods would probably fall under the remit of either the SPS or the TBT Agreement. There is a debate over which Agreement would be most appropriate. Fundamentally this comes down to a disagreement over the perceived safety of GM foods. If they are considered safe, then the evaluation of any trade measures taken to restrict market access or to require labelling of GM foods would fall under the remit of the TBT Agreement. If the concern relates to a scientific appraisal of product safety, then the SPS Agreement applies and a risk analysis is conducted. Clear classification of the labelling of GM foods as an SPS or a TBT issue has yet to be forthcoming because there is no agreement on their safety (Caswell, 1999). The lack of clarity on this issue only serves to muddy the waters of international commerce, creating uncertainty for firms – something the GATT/WTO was created to avoid.

The controversy over the labelling of GM foods is symptomatic of a wider problem within the international trading environment. Consumer preferences are highly differentiated and, increasingly, 'process attributes' are of importance in consumer purchase decisions. We can think of a product as consisting of a series of attributes or characteristics. For food products these might include tangible product attributes such as taste, texture, colour, fat content, price, package size, etc. However, it might also include intangible process attributes which relate to the means by which the product was produced, e.g. animal welfare friendly, environmentally friendly, GMO-free, etc. A cornerstone of the WTO is that countries cannot impose national standards on the processes used to produce foreign products to prevent discrimination between nations in the use of trade measures.

Another core principle of the WTO is the Most-Favoured Nation Principle that requires countries to treat 'like' products from all countries in the same manner. The USA and Canada would argue that, based on currently available scientific evidence, GM and non-GM rape (canola) products are 'substantially equivalent'. However, if the presence or absence of a GMO causes some consumers to consider them to be completely different products, not substitutes in any sense of the word, then are they still 'like' products? On the other hand, accepting that these are not 'like' products could open the floodgates to misuse of trade restricting measures based on dubious and difficult to substantiate consumer concerns.

At the heart of this issue is a fundamental shift in who is asking for protection from imports. The underlying premise of the GATT/WTO is to reduce the opportunities and incentives for countries to restrict trade in response to demands for protection by domestic producers. What is clear in the case of GMOs, is that it is *consumer and environmental interest groups*, not producer interest groups, who are asking for protection from what are perceived to be harmful imports. This difference is critical.[2] The fundamental principles of the GATT/WTO, its institutional structure and its ancillary Agreements are all oriented towards reducing the threat of producer-based protectionism. This leaves the system ill-equipped to deal with demands for trade restricting measures from other groups.

What is the appropriate institutional framework within which to deal with these issues? In 1998, then EU agricultural commissioner, Franz Fischler suggested that the EU would like to renegotiate the SPS Agreement in order to allow trade restrictions in response to consumer preferences. The US government was quick to denounce what it saw as a highly disturbing proposal that threatened the achievements made in establishing the SPS Agreement. There is merit in both viewpoints. The EU was simply responding to the need to recognize that consumer preferences differ across countries and that these preferences must be given credence. On the other hand, the US was correct in rejecting meddling with the SPS Agreement to accommodate consumer preferences.

The SPS Agreement deals specifically with the scientifically verifiable health and safety of products. It is a very important mechanism with which to deal with the deliberate misuse of health and safety related trade restrictions to protect domestic producers. The problem with GM foods is that there is not enough information to make this a health and safety issue. The balance of scientific evidence to date suggests that the GM foods which have been approved within the EU, the USA and Canada are safe. However, this does not change the fact that some consumers do not wish to consume GM food for a variety of perceived food safety and ethical reasons. In the absence of methods to segregate and identify GM and non-GM food products, consumers will be unable to distinguish between the two. This is an *information* problem, not a health problem.

To ignore the legitimacy of consumer preferences as a source of protectionist pressure is, however, short-sighted. Moreover, it places the entire WTO system at risk. The WTO remains a political compromise, a set of voluntarily agreed to 'rules' by which international trade is governed. It represents a balance between assuring

[2] Calls to limit or halt the approval of GM products and for mandatory labelling of GM foods in the EU have come from a wide variety of consumer and environmental groups. Noticeably absent from this clamour for regulation have been EU agricultural producer interests, many of whom would stand to benefit from the agronomic advantages which the new input-trait GM crops confer.

a stable international commercial environment for firms and the desire of politicians to respond to domestic political pressure for protection – pressure that historically has come from producers. This is the political reality of the WTO. Increasingly, however, we must recognize that a demand for protection from imports might come from other groups in society such as consumer groups, environmental groups, etc. To deny politicians the option of responding to political pressure from these groups denies the political reality of the WTO as a compromise between these conflicting pressures. The basic premise on which the WTO is based still holds, i.e. that the costs of exercising a political 'out' from one's WTO commitments should be sufficiently large to deter abuse of this option (Perdikis *et al.*, 1999).

The WTO has a central role to play in resolving international trade disputes over the products of biotechnology. This is a role fraught with many challenges. A key challenge lies in adapting the trade liberalization framework to allow explicitly for the notion that, in addition to traditional demands for protection from producer interest groups, consumers may demand protection from imports. In a world of imperfect information in which quality signals (e.g. product labelling, certification, etc.) are necessary to inform consumers of the underlying attributes of a product, these may be legitimate calls for protection. An even greater challenge will lie in incorporating consumer interests explicitly into existing or new WTO agreements without the abuse of these changes by protectionist producer interests. While the WTO retains its central role, two other international institutions also have important roles to play in the trade environment for GM food: the Codex and the Cartagena Biosafety Protocol (CBP).

The Codex Alimentarius Commission

Established in 1961 by the UN Food and Agriculture Organization, the objective of Codex is to encourage the formulation and harmonization of food standards worldwide through consensus-building discussion and negotiation. The adoption of Codex standards as the arbiter of food safety issues by the SPS Agreement has increased the impact of its standards. Two aspects of the Codex are important with respect to GM Foods. The Codex Committee on Food Labelling is host to international discussions on the rules for food labelling, including biotechnology. Further, in 1999 an 'Ad Hoc Intergovernmental Task Force on Foods Derived from Biotechnology' was established to develop standards for GM foods. Discussions will focus on scientific evidence, risk analysis and 'other legitimate factors relevant to the health of consumers and the promotion of fair trade practices' (Codex Alimentarius Commission, 1999, p. 53). The 'other legitimate factors' have subsequently been identified as: ethical, religious and cultural considerations, consumer concerns/interest, food security, enforcement capacity, environmental risk, facilitating international trade and food diversity (Codex Alimentarius Commission, 2000). The Task Force is due to complete its work by 2003 and will coordinate as necessary with other Codex Committees, such as the Committee on Labelling. With trade disputes over GM foods looming, it remains to be seen whether the recommendations of the Task Force will be sufficiently timely to deal effectively with trade tensions in this area (Hobbs, 2001).

The CBP

In January 2000, 138 countries agreed to the CBP dealing with the safe transfer, handling and use of 'living modified organisms' resulting from biotechnology. The Protocol covers both organisms that are intended for release into the environment and those destined for consumption. The CBP contains a number of clauses that are in direct contradiction to existing WTO commitments (Isaac *et al.*, 2002). Under the agreement, international commodity shipments which 'may contain' genetically modified organisms must be labelled – a process label. It accepts the 'precautionary principle' meaning that 'scientific principles' are no longer the arbiter of trade rules and it explicitly allows socio-economic impacts as a reason to impose trade barriers. The CBP has no dispute mechanism meaning importers regulations cannot be challenged. In short, it is open for capture by protectionist interests. The CBP must be ratified by 50 countries before it come into force. The USA is not a party to the CBP.

This raises the important question as to whether the CBP or the WTO will take precedence. It would appear that the CBP has pre-empted a WTO decision on this issue. However, a possible 'out' is provided to countries by a 'savings clause' within the CBP which states that the Protocol does not override countries' rights and obligations under other international agreements, such as the WTO (Isaac *et al.*, 2002). If ratified, the parallel but conflicting WTO and CBP will leave trade law pertaining to biotechnology in a considerable muddle. Certainly, there is little transparency for those considering investing in biotechnology.

Conclusions

The frontlines of international trade are likely to be the place at which the clash over diverging national policies towards GM foods and differences in consumer preferences is most keenly felt. This presents a challenge to the international trading system. The challenge lies in reorienting the WTO to deal with pressure for protection from imports from interest groups in society other than domestic producer groups. The core principles of the WTO remain important and the basic premise that there are net gains in economic welfare from freer trade remains valid. Open discussion and debate over the various options of labelling and approval for GM and GM-free foods is extremely important. As consumers are not a homogeneous mass either within or across borders, it behoves decision makers to accommodate these differences in international trading rules while preserving the integrity of the international trading system. These are the challenges for the Doha Round negotiators.

References

Caswell, J. (1999) An evaluation of risk analysis as applied to agricultural biotechnology (with a case study of GMO labeling). *Transitions in Agricultural Biotechnology: Economics of Strategy and Policy*, NE-165 Conference Proceedings, Washington, DC.

Codex Alimentarius Commission (1999) *Codex Alimentarius Commission 23rd Session Report*. FAO Headquarters, Rome, 28 June–3 July.

Codex Alimentarius Commission (2000) *Report of the First session of the Codex Ad Hoc Intergovernmental Task Force on Foods Derived from Biotechnology*. UN/FAO Headquarters, Rome, April. Retrieved from the worldwide web 3 July 2000. ftp://ftp.fao.org/codex/ALINORM01/al01_34e.pdf

Gaisford, J.D., Hobbs, J.E., Kerr, W.A., Perdikis, N. and Plunkett, M.D. (2001) *The Economics of Biotechnology*. Edward Elgar, Cheltenham, UK.

Hobbs, J.E. (2001) Labelling and consumer issues in international trade. In: Michelmann, H.J., Rude, J., Stabler, J. and Storey, G. (eds) *Globalization and Agricultural Trade Policy*, Lynne Rienner Publishers, Boulder, pp. 269–285.

Hobbs, J.E. and Plunkett, M.D. (1999) Genetically modified foods: consumer issues and the role of information asymmetry. *Canadian Journal of Agricultural Economics* 47(4), 445–455.

Isaac, G.E., Phillipson, M. and Kerr, W.A. (2002) *International Regulation of Trade in the Products of Biotechnology*. Estey Centre Research Paper No. 2, Estey Centre For Law and Economics in International Trade, Saskatoon, Canada.

Kerr, W.A. (1999) International trade in transgenic food products: a new focus for agricultural trade disputes. *The World Economy* 22(2), 245–259.

Knight, F. (1921) *Risk, Uncertainty and Profit*. Houghton Mifflin, Boston, Massachusetts.

Perdikis, N. (2000) A conflict of legitimate concerns or pandering to vested interests? Conflicting attitudes towards the regulation of trade in genetically modified goods – the EU and the US. *The Estey Centre Journal of International Law and Trade Policy* 1(1), 51–65. http://www.esteyjournal.com

Perdikis, N., Kerr, W.A. and Hobbs, J.E. (1999) Can the WTO/GATT Agreements on Sanitary and Phyto-Sanitary Measures and Technical Barriers to Trade be regnegotiated to accommodate agricultural biotechnology? *Transitions in Agricultural Biotechnology: Economics of Strategy and Policy*, NE-165 Conference Proceedings, Washington, DC.

Roberts, D. (1998) Preliminary assessment of the effects of the WTO Agreement on Sanitary and Phytosanitary Trade Regulations. *Journal of International Economic Law* 1, 377–504.

6 The Rationale Behind WTO Agreements and Agricultural GMO Controversy[1]

Roberto Esposti[1] and Alessandro Sorrentino[2]
[1]*Dipartimento di Economia, Università Politecnica delle Marche, Ancona, Italy;*
[2]*Dipartimento di Studi Aziendali, Tecnologici e Quantitativi, Università della Tuscia (Viterbo), Italy*

Abstract

This chapter deals with the trade implications of the different policy and regulatory options currently adopted for managing the main issues raised by the recent large diffusion of genetically modified (GM) crops. The chapter focuses on the World Trade Organization (WTO) negotiations and the conflicts that emerged in this context with respect to the GMO trade. The analysis is based on the recent theoretical contributions on the WTO agreements as Pareto-enhancing devices. The chapter analyses under which conditions this approach to the WTO negotiations can maintain its requisites in the case of trade in genetically modified organisms. In other words, under which conditions negotiating countries can find a satisfactory agreement for trading GM products.

Introduction

The analysis proposed in this chapter aims to show that, under some specific conditions, the World Trade Organization (WTO) can represent an appropriate institutional context for negotiating the genetically modified organisms (GMOs) regulations along with international trade liberalization. This negotiation should conciliate two opposite attitudes, here represented by USA and EU respectively: a strongly promoting approach towards commercial development and diffusion of GM-crops; a strongly preventive regulation discriminating the GM from the traditional crops (Paarlberg, 2000).

The basic idea is to extend the Bagwell and Staiger (1999, 2001a,b) analysis to the case of GMO trade. In principle, a Pareto-enhancing multilateral agreement can be found by allowing both GMO free trade and unilateral domestic regulations (for instance, the mandatory labelling in the EU), whose effects on the terms-of-trade (the 'trade effects') are compensated for by appropriate import-tariffs reduction on the GM-free analogous products. Such a solution, although consistent with the WTO principles and rules, can face some difficulties in the current WTO practice; therefore, how actually to implement this kind of compensation remains an open issue.

[1] This paper is part of the INEA Research Project: 'Governing the Agricultural Biotech Revolution: Policies and Institutions'. The authors are listed alphabetically and sections may be attributed as follows: 'Introduction' and 'Problems with GM-crops and Possible Solutions' to Sorrentino; 'The Rationale Behind the WTO Agreements, and the GMO Controversy' and 'Some Final Considerations' to Esposti.

The Rationale Behind the WTO Agreements and the GMO Controversy

Substantial differences in approaching the GMO regulation in USA and EU eventually generated a trade dispute. Though the regulations on GMO cannot be expressly trade measures, major differences in regulation eventually produce different conditions of access to the international markets. Inevitably, therefore, the GMO controversy between USA and EU is essentially a trade dispute; thus, it may naturally be negotiated in those international institutions aiming to settle these international trade disputes to eventually increase the overall market access.

Pragmatically, not only involving the General Agreement on Tariffs and Trade (GATT/WTO) is inevitable, but it might also be the most appropriate way to tackle the GMO dispute, since some relevant Agreements already exist (Sanitary and PhytoSanitary (SPS), Technical Barriers to Trade (TBT), etc.) to allow approaching the negotiation on already accepted and established framework and device. In principle, however, though we may accept that issues regarding GM-crops in the international context turn into trade disputes, we can seriously question whether the GATT/WTO existing agreements, as well as the WTO as institution, are the appropriate background in which the dispute can be rationally settled. In general and abstract terms, the starting point is to understand what is the rationale behind the WTO agreements, that is why their principles and rules are multilaterally accepted as efficient devices. Once this first point is established, the problem becomes to understand if this rationale holds also in the case of the GM-crops dispute. The following sections will deal in sequence with these two issues.

Recent contributions by Bagwell and Staiger (1999, 2001a,b) have substantially improved the representation and comprehension of the rationale underlying the WTO agreements. In particular, Bagwell and Staiger (2001b) summarize their approach to WTO agreements by admitting negotiations on both tariffs and national standards and regulations. By working with a two-countries–two-commodities perfectly competitive general equilibrium trade model, they demonstrate how any country finds it efficient to define its own terms-of-trade with a bilateral tariffs reduction agreement rather than relying on unilateral decisions. Essentially, the trade agreements allow a cooperative equilibrium of terms-of-trade which is always preferable, for both countries, to possible non-cooperative equilibria.

However, this approach relies on two basic assumptions. First, it is assumed that in this game model, any country pay-off is defined by some measure of the country welfare. In other words, the assumption is that any country bases its own trade policy on no other political motivation than to maximize internal welfare. Second, the model assumes that the national welfare is only affected by its own terms-of-trade. In other words, interaction among countries occurs only through the international market and no other non-pecuniary externality is admitted, so that counties have no other reason to worry about others' behaviour than the effects on their own terms-of-trade.[2]

On this basis, the Bagwell–Staiger (2001b) model states that bilateral (multilateral) negotiations define respective import-demand and export-supply curves that allow the highest possible market access to all countries thus allowing the best (in welfare terms) terms-of-trade. In this sense, the WTO agreements are Pareto-enhancing devices and their role is then to remove the inefficiency, so that member governments can enjoy greater welfare (Bagwell and Staiger, 1999). Accordingly, the bilateral (multilateral) trade agreements reduce the effects of negative externalities of unilateral tariffs by defining a sort of optimal right on that common good represented by free market access.

---

[2] In the GMO case unilateral regulations can have significant non-pecuniary externalities. That is, they can affect other countries' welfare not via the impact on the terms-of-trade. For GM-crops, non-pecuniary external effects mainly refer to those global environmental commons that can be influenced by the regulation: for example, a precautionary approach can largely protect biodiversity compared with a promotional approach. Therefore, a different internal regulation can differently make this global common available also to the foreign consumers/stakeholders.

However, what may really undermine the above-depicted approach, is admitting that the consumer's utility, and consequently the national welfare, is not only dependent on prices and on the terms-of-trade. In particular, environmental, labour and health standards can affect consumer welfare as well. In this case, we may doubt that the WTO's effort for securing market access for exporters is really enough to make the multilateral agreements act as Pareto-enhancing devices. In fact, focusing only on terms-of-trade could eventually lead each government to resist raising those labour and environmental standards applied to national producers and/or consumers ('regulatory chill') – and perhaps even to lower these standards ('race to the bottom') – that would otherwise impose on the base of its own internal optimal policy mix (Bagwell and Staiger, 2001a). In this case, the free trade may not rest on the efficiency frontier for any country.

According to Bagwell and Staiger (2001a,b), however, this objection does not affect the main characters of the WTO agreements as Pareto-enhancing devices. In their view, the central purpose of the WTO rules is to create a 'negotiating forum' where member governments can voluntarily exchange market access commitments. Maintaining the assumption of no non-pecuniary externalities, the multilateral agreements can still allow the maximization of the national welfare also under these non-trade concerns.

The key point of their argument is that WTO allows the definition of efficient levels of market access in any case. How this market access level is defined is another sort of problem. In particular, the same optimal level of market access can be granted by the negotiating countries with different mixes of trade policies (tariffs) and domestic standards/regulations. Any negotiating country can define its own optimal mix while granting the optimal multilateral market access level. For example, by increasing its domestic standards as well as the import tariffs on the same commodity, a negotiating country can maintain the exporters access rights on its own markets. On the contrary, market access negotiations only targeted on tariffs cannot achieve efficient policy outcomes. Exclusive negotiations on tariffs would introduce an incentive to distort one's domestic standards to reclaim unilaterally a portion of the market access that the negotiated tariff reduction has granted.

This incentive underlying the WTO negotiation on tariffs would introduce inefficiencies into domestic standards choices. However, the WTO principles and rules, more than the WTO practice, allow for the combination of any mix of tariffs reduction and domestic standards. In principle, governments may first use tariff negotiations to achieve the efficient levels of market access; then, with the prospect of so called 'non-violation complaints' (admitted in GATT Article XXIII) securing market access at the negotiated level, they can unilaterally set the mix of tariffs and domestic standards that secure the agreed market access to other negotiators. According to this argument, WTO rules maintain the capacity to act as an efficiency enhancing device, also when any government wishes to fix internal standards according to its internal optimal mix.

Problems with GM-crops and Possible Solutions

This interpretation of the rationale driving the WTO negotiations could apparently be extended also to the above-mentioned GMO trade controversy between the USA and the EU. After all, as suggested by Anderson and Nielsen (2001), this controversy essentially depends on the following 'clash of rights': the right of the EU to set its own standards and regulation for GM-crops trade according to the specific alleged preferences of the European consumers; the market access right of the US products in the EU markets. Following the Bagwell and Staiger analysis, this clash of rights should not exclude the chance of a 'free trade' agreement for GM-crops in the WTO context. On the contrary, the WTO would just represent the appropriate device since, as described, it can efficiently manage this kind of 'clash of rights'. Eventually, the problem in the Bagwell–Staiger perspective would be to increase the use of tariffs-standards compensation in the WTO negotiation practice.

However, applying this kind of analysis to the GMO controversy reveals some more issues that make it much more intricate. In particular, there are three more key issues to be considered. First, as suggested by Sheldon (2002), the current EU regulation is based on the assumption that the GM and the conventional crops are different products. Since they are not like goods, the Bagwell–Staiger approach in terms of rebalancing more restrictive internal standards with higher tariffs does not work as such, since this balance should maintain the market access, and therefore the terms-of-trade, for the same product invariant. By skipping the substantial equivalence, this rebalancing is, at least in the strict sense, not affordable. A second related issue is that the EU regulations on GM-crops cannot be regarded as an internal technical standard. It applies not only to internal products but also to imported products. Consequently, this regulation does not reduce the EU terms-of-trade. From the US perspective, it rather reduces imported products competitiveness thus negatively affecting the US terms-of-trade. In the end, it acts like a protectionist measure that cannot be re-balanced along the line depicted by Bagwell and Staiger.

As a matter of fact, the process of GMO regulation in the EU becomes a WTO issue, that is trade distorting, not because the EU 'standards' disadvantage the internal production but because they force the imported products to increase their production costs. By imposing mandatory labelling on GMO products (Directive 18/2001), the EU implicitly increases the production costs of both GM and GM-free US crops; the former as labelling costs, the latter as identity preservation costs (Laplan and Moschini, 2001). Since the EU production of GM-crops is substantially zero, this regulation eventually affects the US production costs and not the EU ones. In other words, this is not a case where domestic standards are prevented by the risk of a reduction in their own terms-of-trade. It is rather a case in which domestic-border standards impose a reduction in the terms-of-trade of the imported products, thus eventually generating a 'race to the top' process of regulation.

The third issue concerning the GMO trade controversy with respect to the Bagwell–Staiger approach is related to the exclusion of non-pecuniary externalities from their analysis. As mentioned, this assumption restricts the international effects of domestic regulations only to the terms-of-trade. According to Bagwell and Staiger (2001b), domestic labour and environmental standards could also raise a set of humanitarian/global commons issues (international non-pecuniary externalities) that are not fundamentally market access issues. Possible global commons concerns should be handled more properly by other international institutions than WTO. In the case of GM-crops, this assumption can be critical. In particular, if we admit the EU regulation aims to increase the internal consumers' welfare by allowing them to express their preferences for GM-free products, we should also consider that this preference might be linked to global commons. If the consumers prefer the GM-free products because they think these do not harm the 'global' environment (for instance biodiversity), their welfare is also affected by the US internal regulation since this affects the 'global' environment as well. Therefore, non-pecuniary externalities cannot be ruled out from the analysis whenever they can affect the optimal internal combination of tariffs and standards/regulations. In this case, indirectly, they also have a trade effect and eventually affect the capacity of the WTO rules to achieve both optimal market access levels and optimal internal policy mix between tariffs and regulation.

The basic intuition of the Bagwell and Staiger analysis is that the WTO negotiations remain efficient also when domestic regulations are introduced, since they still allow appropriate combination of regulations and tariffs reduction that maintains the same level of market access. We now try to show how this efficient mix of policies can be afforded in the case of EU regulation on GMO trade thus considering GM and GM-free crops as different products, applying the regulation to all (domestic and imported) products and assuming that these regulations are costly. Then we will also consider what happens when we admit non-pecuniary externalities of domestic regulations.

Let us start from the Bagwell and Stiger (2001a) general representation of the government objectives and consider a bilateral trade agreement. As mentioned, the assumption is that both negotiating governments aim to maximize their internal welfare. This welfare depends on three key-variables: the level of internal prices (*P*), the terms-of-trade (*T*) and the level of internal standards (*S*). To make this point clear, let us consider as an example an agricultural commodity such as maize and USA and EU as the two negotiating countries. The respective welfare can be depicted as follows:

US internal welfare: $W^{USA}[S^{USA}; P^{USA}(\tau^{USA}, T^{USA}); T^{USA}(\tau^{USA}, \tau^{EU})]$
EU internal welfare: $W^{EU}[S^{EU}; P^{EU}(\tau^{EU}, T^{USA}); T^{EU}(\tau^{USA}, \tau^{EU})]$

where $T^{EU} = 1/T^{USA}$. It is also generally assumed that: $\partial W/\partial S > 0$, $\partial W/\partial P < 0$, $\partial W/\partial T > 0$. Considering US as net exporter of maize and EU as net importer, the US terms-of-trade in the case of maize is given by the ratio between the maize world price and the world price of one normal good US import from EU and that we will assume constant over the analysis. Therefore, changes in the terms-of-trade only occur if the maize world prices change. If this increases, the terms-of-trade T for USA will increase, while the terms-of-trade $1/T$ for EU will decrease. Both countries will increase their welfare if the respective terms-of-trade increase. Moreover, the assumption is that the internal standard S is established to increase the welfare and also that the national welfare is inversely linked to maize internal price P. This internal price, in turn, depends on either the terms-of-trade, that is on the maize world price, and import tariff τ.

According to this simple representation, an internal policy is given by some mix of S and τ; this apparently does not affect the welfare of the foreign country. However, the linkage between the internal price and the world price, therefore the terms-of-trade, transmits indirectly, i.e. through the world market, this internal mix to the foreign country eventually affecting its welfare. In this kind of analysis, the only external effects of internal policy mix are changes in the world prices (at this stage, again, non-pecuniary externalities are not admitted). Efficient negotiated agreements are therefore combinations of the internal instruments (S and τ) that make internal welfare increase for both USA and EU. In traditional tariffs-reduction agreements, the idea is quite straightforward (Bagwell and Staiger, 1999). Both USA and EU can reduce their import tariffs on the respective import goods,[3] thus reducing the world and internal prices for both but leaving unchanged the terms-of-trade (Case 1 in Table 6.1); thus the welfare increases for both. This is also why the final gain of this kind of agreement should be an increased consumers surplus.

In this framework, it is also straightforward to understand the implication of increasing the internal standards according to Bagwell and Staiger (2001a,b) (Case 2 in Table 6.1). Any country aims to increase the internal standard S since this improves the national welfare by increasing, for instance, the consumer food safety, the environmental safety, etc. However, an increase in internal standards of the importing country implies, implicitly or explicitly, an increase of the internal production costs, thus increasing the competitiveness, i.e. the market access, of the imported maize. This, in turn, leads to a supply reduction in the world market and, therefore, an increase in the world price. These effects on the world market imply a lower terms-of-trade for the importing country, thus a welfare reduction. The risk is that, eventually, we observe $dW/dS < 0$, i.e. an overall reduction in importing country welfare, since the decrease of the terms-of-trade overcompensates the positive welfare effect of standards. This is basically what is meant by 'regulatory chill': any country is prevented from choosing the proper internal standards by the consequent reduction in its own terms-of-trade. However, this reduction in terms-of-trade due to higher internal standards can be rebalanced by an appropriate increase in import tariff τ. For any country importing maize, increasing the import tariff determines better terms-of-trade, due to the consequent reduction of the maize world price. It follows that there can be a set of different combinations of S^{EU} and τ^{EU} providing the same maize world price, therefore leaving the terms-of-trade unchanged. This is the basic idea of introducing the standards in the WTO negotiations: any country can unilaterally define its optimal combination of standards and import tariffs provided that the terms-of-trade are not affected. Under this condition, the other negotiating country will accept the unilateral decision on standards and import tariffs since

[3] Both are assumed to be normal goods and we are also assuming that involved countries actually carry out a multi-product negotiation.

Table 6.1. Conditions and mechanisms for efficiency in WTO negotiations.

Case	Required conditions (assumptions)	Mechanisms
1. Promoting multilateral market access (Bagwell and Staiger, 1999)	• Domestic welfare maximization as government only objective	Multilateral tariffs reduction
2. Promoting multilateral market access and allowing domestic regulations (Bagwell and Staiger, 2001a,b)	• Domestic welfare maximization as government only objective • Tariffs and regulations refer to like goods • Regulation only applies to domestic production • Non-pecuniary externalities of regulations (global commons) excluded	Balancing domestic regulation with tariffs increase: **1.** Finding the internal optimal mix between regulations and tariffs (avoiding 'regulatory chill') **2.** Maintaining the same level of market access (terms-of-trade)
3. Allowing free trade and the EU regulation (mandatory labelling and traceability) on GM products	• Domestic welfare maximization as government only objective • Skipping the substantial equivalence principle (non-like goods) • The EU imposes a (large enough) import tariff on the commodity • GM products enough strictly inferior for EU consumers with respect to the GM-free ones • EU self-sufficiency ratio significantly > 0 • Non-pecuniary externalities of regulations (global commons) excluded	Balancing EU mandatory labelling with reduction of tariffs on the GM-free product: **1.** Tariff (τ) reduction should compensate the tariff-effect of the segregation costs (ΔC) **2.** The internal prices ratio P_{FREE}/P_{GM} determines a welfare increase in the EU
4. Allowing free trade as well as global regulations (global mandatory labelling + traceability) on GM products	• Domestic welfare maximization as government only objective • Skipping the substantial equivalence assumption (non-like goods) • Non-pecunairy externalities of regulations (global commons) admitted and relevant • GM products enough strictly inferior for all (EU–USA) consumers with respect to the GM-free ones	**1.** Terms-of-trade increase for US does not compensate the welfare loss due to internal prices increase **2.** The internal prices ratio P_{FREE}/P_{GM} determines a welfare increase for the world (EU + USA) consumers **3.** Importing country (EU) tariff reduction could further increase the US terms-of-trade

they will leave unchanged the agreed market access levels.

Does this kind of analysis apply to the EU recent regulation (Directive 18/2001) on GM-crops trade (Case 3 in Table 6.1)? This regulation imposes mandatory labelling and traceability of GMOs for both internal and imported products and refers to those commodities that can be traded both in conventional (GM-free) and GM form, as in the case of maize. As mentioned, two main issues may arise in interpreting this case as a traditional unilateral internal standard. First, this regulation implies market discrimination of the GM-free and GM maize. They become two distinct products within the EU while, if the regulation is not adopted worldwide, there will be still only one maize world price, therefore only one term-of-trade. Second, this regulation also applies to imported products and, consequently, its costs have also to be borne by the foreign maize producers.

It must be emphasized that the current EU regulation implies higher production costs (ΔC) for both GM product (labelling costs) and GM-free product (identity preservation costs) (Lapan and Moschini, 2001). Whatever is the initial mix of the two different types of maize, this regulation will increase the internal price for both. Since it applies also to imported maize, this regulation eventually acts as a further import-tariff on both products. This clearly indicates that the EU regulation on GM-crops does not increase the competitiveness of the imported product, and therefore does not increase the market access rights for US products. Consequently, no compensation with import tariff increase does actually make sense, the standard itself behaving as a tariff increase.

Since the EU regulation aims to segregate the GM and GM-free maize, the consequent internal welfare gain exists only if the EU consumers strictly prefer the GM-free product.[4] Since the unit production costs of the GM-free maize as well as the consumers willingness to pay are greater, the GM-free maize price (P_{FREE}) will be greater than the GM maize price (P_{GM}). Therefore to make the regulation welfare-increasing in the EU it must be ($P_{FREE} > P_{GM}$). However, this welfare gain from the market segregation, as well as the maize price increase due to the segregation costs (ΔC), is not transmitted in the world market, where the two products are not discriminated. The maize world market price, and therefore the US terms-of-trade, will be only affected by the 'increase in tariff' effect of the EU regulation. Therefore, the world price and the terms-of-trade will decrease with a consequent reduction in the US welfare.

The main implication of this analysis, is that EU regulation on GM maize is not a measure to be compensated by an import tariff increase that leaves the terms-of-trade unchanged. On the contrary, to maintain the same terms-of-trade, the EU will have to compensate its regulation with a large enough reduction in maize import tariff. Only in this case, the USA will accept the unilateral EU regulation together with a bilateral agreement on tariffs in the WTO negotiations. How should this compensatory tariff reduction be accomplished and how strong should it be? In principle, tariff reduction could regard either GM and GM-free maize since in both cases the consequence for the USA would be an increase in its terms-of-trade. However, as it is typically proposed (Sheldon, 2002), the tariff reduction should be applied to the GM-free products. As mentioned, the basic assumption of this analysis is that the European consumers strictly prefer the GM-free product and this makes the market segregation a welfare-increasing policy. In this case, reducing the import tariff on GM-free product would reduce the ratio P_{FREE}/P_{GM} thus causing a further positive welfare effect in the EU.

In any case, there are two important conditions for this compensation to work. First, there must be an initial import-tariff on the commodity and this tariff should be large enough to make its reduction significant.[5] Second, this tariff reduction should be strong enough to restore the initial terms-of-trade for the USA; that is, the tariff reduction has to increase the import demand, thus reducing the residual world market supply and, in turn, increasing the maize world price. This effect on the world market has to be strong enough to restore the initial world price for the undifferentiated maize. This compensatory tariff reduction should re-balance the 'increase in tariff' effect induced by the mandatory labelling-identity preservation costs (ΔC), and therefore has to be proportioned on them.

In short, the EU should compensate the higher production costs imposed on the US maize by increasing the market access to the EU markets. If the tariff reduction is applied to GM-free maize, it means that the EU regulation should be compensated by an increase of the US share in the EU maize market. In fact, the internal price increase, caused by the segregation costs, is expected to reduce the whole EU internal demand of maize. If the US share in the EU market were maintained at the initial level, this reduction would imply a lower demand for imported maize, thus still inducing an excess supply in the residual world market compared with the pre-regulation

[4] It means that if the prices of GM and GM-free products are equal, the consumers will always consume only the GM-free one.

[5] EU is largely a net importer for soybean and GM-soybean production is particularly relevant in the USA. However, the proposed solution cannot be referred to the GM-soybean case since currently the EU does not apply any import tariff to soybean. On the contrary, the EU imposes a tariff on the imported maize. Though the EU is a maize importer, the self-sufficiency is quite high (currently about 90%) and GM-maize production is also quite significant in the USA, though not as much as for soybean. Therefore, the maize case seems to have the appropriate requisites to be used as an example for our analysis.

period.[6] Therefore, if the compensatory tariff reduction has to fully 'drain' this excess supply, it also must increase the market share of the imported (US) product (Fig. 6.1).[7] As mentioned, the increase in this share should be proportional to the (ΔC) implied by the regulation. Actually, even if we assume that these additional costs do not affect the internal US market price at all but only refer to the exported product, increasing the access of the US maize into the EU market is still needed to maintain the initial US terms-of-trade. It also follows, however, that to make this compensatory tariff reduction acceptable for the USA, the EU self-sufficiency rate for the GM-free maize has to be large enough to allow a large enough compensatory 'transfer' of the market share to the USA.

Under the above conditions, summarized in Table 6.1, it would be possible within the WTO principles and rules and following the Bagwell and Staiger (2001a) analysis, to find Pareto-enhancing agreements between EU and USA, saving at the same time the right of unilateral internal regulation (thus avoiding 'regulatory chill') and the right of exporting countries to freely trade GM-crops. It must be emphasized that this possible negotiable arrangement essentially depends on the EU consumer preferences, in particular on her/his concerns about the GM-crops. The extent of her/his preference for the GM-free product also affects the extent of the EU welfare gain from the segregation, therefore defines to what extent associated costs (ΔC) are acceptable. On the one hand, these costs make the maize price increase in the EU, thus reducing the welfare gain (since welfare function depends on the internal price). On the other hand, it also affects the required level of the compensatory tariff reduction. Consequently, although the WTO agreements apparently do not directly admit consumer concerns, the extent to which the EU consumers prefer the GM-free crops indirectly but largely affects the eventual WTO solution of the controversy.

There is also a further point to be considered about the EU consumer concerns and it

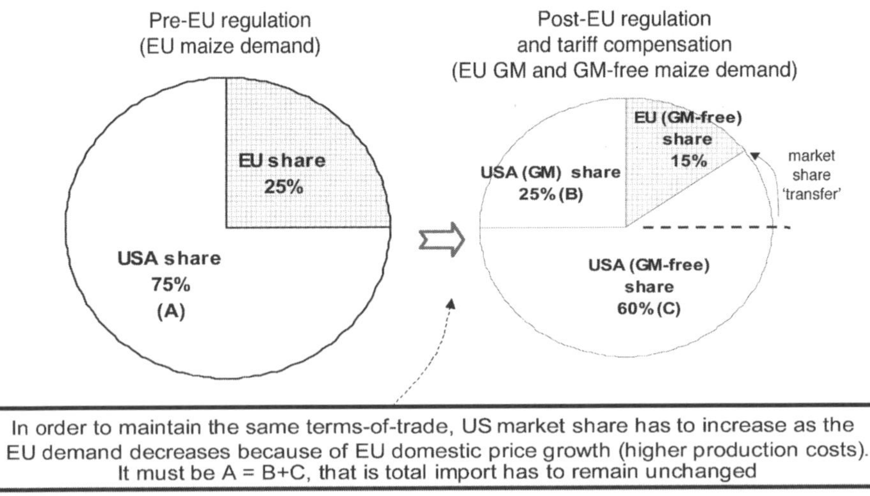

Fig. 6.1. Efficient (Pareto enhancing) WTO negotiation on GM-crops: the example of maize (percentages are imaginary).

[6] It should be also mentioned that the segregation costs could also significantly reduce the internal supply. A detailed analysis of the effect of the labelling-identity preservation costs on the agricultural supply is a complex issue (Bullock and Desquilbet, 2002) and is beyond the scope of this chapter. However, we do not rule out this possibility; we simply assume that the reduction in internal supply does not compensate for the supply increase in the world market induced by the regulation.

[7] As mentioned and unlike the figure, the EU self-sufficiency for maize is not much lower than 1. The figure just wants to emphasize how low self-sufficiency can be binding in compensating restrictive regulations with tariff reduction.

refers to the aforementioned third issue raised by the GM-crop trade. It might be the case that the EU consumer preference does not refer only to the internal regulation. She/he prefers the GM-free products not only for her/his own food safety or 'local' environmental concerns, but also for the global environmental concerns raised by the GM-crops (for instance reduction of the biodiversity worldwide). Under this hypothesis, and following the usual notation, the EU welfare should be depicted as $W^{EU}[S^{USA}; S^{EU}; P^{EU}(\tau^{EU}, T^{USA}); T(\tau^{EU}, \tau^{USA})]$. In particular, if the consumers express global concerns, global standards (or global regulation), rather than the local ones, are relevant for her/his welfare. In short, it would be $\partial W/\partial S^{EU} \cong 0$ and $\partial W^2/\partial S^{EU}\partial S^{USA} > 0$.

These are the non-pecuniary externalities ruled out by Bagwell and Staiger (2001a) as non-market issues. However, under this hypothesis, 'regulatory chill' could arise not with direct reference to market access issues, but as a consequence of the cross-countries effect of different regulatory frameworks. In particular, the optimal mix of regulation and tariffs for the EU cannot be pursued if there is not an analogous regulation in the other countries (global regulation). In this case (Case 4 in Table 6.1), the WTO should not only promote multilateral market access rights, allowing unilateral standards at once, but also promote multilateral regulations.

The issue becomes whether or not the WTO principles and rules can make the EU regulation for GM-crops be applied also in USA, again by introducing some compensatory arrangement to make the agreement Pareto enhancing. Extending the EU regulation scheme would essentially mean imposing the associated costs (ΔC) also to the US internal production. In this case, the US welfare is negatively affected by the increase of the internal prices and this welfare reduction, for normal goods, would never be fully compensated for by the expected increase in the terms-of-trade (if the regulation were applied globally, the maize world price would, in fact, increase). It follows, that the only condition to make this global regulation affordable (and Pareto enhancing) is that the GM-free maize has to be globally strictly preferred, that is by both US and EU consumers.[8] This holds particularly if this preference largely depends on the consumers' global concerns either in EU and USA, so that the regulation can produce relevant positive non-pecuniary externalities from both side.

Some Final Considerations

Within the WTO, the SPS and TBT Agreements are the institutional frameworks where a multilateral arrangement on GMO trade could be found. In the interpretation of Heumueller and Josling (2001), the SPS Agreement should be concerned with those trade restrictive measures whose only objective is to protect the health and sanitary conditions of consumers, plants and animals. On the contrary, the TBT Agreement would better fit domestic regulations aimed at multiple objectives specifically concerned with environmental safety and protection, consumers' right to free choice or other possible ethical-religious issues.

To admit a positive solution of the GMO controversy, the WTO agreements would, in principle, have to work as a 'negotiating forum' for 'trading', on a multilateral base, market access rights as well as unilateral domestic regulations. In practice, however, within the SPS and TBT Agreements, the main driving concern is currently to avoid the use of internal regulations as 'hidden' forms of protectionism. So that, rather than being 'negotiating forums', such agreements become essentially 'codes of fair behaviour' accepted by any country in selecting admitted unilateral standards and regulations.

Besides this discrepancy between the principles and the practice, there are also real differences between the SPS and TBT when referred to the case of GMO trade and regulation. If the non-pecuniary externalities are excluded, the most common reason for an internal regulation discriminating the GM and GM-free products is the alleged uncertainty about possible sanitary effects, still widely unknown and, therefore, not

[8] To avoid an overall welfare reduction for the USA, a further adjustment of the US terms-of-trade could still be needed and afforded through a further tariff reduction in the EU on the GM-free product (see Case 4 in Table 6.1).

demonstrable. These concerns could be hardly admitted within the SPS Agreement, as in such agreement the approach to any internal regulation is strictly science based and does not admit any uncertainty in the scientific evidence about the risk.[9]

Under this perspective, and according to the actual practice, the TBT Agreement seems to provide more chances for negotiating a solution on the GMO trade controversy. Unlike the SPS, unilateral regulations under the TBT agreement can be supported by different kinds of motivation (even strong ethical-religious or socio-political justifications) and they have not to be fully justified only on a strictly scientific ground; GM food and crops could pertain to similar cases.

The original institutional characters of the TBT Agreement makes it much closer to the idea of 'negotiating forum' than the SPS Agreement. Thus, it could represent the appropriate institutional context in which to establish the Pareto-enhancing trade-off between unilateral regulations on GMO and tariffs reduction on the GM-free products. This can happen whenever possible concerns about the non-pecuniary externalities of the regulations were excluded from the negotiation, that is when the consumers' demand for a regulation only involves local/national standards; this is mainly the case of concerns about food safety. However, if the welfare gain for the consumers is linked to a global common (for instance conservation of biodiversity), what really becomes relevant for consumers is the global standard, that is regulation holding worldwide, since the commons they want to be protected cannot be provided, in any case, only at the national level: the welfare of the EU consumers is affected by both the EU and US regulation and vice versa. A Pareto-enhancing agreement on GMO trade involving this kind of non-pecuniary externality would require quite restrictive conditions and is hardly affordable even in the WTO/TBT context.

In this chapter, we propose some possible solutions to the potential controversies generated by GMO trade. These solutions require a 'negotiating forum' where to trade multilateral market access rights versus unilateral internal regulations and standards. Other multilateral environmental agreements (for example the Cartagena Biosafety Protocol) and all those international institutions whose task is to settle international standards, though more consistent in dealing with global common issues (Phillips and Kerr, 2000), do not fully meet this condition since they essentially aim to fix a common shared practice and are not 'negotiating forums'. On the contrary, the WTO agreements, and particularly the TBT Agreement, could satisfy this requirement. If negotiating countries do not spontaneously find shared global standards to finally solve the GMO controversy, the WTO negotiations (especially within the TBT Agreement) can still allow an efficient combination of domestic regulations and market access rights.

References

Anderson, K. and Nielsen, C.P. (2001) GMOs, the SPS agreement and the WTO. In: Anderson, K., McRae, C. and Wilson, D. (eds) *The Economics of Quarantine and the SPS Agreement*. Centre for International Economic Studies and AFFA Biosecurity Australia, Australia.

Bagwell, K. and Staiger, R.W. (1999) An economic theory of GATT. *American Economic Review* 89, 779–795.

Bagwell, K. and Staiger, R.W. (2001a) Domestic policies, national sovereignty and international economic institutions. *Quarterly Journal of Economics* 116, 519–562.

Bagwell, K. and Staiger, R.W. (2001b) The WTO as a mechanism for securing market access property rights: implications for global labour and environmental issues. *Journal of Economic Perspectives* 15(3), 69–88.

Bullock, D.S. and Desquilbet, M. (2002) The economics of non-GMO segregation and identity preservation. *Food Policy* 27, 81–99.

Heumueller, D. and Josling, T. (2001) Trade restrictions on Genetically Engineered Foods: the application of the TBT agreement. Paper presented at 5th International Conference on the Economics of Biotechnology, *International*

[9] Actually, Article 5.7 of the agreement on the SPS measures admits cases where the scientific evidence is only partial. This case, however, is considered an exception to the general rule and is exclusively transitory. It only allows temporary unilateral measures to be introduced in negotiated agreements.

Consortium on Agricultural Biotechnology Research, Ravello, Italy, 15–18 June.

Josling, T. (2001) International institutions, World Trade rules, and GMO's. In: Nelson, C. (ed.) *Genetically Modified Organisms in Agriculture*, Academic Press, New York, pp. 117–130.

Laplan, H.E. and Moschini, G. (2001) GMO labelling and trade: consumer protection or (just) protectionism. Paper presented at the Conference on *Globalization, Biotechnology and Trade*, International Agricultural Trade Research Consortium Annual Meeting, Tucson, Arizona, 14 December.

[OECD] Organisation for Economic Co-operation and Development (2000) *Summary Report. Conference conclusions and recommendations on Agricultural Knowledge System (AKS) addressing food safety and environmental issues.* II conference of directors and representatives of the AKS, 10–13 January, Paris.

Paarlberg, R.L. (2000) *Governing the GM-crop revolution.* Food, Agriculture and the Environment Discussion Paper 33, IFPRI.

Phillips, P.W.B. and Kerr, W.A. (2000) The WTO versus the Biosafety Protocol for trade in Genetically Modified Organisms. *Journal of World Trade* 34(4), 63–75.

Sheldon, I.M. (2002) Regulation of biotechnology: will we ever 'freely' trade GMOs? *European Review of Agricultural Economics* 29(1), 155–176.

7 Trade Restrictions on Genetically Engineered Foods: the Application of the TBT Agreement

Dirk Heumueller[1] and Tim Josling[2]
[1]School of Law; [2]Institute for International Studies, Stanford University, California, Stanford, USA

Abstract

A growing number of consumers want to know more about how their food is produced, in particular whether or not it has been subject to genetic engineering. Labelling requirements, pre-market approvals, and even bans are on the agenda of many countries. Domestic policy makers argue that they restrict trade to manage risks that might be linked with these foods. However, within the World Trade Organization (WTO) system, trade restrictions must meet the conditions of the General Agreement on Tariffs and Trade (GATT), Sanitary and Phyto-sanitary (SPS) and Technical Barriers to Trade (TBT) Agreements. Considering the multifaceted character of the issue of genetically engineered foods, this chapter suggests that the TBT Agreement will often provide the measure.

The chapter analyses the role of the TBT Agreement within the WTO system with respect to trade restrictions on genetically engineered foods, in particular those related to labelling. The TBT Agreement does not cover import bans, which are come under the provisions of the GATT (94). It does not apply to sanitary and phytosanitary measures that are introduced for plant, animal or human health reasons alone, which are exclusively covered by the SPS Agreement. The TBT Agreement covers all technical regulations that are not covered by the SPS Agreement, including measures with multiple objectives even if one of those objectives is related to health.

With respect to the issue of labelling, the chapter examines whether the TBT Agreement would allow countries to justify broad labelling regimes if these correspond with the needs and preferences of consumers. The problem usually raised with respect to the flexible use of precaution and labelling is that it quickly leads to a misuse of such regulations to protect domestic suppliers.

The chapter concludes with suggestions of ways to require sufficient empirical evidence regarding what people want and why, so as to avoid the misuse of the substantive TBT rules as disguised protectionism.

Introduction

Using modern genetic engineering in food production is relatively new and its regulation is still evolving.[1] While genetically engineered foods (GEF) enjoy relatively free market access in the USA, the EU, Japan, Australia, and others impose labelling requirements, strict pre-market approvals, and even bans on GEF. As a result the regulation of trade in GEF is a 'hot' issue in legal and policy debates.

The World Trade Organization (WTO) does not specifically regulate GEF but rather constrains countries in restricting trade in generally in certain ways. Three parts of the WTO are particularly relevant to the issue: the General Agreement on Tariffs and Trade (GATT) (94), which discourages discrimination, limits the ability of governments to impose border restrictions and imposes the equal treatment of foreign and domestic goods; the Sanitary and Phyto-sanitary (SPS) Agreement, which regulates trade restrictions for the purpose of protecting animal, plant and human health; and the Technical Barriers to Trade (TBT) Agreement, which attempts to limit the trade impacts of standards and other technical regulations, which can become trade barriers. Much of the discussion has been focused on the impact of a science-based approach to GEF within the SPS Agreement. This chapter suggests that the TBT Agreement along with the GATT (94) will often be the relevant legal text for judging domestic trade restrictions on GEF.[2]

Two questions are vital. First, what is the relevance of the TBT Agreement for trade restrictions on GEF? Second, to what extent does the TBT Agreement allow a science test and a comprehensive risk assessment that includes the 'precautionary principle' as well as consumer rights, public morals and even human rights?

The Relevance of the TBT for GEF

The TBT Agreement covers trade barriers resulting from 'technical regulations' (mandatory requirements) and 'standards' (voluntary regimes) with respect to 'all products' as well as conformity assessment procedures (e.g. certification). It expressly includes labelling requirements relating to 'product characteristics or their related processes and production methods'.[3] This chapter will focus on technical regulations rather than standards and conformity assessment, and on the issue of labelling in particular.

SPS and TBT and GEF

The TBT Agreement, like the SPS Agreement, aims at 'distinguishing genuine non-protectionist measures from disguised trade protection'.[4] However, the drafters of the TBT Agreement expressly exempted SPS measures.[5] Those measures are subject to the SPS Agreement. However, a sanitary or phytosanitary measure requires per definition a direct connection between the import of a particular food or feedstuff and the spread of a specific disease or pest. A labelling requirement, to fall under the SPS Agreement, has to be 'directly related' to food safety.[6] Abstract and theoretical concerns

---

[1] *See* Nelson, G.C. (ed.) (2001) *Genetically Modified Organisms in Agriculture: Economics and Politics*. Academic Press (providing an overview of the controversy).

[2] *See also* Buckingham, D.E. and Phillips, P.W.B. (2001) Issues and options for the multilateral regulation of GM foods. (2) 1 *The Estey Centre Journal of International Law and Trade Policy* 2(1), 178–189.

[3] *See* No. 2 of Annex 1 of the TBT Agreement, Article 1.3 of the TBT Agreement.

[4] *See* Article 2.2 and clause 5 of the preamble of the TBT Agreement.

[5] Article 1.5 of the TBT Agreement states:

'The provisions of this Agreement do not apply to sanitary and phytosanitary measures as defined in Annex A of the Agreement on the Application of Sanitary and Phytosanitary Measures'.

[6] No. 1 of Annex A of the SPS Agreement states:

'. . . Sanitary or phytosanitary measures include all relevant laws, decrees, regulations, requirements and procedures including, *inter alia*, end product criteria; processes and production methods; testing, inspection, certification and approval procedures; quarantine treatments including relevant requirements associated with

continued

are apparently not covered by the SPS Agreement. Clearly, a labelling regime that *only* aims at preventing people from direct and concrete health risks, such as potentially allergic reactions, cancer, or damage to the human immune system, has to be examined under the SPS and not under the TBT Agreement.[7] Measures that lack this direct link are not covered by the SPS Agreement.

Which of the agreements would constrain labelling measures that respond to concerns about links between GEF and a person's 'right to know' and 'to choose'; questions of food safety, sustainable agricultural development and ecosystem use; democratic decision making; the power of multinationals, and deeper ethical/religious considerations?[8] Decision making in the WTO so far indicates a case-by-case judgement, depending on the reasons and the evidence presented in favour of a trade restriction.[9] As opposed to the SPS Agreement, the TBT Agreement refers in Article 2.2 to various objectives and even suggests an open-ended catalogue of policy aims. Thus, the TBT Agreement is clearly intended to apply to measures that address multiple ends. Correspondingly, one should interpret Article 1.4 of the SPS Agreement[10] and Article 1.5 of the TBT Agreement as implying that domestic trade restrictions with multiple ends have to be examined under the TBT Agreement or the GATT. This would be the case if an importing country imposed a labelling restriction because of general fears about consumer reactions not related to specific health concerns.

GATT and TBT on GEF

According to Article XVI (3) of the WTO Agreement and a general interpretative note, the TBT Agreement has to take priority over the GATT (94) to the extent that a measure in question is covered by both agreements.[11] A country can impose a trade-related regulation in a number of ways. The regulation can be in the form of a simple prohibition on imports. However, complete import bans fall outside the scope of the TBT Agreement since they are neither a technical regulation nor a standard. They have to be judged under the GATT (94) or the SPS Agreement which do allow a quantitative restriction (such as a ban) on imports. But many technical regulations require licences for importation with various constraints attached to the issue of the import permit. If the licences are freely available (i.e. the quantity is not controlled) the regulation presumably falls within the purview of the TBT Agreement. Imports may have to be labelled or packaged in particular ways. These regulations are clearly also within the scope of the TBT Agreement. But in fact, many TBT issues reside in the grey area between an outright ban and a simple non-restrictive label, and include conditions put on the issue of a licence.

Limiting Trade Restrictions on GEF via the TBT Agreement

The TBT Agreement is on the face of it quite permissive. The Agreement does not limit

the transport of animals or plants, or with the materials necessary for their survival during transport; provisions on relevant statistical methods, sampling procedures and methods of risk assessment; and packaging and labelling requirements directly related to food safety'.

[7] *See* Report of the WTO Panel on EC measures concerning meat and meat products (Hormones), WL 569984 DS26/R/USA, 18 August, 1997 at 190.

[8] Lay Panel Report, first Australian consensus conference – gene technology in the food chain http://www.austmus.gov.au/pdf/layreport.pdf

[9] *See* Canada's submission in the Report of the WTO Panel on EC measures concerning meat and meat products (Hormones), complaint by Canada, WL 561677 WT/DS48/R/CAN, 18 August, 1997 at sec. 318.

[10] Article 1.4 of the SPS Agreement states:

'Nothing in this Agreement shall affect the rights of Members under the Agreement on Technical Barriers to Trade with respect to measures not within the scope of this Agreement'.

[11] WTO Agreement, Apr. 15, 1994, Final Act Embodying the Results of the Uruguay Round of Multilateral Trade Negotiations 7, 33 I.L.M. 1125, Annex 1 A, *Multilateral Agreements on Trade in Goods* – general interpretative rule to Annex 1A.

a party's right to impose domestic trade restrictions when pursuing a legitimate goal in a non-protectionist way. The key provisions of the TBT Agreement are Articles 2.1 and 2.2, which define the basic concepts of legitimate goals and non-protectionist actions related to technical regulations promulgated by central government bodies. Other Articles deal with standards and conformity assessments and by the actions of other (sub-federal and local) government agencies. This section looks at the significance of these Articles, considering specifically the key terms used, the way in which a panel might interpret these terms and the implications of this interpretation.

GEF and Article 2.1 of the TBT Agreement

Terms. For already imported products, Article 2.1 restates the national treatment and most favoured nation obligation of the GATT, two of its key non-discrimination principles:

> Members shall ensure that in respect of technical regulations, products imported from the territory of any Member shall be accorded treatment no less favorable than that accorded to like products of national origin and to like products originating in any other country.

Assessment. Looking at labelling regimes for GEF two points seem to be critical: what are the 'like products' in question, and what is the meaning of 'less favorable'?

The GATT has historically judged the 'likeness' of products on a case-by-case basis considering 'the product's end-uses in a given market; consumers' tastes and habits, which change from country to country; the product's properties, nature and quality'.[12] Usually, 'chemically-identical' products are considered as 'like products'.[13] According to this measure, soy oil made of genetically engineered beans would probably be considered as like-product to non-genetically engineered (but chemically identical) domestic oil. However, one may get a different result if one points to consumer's preferences as does a recent WTO judgement.[14]

To demonstrate a less favourable product treatment can be tricky for *de facto* discrimination. Assume the importing country completely prohibits domestic manufacturers from using genetically engineered crops, so that there would be no domestic genetically engineered soy oil. Nothing in the WTO constrains a country from taking such a step. A domestic licensing and/or labelling requirement for genetically engineered soy oil (e.g. 'made of genetically engineered soy beans') may be judged as less favourable treatment if no such labelling is required for domestic conventional soy oil. Moreover, how would one characterize the situation, given that exporters are producing genetically engineered soybeans while domestic industry is not? The label may be 'less favourable' to the import without any overt discrimination.

Implications. The previous examples suggested that applying the usual measures of likeness and less favourable treatment can potentially clash with the domestic politics of GEF regulation. Both the reality of different consumer perceptions and the fact that most regulations have a differential impact on domestic and imported goods suggests that it will be difficult to apply Article 2.1 in a non-controversial manner.

GEF and Article 2.2 of the TBT Agreement

Complementary to Article 2.1, Article 2.2 of the TBT Agreement applies to measures that affect the actual importation of products. Article 2.2 reads:

> Members shall ensure that technical regulations are not prepared, adopted or applied with a view to or with the effect of creating unnecessary obstacles to international trade. For this purpose, technical regulations shall not be

[12] *See* WTO Panel Report on USA: Standards for Reformulated and Conventional Gasoline, WT/DS2/R.29 at 34, para. 6.8, reprinted in 35 I.L.M. 274 (29 January, 1996).

[13] *See* id. at para. 6.9.

[14] *See* WTO Appellate Body Report on *European Communities – Measures Affecting Asbestos and Asbestos-containing Products*, WT/DS135/AB/R para. 72–75 (12 March, 2001) at para. 122–124, 130 (stressing the role of 'consumers' tastes and habits' to be significant even with respect to a regulatory suppressed 'latent demand').

> more trade-restrictive than necessary to fulfil a legitimate objective, taking account of the risks non-fulfillment would create. Such legitimate objectives are, *inter alia*: national security requirements; the prevention of deceptive practices; protection of human health or safety, animal or plant life or health, or the environment. In assessing such risks, relevant elements of consideration are, *inter alia*: available scientific and technical information, related processing technology or intended end-uses of products.

According to this provision, for example, a governmental labelling regime could be justified under two conditions: (i) the labelling regime aims to fulfil a legitimate objective; and (ii) there is no other less trade restrictive measure available to fulfil the legitimate objective.

Expressly listed objectives for measures that restrict trade in GEF

The range of legitimate objectives is both wide and open-ended ('*inter alia*'). Of particular interest on the issue of GEF are the objectives of preventing deceptive practices, 'protection of human health and safety, animal or plant life or health', and protection of the environment.

PREVENTING DECEPTIVE PRACTICES If people care about a product characteristic that is not easily observable, such as the use of genetic engineering today, it is fair to assume a potential for fraud within a market.

Terms. The term 'deceptive practice' lacks a precise definition in international law.[15] A reasonable starting point to bring some light to the term deceptive practices seems to be that it is deceptive if one makes a demonstrably false claim about a material fact in connection with a food product. For example, if maize for popcorn is labelled 'not genetically engineered' and it is in fact *Bt*-maize, there should be little doubt that this label is deceptive. The stated circumstance is somehow provable in an 'objective' way.

Things become more complex if one refers to the ability of a label to influence consumers' beliefs. A label might produce an inaccurate belief about a material fact of a food. For example, if a soy oil might be labelled 'GMO-free' if it is in fact free of detectable genetically engineered material at the time of sale. But the oil could have been made from genetically modified soybeans and the process of refining has destroyed the DNA. The claim made by the label is not false unless the law defines GMO-free more broadly.[16] However, a consumer might get the false impression that the soy oil is made of non-genetically modified soybeans.

A consumer might be led to assume that the GMO-free labelled food is 'somehow safer or better than its genetically manufactured counterpart, or that the use of genetic engineering techniques adversely affects the character, quality, or nature of the food'.[17] In this case, the GMO-free label is true but misleading. Thus, even truthful information could mislead the consumer by producing an inaccurate belief about non-material facts.

Assessment. A selling practice can only be qualified as 'deceptive' if the provided information affects consumers' choice.[18] In short, whether or not a practice might be seen to be deceptive heavily depends on the particular product in question, the knowledge and preferences of consumers, their number, and reasons for their impressions. These factors can vary from society to society and/or country to country and are subject to empirical evaluation.

The objective of the prevention of deceptive practices is at the core of the workings of a modern market system. To argue that labelling GEF is necessary to achieving such an objective, a member could have to provide evidence through empirical data that a significant number of consumers are actually misled and that simple educational campaigns alone would not have

[15] *See* unpublished GATT Panel report on USA – Restrictions on Imports of Tuna, DS21/R – 39s/155 at p. 76–77 (3 September, 1991), 30 I.L.M. 1991 at pp. 1594 et seq.

[16] *See* Stecklow, S. (1999) 'Genetically modified' on the labels means . . . Well, it's hard to say. *The Wall Street Journal* 26 October, at 1, A 14.

[17] Degnan, F.H. (1997) The food label and the right-to-know. *Food and Drug Law Journal* 52, 49, 59–60.

[18] *See* Unpublished GATT Panel report on USA – Restrictions on Imports of Tuna, DS21/R – 39s/155 at p. 76 (3 September, 1991).

the desired effect that consumers make better informed purchases that reflect their preferences. By contrast, defending such a ban would be much more difficult since it reduces consumers' 'pool of choice' and is in principle more trade restrictive.

Implications. The analysis reveals that the point of reference that determines when a trade practice could be qualified to be deceptive is the consumer's level of understanding. In practice, a WTO panel would need to refer to case-by-case empirical evidence.

PROTECTION OF HUMAN HEALTH AND SAFETY AND THE ENVIRONMENT Labels for GEF are often justified as a way of allowing consumers to avoid such foods if they have concerns about their long-term health impacts. Such a labelling requirement could be based on the lack of long-term experience with the product. To what extent does Article 2.2 of the TBT Agreement allow the basing of a trade restrictive measure on such hypothetical concerns?

Terms. With regard to specific health concerns, the SPS Agreement clearly requires a sufficient scientific basis for the measure in question. The requirement of 'scientific evidence' to support domestic regulations has been subject to intensive debate within the context of the SPS Agreement. In essence, it requires 'a rational relationship between the measure and the risk assessment'.[19] '[T]he absence of an actual causal link' between the measure in question and the risk that is intended to counteract is seen to be 'a strong indication of the absence of a rational relationship'[20] while a 'proper risk assessment . . . must evaluate the 'likelihood', i.e. the 'probability', of entry, establishment or spread'.[21]

Article 2.2.4 of the TBT Agreement lists available scientific information only as an element of its risk assessment. This is as close as the TBT Agreement gets to mandating scientific risk assessment.[22] The TBT-risk assessment is not limited to a requirement 'based on scientific principles [. . .] and not maintained without sufficient evidence' as provided in Article 2.2 of the SPS Agreement. On the contrary, the catalogue of relevant criteria is opened for other aspects than scientific judgement that might be relevant in any specific case. This seems to be consistent considering the scope of legitimate objectives that might be targeted with a technical regulation. It would not make much sense, for instance, to require a risk assessment (based on natural sciences) regarding the legitimate objective of 'preventing deceptive practices'. The TBT Agreement also has no special rule for provisional measures in case of insufficient scientific knowledge/scientific uncertainty as contained in Article 5.7 of the SPS Agreement: Article 2.2 of the TBT Agreement also has to cover such issues. Considering the wider objective of environmental protection, the TBT Agreement would also appear to be less demanding in terms of a direct link between the measure used and its scientifically provable effects on attaining an objective.

Assessment. A WTO panel would have to judge whether the current knowledge provides some hints that support a special treatment of a GEF with respect to human health.

Assume that a country requires all processed foods, such as cornflakes, to be labelled 'made of genetically engineered maize' if the food (testably) contains such maize. Presenting an 'actual causal link' between the labelling requirement and the risk for human health as required by the SPS Agreement appears to be difficult. Under the SPS Agreement, the importing country would probably have to refer to Article 5.7 of the SPS Agreement on grounds of the lack of scientific knowledge (though the EU did not in the beef-hormone case).[23] In the case

[19] Report of the WTO Appellate Body on Japan – measures affecting Agricultural Products, WT/DS76/AB/R (99-0668), 22 February, 1999 at 73–76.

[20] Report of the WTO Appellate Body on Japan – measures affecting Agricultural Products, WT/DS76/AB/R (99-0668), 22 February, 1999 at 83.

[21] Report of the Appellate Body on Australia – measures affecting importation of salmon WT/DS18/AB/R (98-4035); AB-1998-5 20 October, 1998 at para. 123.

[22] The whole TBT Agreement uses the term 'risks' only three times: in Articles 2.2.2, 2.2.4, 5.1.2.

[23] *See* World Trade Organization (1999) *Seattle: What's at Stake? Resource Booklet for the Seattle Ministerial Meeting*. At 40 and 41.

of the TBT Agreement it may be enough to show that the label is fulfilling another legitimate objective and/or multiple objectives and that scientific and technical information has been used in assessing its appropriateness. Following the beef-hormone case, a court may conclude that a labelling requirement for provably genetically engineered foods cannot be based on the objective of protecting human health since those foods do not significantly differ from conventional foods in their nutritional value.

Implications. If a measure relates to natural science-based concerns, available scientific information needs to be considered within the TBT-risk assessment; purely hypothetical concerns are probably not sufficient.

. . . 'not be more trade-restrictive than necessary'

Terms. The TBT Agreement does not expressly define the content of its 'necessary' test. The exact meaning of the terms itself is not clear and can vary depending upon the type of measure and the rule in question.[24]

Assessment. In the SPS Agreement, a footnote clarifies the 'not more trade-restrictive' test by explicitly referring to a 'reasonably available' standard regarding 'technical and economic feasibility of an alternative'. Despite the fact that the parties themselves determine the 'appropriate level of protection', 'not more', is not understood to require that the state has chosen the 'optimal' and strictly 'least trade restrictive' alternative among all of the available measures. The alternative measure must be 'significantly less restrictive to trade'. This implies a not too rigorous cost–benefit test that examines possible alternatives.

The TBT Agreement also entitles domestic regimes to determine the appropriate level of protection according to their needs while it additionally entitles them to choose a legitimate objective to pursue.[25] Given the purpose of the 'necessary' test, namely making specific the broad and overall objective of the TBT Agreement to sort out protectionist from non-protectionist measures, there may be no need for a stricter standard compared with the SPS Agreement.

Implications. The TBT Agreement 'only' requires some kind of not too rigorous cost–benefit/balancing[26] judgement examining possible alternatives and their feasibility.

. . . 'risks non-fulfillment would create'

Terms. Unless one assumes editorial thoughtlessness and useless wordiness, the phrase 'taking account of the risks that non-fulfilment would create' should have meaning beyond the usual understanding of the 'necessary' test. It requires evaluating possible alternatives compared with the effects of the measure in dispute and its non-fulfilment.

Assessment. Three points are crucial for understanding the potential meaning of this provision: the express mention of alternatives; the place of science and economic reasoning; and the consideration of other factors that enter into a risk assessment.

Article 2.6–2.8 of the TBT Agreement provide a first guide on the issue of alternatives. Domestic decision makers should take into account efforts of international standardization (Article 2.6), recognize where possible foreign standards to be equivalent (Article 2.7), and base technical regulations on product requirements rather than on descriptive characteristics wherever appropriate (Article 2.8). Although these provisions constitute separate obligations under the TBT Agreement, they are nevertheless essential elements of the 'risks of non-fulfilment' test.

The reference to 'available scientific and technical information' suggests that magnitude and probability of the risk in question become relevant. The additional reference to the 'risks of non-fulfilment' goes beyond the risks directly related to the objective, pointing to how the specific measure affects the whole socio-economic system. Since domestic decision makers assess these risks in the first place, the phrase could be

[24] *See* WTO Panel Report on USA: Standards for Reformulated and Conventional Gasoline, WT/DS2/R, 35 I.L.M. 274, para. 120 (adopted on 29 January, 1996).

[25] This is implied in clauses 6 and 7 of the preamble of the TBT Agreement.

[26] *See* Sykes, A.O. (1995) *Product Standards for Internationally Integrated Goods Markets*. Brookings Institute, Washington, DC, 78–79.

seen as providing additional room for domestic discretion.

With respect to other factors to be taken into account, more scope for disagreement remains. According to the nature of risk, risk assessment, and risk management, it is possible that judgements will vary depending on who assesses the risk and under which premises.[27] A country that prefers to stay on the 'safe' side will probably argue that more evidence may be required to be sure that the risks are acceptable. The argument relates to the so-called precautionary principle. How does this fit into the 'risks of non-fulfilment' test of the TBT Agreement? The WTO Agreements – including the TBT Agreement – do not explicitly use the term 'precautionary principle'. However, Article 2.2 of the TBT Agreement links the term 'risks' with 'assessing'. One may understand this language as referring to a concept of risk and risk assessment as it is used and defined within the frame of the WTO and in international law.[28] Since 'available scientific and technical information' is only one element of the TBT-risk assessment, the phrase should be read as referring to the broad process of political decision making in the field of risk and scientific uncertainty.

Implications. The phrase may provide additional room for domestic discretion. It may also accommodate precautionary measures on GEF. The Cartagena Biosafety Protocol (CBP) substantiates the precautionary principle for living genetically modified organisms (LMOs), including when they enter into international trade: it could be seen as providing a frame for essential trade regulations related to LMOs while the WTO agreements provide the general non-discrimination rules, additional procedural requirements, and a dispute settlement regime. Whether there could be a similar division of labour for other GEF products is the source of some contention.

Protecting public morals, the consumer right to know and human rights

Terms. Countries may wish to make use of the open-ended nature of Article 2.2.3 ('*inter alia*') to claim other objectives, such as the 'public morals' exception stated in Article XX(a) of the GATT(94).

Assessment. So far, the 'public morals' exception has had a relatively broad coverage without so far undermining the WTO system as such.[29] Israel, e.g. prohibits importing non-kosher meat products. The exception may also provide the measure for judging domestic regulations based on consumer rights or human rights.

The stated 'right to know' and 'to choose' might be linked to existing and evolving human rights of self-determination, aspects of democratic participation and a right to preserve cultural identity. The ultimate goals stated in the preamble of the WTO Agreement are raising of the living standards, sustainable development, and the actual needs and concerns of people. Given the set of the concerns of many people, it would probably be far too narrow and contradict the spirit of the preamble to define 'raising standards of living' only in terms of economic growth and monetarily measurable gains. At the risk of oversimplifying: if people seriously care about something, then it should be taken into account. The TBT Agreement refers implicitly to actual consumer preferences when accepting the objective of preventing deceptive practices. It would seem consistent to recognize such preferences in other parts of the Agreement. The resulting diversity could even be seen as efficiency enhancing.[30]

With respect to GEF, the real struggle today is whether comprehensive and mandatory labelling may be deemed necessary, rather than

[27] *See* submission of the European Communities in Report of the Appellate Body on EC measures concerning meat and meat products (Hormones), WL 25520 (WTO) at para.121 (16 January, 1998).

[28] *See* Howse, R. (1998) The turtles panel – another environmental disaster in Geneva. *Journal of World Trade* 32(5), 73, 94–95 (explaining Article 31(3)(c) of the Vienna convention).

[29] *See* GATT Panel on USA – Measures Affecting Alcoholic and Malt Beverages, June 19, 1992, GATT B.I.S.D. (39th Supp.) at 206, para. 3.125 (1993) (The USA argued that a different tax treatment of high and low alcohol beer could be justified by Article XX(a) of GATT.)

[30] *See* Esty, D. and Geradin, D. (1998) Environmental protection and international competitiveness – a conceptual framework. *Journal of World Trade* 32(3), 4, 11 and 46.

a voluntary regime.[31] Economic analysis may often provide support for various positions. For example, Howard Beales recently concluded that special mandatory labelling for genetically modified organisms is unjustified due to its costs while voluntary labelling provides the regulatory means of choice for addressing consumers' demands on information.[32] On the other hand, Lydia Zepeda explains how the various alternatives relate to the distribution of costs, income-dependent choice ability of consumers, and international competitiveness, and suggests that even mandatory labelling of GEF can make sense.[33] Here, a panel may limit its judgement to aspects of system conformity and procedural equal treatment and transparency instead of judging a regime based on its material substance.

Implications. Based on the TBT Agreement, a domestic measure that tries to tackle the broader set of concerns may not be judged only by asking whether a GEF is 'safe' or different from foods from traditional breeding programmes when assessed from a current scientific perspective.

Closing Pandora's Box

It could be argued that allowing trade-restrictive measures for GEF seems to open a Pandora's box, out of which could flow all manner of protectionist devices masquerading under precaution, the consumer's right to know, and environmental protection. How to close this box and avoid a 'slippery slope' would be a key consideration.

One might think of three possibilities: using tighter 'necessary' test standards by requiring a strict proportionality between measures and objectives; imposing stricter limits on the circle of legitimate objectives; demanding more detailed empirical evidence for judgements.

Applying a stricter balancing or proportionality test goes back to the question of to what extent the broader set of values should be considered. To require a strict proportionality test would mean that the WTO system itself would assess and judge ends. However, the TBT Agreement expressly provides priority to the objectives and the level of protection as set by the Member State; there is no explicit trade-off envisaged between trade and the effectiveness of the means to reach a non-trade goal. Second, if one nevertheless questions this rather clear language, in case of an ambiguity the meaning to be preferred is that which 'interferes less with the . . . supremacy of a [WTO member]'. Third, trade conflicts may arise from differences in the underlying values themselves. Neither science nor any other objective standards are available to give priority to one or the other value. A court might lack a sufficient basis for judgement.[34] Thus, one should avoid a strict proportionality test to close the box.

Nevertheless, the other two ways of drawing the line show more promise: a substantive limit requiring countries to stay within existing international law on the precautionary principle and human rights/civil society concerns; and a procedural limit requiring corresponding empirical evidence when referring to consumers' demands and cultural habits, demonstrating the existence of people's needs. The procedural requirements should provide an equivalent to a scientific justification and compensate the lack of material filters.

The substantive limit essentially means that the WTO integrates itself into the dynamic system of global governance in conformity with its own aims of 'raising standards of living' and 'sustainable development'. It avoids tensions with actual civil society movements as well as with international law outside the WTO, such as the CBP or the UN Charter on Human Rights. This does not mean *carte*

[31] *See* Phillips, P.W.B. and Isaac, G. (1998) GMO labelling: threat or opportunity? *AgBioForum* 1(1), 25–30.

[32] *See* Beales, J.H. (2000) Modification and consumer information: modern biotechnology and the regulation of information. *Food & Drug Law Journal* 55, 105, 113.

[33] *See* Zepeda, L. (2001) 'Don't ask, don't tell: US GM food labelling policy'. Paper presented at the American Association for Advanced Science, San Francisco (18 February, 2001).

[34] *See also* Esty, D. and Geradin, D. (1998) Environmental protection and international competitiveness – a conceptual framework. *Journal of World Trade* 32(3), 4, 45. (stating: 'Whose values should be used? Valuation is, to some degree, an inescapably political exercise'.)

blanche for 'consumer' advocates. For example, an objective of 'consumer protection' which implied 'spoon-feeding' its citizens by the state also in the long run may lack a basis in human rights law since it tends to counteract the individual and group rights approach of human rights. On the other hand, an objective of recognizing the 'consumer right to know' may well be based on internationally acknowledged human rights.

Requiring more supportive empirical evidence on the procedural side, regarding what people want and why, can provide the necessary compensation for the possibility of greater trade restrictions. With respect to domestic labelling regimes for GEF, a state should provide proof that education, the forces of the free market, or a voluntary labelling regime would not meet public concerns. Furthermore, the material and procedural transparency requirements should be taken seriously and strictly. A WTO panel may want to refer to international institutions for necessary expertise, such as an international panel on GEF.

Conclusions

The TBT Agreement will often provide the relevant measure for judging labelling regimes for GEF. The range of legitimate objectives could include those that are mentioned in the GATT, such as 'public morals', and those that have been acknowledged in international law, which may include human rights law. When judging trade disputes, procedural constraints on transparency and the evidence required should compensate for the substantive openness and flexibility.

8 Environmental Liability and Research and Development in Biotechnology: a Real Options Approach

Odin Knudsen[1] and Pasquale L. Scandizzo[2]
[1]The World Bank, Washington, DC, USA; [2]University of Rome 'Tor Vergata', Rome, Italy

Introduction

Traditionally, environmental policy has relied on *ex ante* regulation such as standards, taxation or outright prohibition to inhibit industries from causing damage to the environment. These *ex ante* regulations include emissions and technological standards, taxes on pollution, and prohibitions on dumping. The effect of these forms of regulations on productivity growth and innovation has been controversial. Gollop and Roberts (1983) analysed the effect of restrictions on sulphur dioxide emissions on the rate of productivity growth in the electric power industry and found that productivity growth was reduced significantly. However, another perspective has emerged known as the Porter hypothesis (Porter, 1991; Porter and van der Lind, 1995) where it is argued that environmental regulation should increase competitiveness of firms, force firms to seek innovation that could be both profitable and socially beneficial. In many ways, the Porter hypothesis is a restatement of Hicks (1932) induced-innovation hypothesis where a change in relative prices of the factors of production spur innovation seeking to economize on the use of the factor for which costs have risen.

While some examples support the Porter hypothesis of innovation to meet environmental regulation achieving cost savings (Sinclair-Desgagné, 1999), the controversy remains as it is argued that someone needed to absorb the costs, either managerial or technological to meet the regulations, at least, until innovation increased productivity or economized on the now more costly factor of production. Schmalensee (1994) suggests that while research and development expenditures devoted to environmental compliance may increase with stricter regulation, it may also divert resources from research on more productive technologies. McCain (1978) argues further that firms may be reluctant to innovate as this may subsequently lead to even stricter regulatory standards.

Some empirical work supports the 'win–win' hypothesis of Porter. Lanjouw and Mody (1996) analyse the increased cost of environmental compliance with respect to the patenting of environmental technologies. Using international data, they find with a one to two year lag, patents do increase following higher compliance costs. Jaffee and Palmer (1997) in a panel of manufacturing industries find that compliance expenditures have a significant positive effect on research and development expenditures. But contrary to Lanjouw and Mody, they do not find evidence that this leads to increased successful patent applications.

Research and development in biotechnology presents a new challenge to environmental regulation. For example while *ex ante* regulation

 The Regulation of Agricultural Biotechnology (eds R.E. Evenson and V. Santaniello)

on the use of new varieties in agriculture or on standards on dissemination of new technology can reduce potential adverse environmental consequences, substantially more risks to industry could result from liability damages from introduction of these innovations in biotechnology. This becomes particularly burdensome when the full risks and consequences of the introduction of new biotechnology are unknown.

While many environmental groups would support the strict application of the precautionary principle, whereby innovation would be stopped or slowed until the risks are known, biotechnology offers such potential benefits to health, agriculture and to the environment through reduced inputs that this approach would be mostly unacceptable. Through the use of real options, Knudsen and Scandizzo (2001) evaluate how the risks of biotechnology under a weaker form of the precautionary principle could be taken into account. By using the concept of a social standard whereby society places a threshold on the acceptable loss from biotechnology causing damage, Knudsen and Scandizzo establish some simple rules on evaluating risks of biotechnology both in a static and dynamic case.

With innovation in biotechnology moving at a very rapid rate, the *ex ante* approach to environmental regulation is likely to hold limited promise. Increasingly governments have introduced legislation that holds the perpetrators of new technologies liable for cleaning up the environmental damage they may have caused. This *ex post* form of regulation is designed in principle to provide a motivation for industry to account for the risk of environmental liabilities and thereby have incentives to reduce development of potentially environmentally risky technologies. The tradition of environmental based liability begins with the US Superfund legislation introduced in 1980 (officially the Comprehensive Environmental Response, Compensation, and Liability Act), which was passed after the discovery of severely contaminated sites in the 1970s. In February 2000, the European Commission adopted a White Paper on Environmental Liability that addresses the liability of damage to nature. While there are many similarities to the Superfund legislation, the European liability regime extends the damage to biodiversity both for hazardous and non-hazardous activities, the latter with fault basis – the liability party should be the operator in control of the activity that caused the damage. The approach envisages that compensation would be paid to effectively restore the damage.

The potential costs and ramifications of this form of liability on biotechnological innovation and development are largely unknown. McGuignan (2000) attempts to access the potential economic impact of environmental liability if Europe adopts the recommendations of the White Paper on Environmental Liability based on the experience of the USA with the Superfund, but concludes that too much uncertainty exists on the judicial and legalistic side to estimate the potential costs. Ulph and Valentini (1999) estimate in terms of effectiveness of environmental liability laws, that limited liability will reduce the effects of this form of *ex post* regulation as firms can avoid large damages through bankruptcy. They conclude that liability should be extended to lenders such as banks.

Clearly, this emerging field of environmental regulation through liability and the introduction of development products from biotechnology create substantial uncertainty and risks. In this chapter, we attempt to access how these risks could affect research and development in industries such as biotechnology where the uncertainty is particularly high but the potential returns are also great. We utilize a real options approach whereby research creates the real option for development and subsequently development poses risks in terms of liability, either through expected liability losses or through a risk option of liability.

In the next section, we develop a simple two-stage model of research and development where expected liability enters into the development phase. We evaluate how liability affects the decision on whether to embark on the project and how uncertainty heightens the effects of liability.

The Model

We consider the problem of research and development (R&D) as a multi-stage investment problem under uncertainty with a probability of a liability claim in the development phase. More specifically, we assume that the prospect confronting the investor may be broken down at least into two (and possibly more) stages. In stage 1, a research project is undertaken which 'may' lead to a significant discovery. If this occurs, a second stage of 'development' may be undertaken. However, the development phase results in unknown risks that may be manifested in a liability suit with probability p and potential award of Y. We assume that the discovery process may be represented by a stochastic process of the Brownian motion variety, whereby the value flow generated by R&D project is expressed as follows:

$$dV = \alpha V dt + \sigma V dZ \quad (1)$$

where dZ is random variable with mean $EdZ = 0$ and variance $E(dZ)^2 = dt$ The α represents the drift or trend term and σ the standard deviation of V. The value V could be the underlying asset value of the firm embarking on the innovation. This value could be realized through an Initial Public Offer (IPO) or be implicit, that is, expressing potential value if the shares were traded. In a multiple product firm, this value would be the incremental value of the new research line. It is of course dependent on the market price of the potential product from the R&D.

Given this set up, we assume for simplicity that investment requires an instantaneous and irreversible commitment of resources both in the research (I_r) and in the development phase (I_d). Given that a discovery has occurred, the development option O_d may be evaluated (Dixit and Pindyck, 1994, p. 248) as $O_d = \max(BV^{\beta_1} + CV^{\beta_2}, 0)$, where B and C are constants determined by boundary conditions and β_1 and β_2 are, respectively, the positive and the negative root of the characteristic equation:

$$r - \beta\alpha \frac{\beta}{2}(\beta - 1)\sigma^2 = 0$$

r being an appropriate rate of discount.

The value of the option to develop the resource should decrease with any decrease in the cash flow generated by the resource under development. But the second term on the right hand side of the characteristic equation goes to infinity as y goes to zero. Thus, the constant C can be set to zero.

In order to evaluate B, we use the optimal stopping conditions (in this case they indicate the optimal stopping of the process of waiting before adopting the process). They state (Karlin and Taylor, 1975) that optimal switching from waiting to adoption occurs the first time the process hits the boundary of the continuation region. This requires in turn that the following conditions are satisfied at the switching point V_d:

Value matching:

$$BV^{\beta_1} = \frac{V}{\delta} - I_d - pY \quad (2)$$

Smooth pasting:

$$\beta_1 BV^{\beta_1} = \frac{1}{\delta} \quad (3)$$

In Equations 2 and 3, as we have already specified, V is the expected value of the cash flow from the innovation, I_d is the investment outlay of the development stage, $\delta = r - \alpha$, where r is the market interest rate (or an other appropriate rate of discount), and B and β_1, two parameters that can be determined, respectively, solving the system (Equations 2 and 3) and applying Ito's lemma (Dixit and Pindyck, 1994, p. 4), and pY is the expected liability claim. In Equation 2, the left-hand side represents the option to undertake the development phase, which is non zero, and dominates the net present value (NPV) on the right-hand side in the so called continuation region, i.e. for the values of expected benefits V, for which it is not worth undertaking the project. Once the development phase is undertaken, on the other hand, the option is no more alive, and its value is zero. We may capture this behaviour by defining the so called extended NPV or NPVE as:

$$NPVE = E\left(\max\left(\frac{V}{\delta} - I_d - pY, 0\right)\right)$$
$$= NPV + \max(BV^{\beta_1} - NPV, 0) \quad (3bis)$$

where $NPV = \frac{V}{\delta} - I_d - pY$ is the expected net present value of the project. The relevance of NPVE is that not only does it inform us of project worth, but also of how much is worth *waiting to undertake the project* through committing irreversible resources. An increase in the value of waiting, in fact, may channel resources to alternative uses, including those related to improvement of information and what we can call 'general preparedness' to undertake the project subsequently. Note that NPVE is necessarily greater or equal to NPV. If it is greater, the value of the option to wait prevails over the expected net present value, and the NPVE equals the option value. If it is equal, the option to wait is lower than the NPV, and the latter measure prevails as an indicator of project worth.

Substituting Equation 3 into Equation 2 and solving for V, we obtain the value of the optimum switching point V_d, i.e. the minimum present value of the expected cash flow necessary to prompt the entry into the development phase:

$$\frac{V_{\mathrm{d}}}{\delta} = \frac{\beta_1}{\beta_1 - 1}(I_{\mathrm{d}} + pY) \tag{4}$$

and, by substituting into Equation 3, we obtain the expression for the constant B:

$$\mathrm{B} = \frac{1}{\delta\beta_1}\left[\frac{\beta_1}{\beta_1 - 1}(I_{\mathrm{d}} + pY)\delta\right]^{1-\beta_1} \tag{5}$$

Equation 4 states that at the optimum entry point, the discounted value of V_{d} must exceed the investment costs I_{d} and the expected costs of the liability pY by a factor of $\beta_1/(1 - \beta_1)$ or the uncertainty 'wedge'. From the characteristic equation, it can be shown that β_1 is inversely proportional to σ: as uncertainty increases, the 'wedge' also becomes larger requiring a larger value of V before the development phase is undertaken. Note that the expected liability increases the necessary value V not just by the expected costs pY as in the deterministic case but by the expected costs times the uncertainty wedge $\beta_1/(1 - \beta_1)$.

For the entry in the research stage, on the other hand, we have to consider the value matching condition given by the equality between the option to undertake the project and the difference between the option to go into the second stage and the investment in research:

$$\mathrm{A}V^{\beta_1} = \frac{1}{\delta\beta_1}\left[\frac{\beta_1}{\beta_1 - 1}(I_{\mathrm{d}} + pY)\delta\right]^{1-\beta_1} V^{\beta_1} - I_{\mathrm{r}} \tag{6}$$

The smooth pasting condition of this problem is met only if the constant A of the option value for entering the research stage is the same as the constant B of the option value entering the development stage. Since this would deny value matching, in order for the research stage to be acceptable to the investor, the left hand side of Equation 6 must be lower than the right hand side. Because of Equations 3 and 4, on the other hand,

$$\frac{1}{\delta\beta_1}\left[\frac{\beta_1}{\beta_1 - 1}(I_{\mathrm{d}} + pY)\delta\right]^{1-\beta_1} V^{\beta_1} \leq \frac{V}{\delta} - (I_{\mathrm{r}} + pY) \tag{7}$$

and, by combining with Equation 6, stated as an inequality, we obtain:

$$\mathrm{A}V^{\beta_1} \leq \frac{V}{\delta} - I_{\mathrm{d}} - I_{\mathrm{r}} - pY \tag{8}$$

Thus, if the research and the development stage are separately acceptable, the whole project is acceptable from the start and there is no need to divide it in stages.

Using the equality case of Equation 8 as value matching and adding the smooth pasting condition:

$$\beta_1 \mathrm{A}V^{\beta_1 - 1} = \frac{1}{\delta} \tag{9}$$

Equation 9, upon substitution into Equation 8 yields:

$$\frac{V_{\mathrm{p}}}{\delta} = \frac{\beta_1}{\beta_1 - 1}(I_{\mathrm{d}} + I_{\mathrm{r}} + pY) \tag{10}$$

where V_{p} denotes the minimum value of the expected cash flow that justifies adoption of the whole project. Clearly $V_{\mathrm{p}} > V_{\mathrm{d}}$, i.e. adoption of the whole project implies that the expected cash flow is high enough to justify adoption of the development stage.

From Equations 9 and 10, we can derive the value of the constant A:

$$\mathrm{A} = \frac{1}{\delta\beta_1}\left[\frac{\beta_1}{\beta_1 - 1}(I_{\mathrm{r}} + I_{\mathrm{d}} + pY)\delta\right]^{1-\beta_1} \tag{11}$$

Because $\beta_1 \leq 1$, from Equation 11 and Equation 5, we conclude that $\mathrm{B} \geq \mathrm{A}$, which is compatible with Equation 6. We can now evaluate explicitly the net present value (NPV), extended to take into account the values of the options created or destroyed by the project in each phase (the NPVE). The NPVE for the entire project is shown in Equation 12 (see overleaf).

Equation 12 indicates that, while the usual expected NPV considers only project worth as resulting from the net value flow acquired from immediate project adoption, the NPV in its extended form considers also the alternative value embedded in the project as an option. If the value of such an option is lower than the traditional value of discounted cash flow, then the expected NPV is the proper value to attach to the project. If, instead, the option has a greater value, then it should be considered as the appropriate measure of project worth. The decision to embark on a project would be predicated on the traditional NPV exceeding the option value of waiting.

That is, proceed with the project of research if the value stream is such that:

$$\text{NPV} \geq \frac{1}{\beta_1 \delta}\left(\frac{\beta_1 - 1}{\beta_1 \delta (I_r + I_d + pY)}\right)^{\beta_1 - 1} V^{\beta_1} \quad (13)$$

The NPVE for the development phase may be written similarly as in Equation 14 (see bottom of page), and the NPVE for the research phase will be equal to the difference $W_p - W_d$, i.e. to the additional gain that research would allow with respect to the development phase (see Equation 15 at bottom of page), where O_p and O_d are, respectively, the values of the waiting option for the entire project and the development phase, while NPV_p and NPV_d are the corresponding net present values.

From Equations 13 and 14, we can see that the NPVE of both the entire project and the development phase may defined as the sum of expected NPV and the option value, respectively, of the whole project and the development phase. The option value, on its part, equals either the excess of the waiting option over the corresponding NPV, or zero, whatever is greater. The value of the NPVE for research, defined as the difference between the two NPVEs, as a consequence, may be negative or positive, although the NPVE for the whole project and for the development phase are both non-negative. In particular, if the option to wait is larger than expected net benefits in each respective continuation region, the NPVE of the project at the research stage is necessarily negative, since the value of the waiting option for the project is lower than the value of the waiting option for the development phase. On the other hand, if both waiting options are lower than the respective NPVs, the two option values in Equation 14 are both zero, and the NPVE for research is again negative, since undertaking the whole project produces a greater net benefit than entering just the first part of it.

The NPVE for the research stage may be non-negative only if the first option value is non-zero and the second is zero (i.e. the waiting option is alive for the whole project, but not for the development phase taken in isolation). This can be true only if the development phase, considered separately from the whole project, is different from the development phase as an integral part of the same project. For example, dividing the project into stages changes the timing of the two phases, so that the development phase will not be implementable instantaneously as it would be if integrated into the project. In this case, the development option may be non-zero (i.e. it may be worth waiting before developing) even though waiting does not have value for the whole project.

Conclusions

The addition of the consideration of environmental liability in this two stage example

$$W_p = \frac{V}{\delta} - I_r - I_d - pY + \left(\max\left(\frac{1}{\beta_1 \delta}\left(\frac{\beta_1 - 1}{\beta_1 \delta (I_r + I_d + pY)}\right)^{\beta_1 - 1} V^{\beta_1} - \left(\frac{V}{\delta} - I_r - I_d - pY\right), 0\right)\right) \quad (12)$$

$$W_d = \frac{V}{\delta} - I_d - pY + \left(\max\left(\frac{1}{\beta_1 \delta}\left(\frac{\beta_1 - 1}{\beta_1 \delta (I_d + pY)}\right)^{\beta_1 - 1} V^{\beta_1} - \left(\frac{V}{\delta} - I_d - pY, 0\right)\right)\right) \quad (14)$$

$$W_r = \left\{ \max\left(\frac{1}{\beta_1 \delta}\left[\frac{\beta_1 - 1}{\beta_1 \delta (I_r + I_d + pY)}\right]^{\beta_1 - 1} V^{\beta_1} - \left(\frac{V}{\delta} - I_r - I_d\right), 0\right) - \max\left(\frac{1}{\delta \beta_1}\left[\frac{\beta_1 - 1}{\beta_1 \delta (I_d + pY)}\right]^{\beta_1 - 1} V^{\beta_1} \right)\right\}$$

$$= [\max(O_p = NPV_p, 0) - \max(O_d = NPV_d, 0)] - I_r \quad (15)$$

increases the threshold value of V for embarking upon the research phase of the project by a proportionally higher value than the expected costs of the liability. This difference increases with the level of uncertainty over the returns σ and with the probability p of the liability damage. If we consider research synonymous with innovation, then the threat of environmental liability is to decrease innovation disproportionably more than the liability. But once the threshold value is exceeded and research is undertaken, the development phase is automatic if the development phase can be begun without delay.

Including a delay in embarking on the second phase complicates the model but the intuition of the solution is as follows. The development or second stage of the project can be evaluated as above, with the threshold value of V or the value needed to exercise the option of the development phase determined. This value is lower than the value for beginning the overall project through launching research. But with the delay, the decision to 'destroy' the research option and create the development option depends on how much the value of V could move from the threshold value during the delay in embarking on the development phase. If V moves downward because of market forces on the output value or because the discovery is less valuable, then the development option will not be exercised and the sunk costs of the research investment and the research option value will be lost. Of course, if V stays the same or moves above the threshold value of the development phase, the development phase of project will be undertaken.

If we let V_p^* be the threshold value for launching the research phase of the project, then after time T the expected value of V will be:

$$E(V_T) = V_p^* e^{\alpha T} \tag{16}$$

And the variance of V will be given by:

$$\mathrm{Var}(V_T) = V_p^{*2} e^{2\alpha T}(e^{\sigma^2 T} - 1) \tag{17}$$

As the interval T increases between the launching of the research and the realization of the development phase (T being the time needed for discovery), the uncertainty as measured by the variance of V becomes greater.

The expected value of the development phase is now from (14) (see Equation 18 at bottom of page).

If α is greater than or equal to zero, then the decision to launch the project is the same as in Equation 13. If α is negative, then the decision to invest in the project (the NPV in Equation 12) must be discounted by $e^{\beta_1(\alpha - r)T}$.

References

Dixit, A. and Pindyck, R.S. (1994) *Investment under Uncertainty*. Princeton University Press, Princeton, New Jersey.

Gollop, F.M. and Roberts, M. (1983) Environmental regulations and productivity growth: the case of fossil-fueled electric power generation. *Journal of Political Economy* 91 (August), 654–674.

Hicks, J.R. (1932) *The Theory of Wages*. Macmillan, London.

Jaffee, A.B. and Palmer, K. (1997) Environmental regulation and innovation: a panel data study. *Review of Economics and Statistics* 79(4), 610–619.

Karlin, S. and Taylor, M.T. (1975) *A First Course in Stochastic Processes*. Academic Press, New York.

Knudsen, O. and Scandizzo, P.L. (2001) Evaluating risks of biotechnology: the precautionary principle and the social standard. Paper presented to the *International Consortium on Agricultural Biotechnology Research*, Ravello, Italy, 15–18 June.

Lanjouw, J. and Mody, A. (1996) Innovation and the international diffusion of environmentally responsive technology. *Research Policy* 25, 549–571.

McCain, R.A. (1978) Endogenous bias in technical progress and environmental policy. *American Economic Review* 68, 538–546.

McGuigan, J.S. (2000) *The Potential Economic Impact of Environmental Liability: The American and European Contexts*. European Commission,

$$W_d^* = \frac{V_p^* e^{\alpha T}}{\delta} - I_d - pY + \left(\max\left(\frac{1}{\beta_1 \delta} \left(\frac{\beta_1 - 1}{\beta_1 \delta (I_d + pY)} \right)^{\beta_1 - 1} V_p^{*\beta_1} e^{\beta_1 \alpha T} - \left(\frac{V_p^* e^{\alpha T}}{\delta} - I_d - pY, 0 \right) \right) \right) \tag{18}$$

Environment Directorate, Economic Analysis Unit, Bruxelles.
Porter, E.M. (1991) America's green strategy. *Scientific American* April, 168.
Porter, E.M. and van der Linde, C. (1995) Towards a new conception of the environment competitiveness relationships. *Journal of Economic Perspectives* 9(4), 97–118.
Schmalensee, R. (1994) The costs of environmental protection. *Balancing Economic Growth and Environmental Goals*. Monograph Series on Tax and Environmental Policies and US Economic Growth, American Council for Capital Formation, Washington, DC (May).
Sinclair-Desgagné, B. (1999) *Remarks on Environmental Regulation, Firm Behavior and Innovation*. CIRANO Working Papers. Centre Interuniversitaire de Recherche en Analyse des Organisations (CIRANO), Montreal.
Ulph, A. and Valentini, L. (1999) *Environmental Liability and the Capital Structure of Firms*. University of Southampton Discussion Papers in Economics and Econometrics 0036, Southampton, UK.

9 Should the Public Sector Conduct Genomics R&D?

Anwar Naseem[1] and James F. Oehmke[2]
[1]Department of Agriculture, Food and Resource Economics, Rutgers, The State University of New Jersey, New Brunswick, New Jersey, USA; [2]Department of Agricultural Economics, Michigan State University, East Lansing, Michigan, USA

Abstract

The nature of the observed market structure and R&D competition in genomics research is used as the basis for a comparative analysis of research under a mixed oligopoly and pure oligopoly when the timing of the innovation outcome is uncertain (as in an R&D race), the winner-take-all assumption is relaxed and the profits in later stages are a function of the R&D expenditures of prior stages. The sufficient conditions under which a mixed oligopoly performs more R&D than the pure oligopoly is derived and shown to be a function of (i) the public sector firm's objective function is strictly greater than in the winning state than it is in the losing state; (ii) profits for the winning and losing private firms in the private duopoly are equal, post innovation; and (iii) the objective function of the firms in the mixed duopoly are increasing in research faster than they are for firms in the pure duopoly. It is suggested that when these conditions are met, the public firm can play a role in increasing the level of research in genomics.

Introduction

Over the past two and a half decades biotechnology has revolutionized major portions of the human health and agriculture industries, transforming them into the life sciences industry. Yet the most revolutionary biotechnologies are still in the early stages. Perhaps the most potent of these technologies is genomics – the science of sequencing all the genes of a given species to study the structure, function and evolution of a diverse organism. Genome research, however, is more than biology; it is also about developing better drugs, foods, industrial products, and, in the case of agriculture, improving plant and animal productivity and quality.

A striking feature of genomic research is the significant levels of investment by a few dominant private firms in competition with an equally well-funded public sector that seek to discover and subsequently patent important gene sequences. This observation is made most obvious by the private sector's Celera Genomics Group challenge to the longer-lived and more expensive, publicly funded Human Genome Project (HGP). The Department of Energy and National Institutes of Health started HGP in 1990, at a cost of approximately US$2.2 billion over the course of the project. In 1992 Craig Venter, a scientist with the HGP, left to form his own private company, Celera Genomics, and claimed that the firm could sequence, using a different technique from the HGP, the whole genome in less than 3 years and at a fraction of the cost (approx. US$200 million). Celera's challenge to the publicly funded HGP signalled the start of the race for the sequencing of the human genome, which was joined by numerous other start-up companies looking to capitalize on the potential of genomic research. The competition between Celera and HGP so accelerated sequencing efforts that by late 2000 both projects were essentially complete ahead of schedule.

Although less confrontational, the private sector is also at the forefront of characterizing plant and animal genomes, sometimes subcontracting from the public sector. In 1998, a consortium led by the International Rice Genome Research Program (IRGSP) in Tsukuba, Japan, began efforts to sequence the rice genome. The participants, who included primarily government and research foundation sponsored labs, took a traditional approach to genome mapping known as the 'stepwise sequence analysis' (Bennetzen, 2002). This approach, while expensive and slow, provides the most precise and complete sequence with a goal of 99.99% accuracy.[1] Soon after the IRGSP was initiated, the Monsanto Co. began funding research to sequence the same variety of rice as that by IRGSP. The Monsanto sequencing strategy was slightly different from that of IRGSP, allowing it to sequence more of the genome with less time and cost. However the strategy does not provide enough information for highly accurate assembly (Bennetzen, 2002).[2] Much more recently, the Beijing Genomics Institute (BGI) and Syngenta, a Swiss-based agricultural biotechnology company, independently produced draft sequences of the rice genome by the quickest and least costly method, the 'shotgun sequence analysis'. Syngenta obtained 99.8% sequence accuracy, identifying more than 99% of the genes at 10% of the cost of the IRGSP strategy (Bennetzen, 2002).[3] On 26 January, 2001 Syngenta and Myriad Genetics announced that they had sequenced the rice genome and planned to provide their database to commercial customers, such as seed companies or agricultural biotechnology companies. The competition from the private sector has in turn spurred the IRGSP into advancing its calendar by almost 4 years (to 2004) and increasing its budget. Japan pledged to increase its annual rice genome research to US$60 million in 2000; a threefold increase over the previous year.

The Human and Rice Genome Projects are classic examples of a research race between firms where the objective is to be the first to discover, and subsequently obtain patents, on important gene sequences. However, current racing models do not conform well to the type of behaviour observed in genomics research for two reasons. First, most racing models have assumed that the research race is between profit maximizing private agents (Reinganum, 1989; Sabido, 1994). This is appropriate for a variety of industrial research in which the public sector is not involved. However in the agricultural sector (and particularly in genomic research) there is a dominant public research sector whose objective, it can be argued, is to maximize not profits but welfare.[4] Thus, such research is best characterized as a race between a social welfare maximizing public organization(s) and a profit maximizing private firm(s). In the context of genomics research, a private firm's objective is to patent the genetic code for important proteins and obtain royalties on the patent. Whereas the objective of a publicly funded entity is to promote further innovations (and hence increase welfare) which it does by making the genetic information more widely available, without regard to maximizing royalty revenues. The asymmetric objectives of these two types of

[1] By early 2002, the project had sequenced 15% of the genome.

[2] In 2000, Monsanto abandoned its rice sequencing effort and donated its data to IRGSP.

[3] A useful analogy in comparing the different strategies is to imagine the whole genome as being a large puzzle. Without the knowledge of what the whole genome (puzzle) resembles, the genome is first broken down into pieces (DNA strands) and the individual pieces subsequently sequenced. After the identification (sequencing) of the smaller pieces has occurred the task of determining how the individual pieces relate to each other and where they fall on the map follows (this is akin to putting a puzzle together). While it is easier to sequence smaller DNA strands, it is much more difficult to 'rebuild' the map with so many pieces.

[4] Following the literature on mixed oligopolies, we abstract here from moral hazard and internal organization issues and define a public firm to be an entity whose objective is to maximize social welfare, whereas a private firm would aim to maximize profit.

firms' gives rise to different behaviours from the case wherein all the firms in the analysis are private.

A second reason why earlier patent racing models fail to capture the intricacies of research such as genomics is in their assumption that the value of the prize is exogenously determined (Sabido, 1994). Genomics is not simply the identification of a sequence of genes, but also involves understanding the properties and relationships of the genetic code embodied in those genes. How accurately and in what manner the genetic sequence has been identified has bearing on the ease of interpreting the functions of the genes in the later stages. The amount of research done and the method employed in the sequencing stage can also influence later stages of genomics research if, for example, one assumes that more expenditure in the sequencing stage will, on average, lower the cost of doing research in the later stages. For example, the scientific knowledge and tools used in the sequencing the plant DNA have the potential of lowering the cost of breeding varieties with agronomically desirable traits. These cost reductions primarily result in more precision in transferring desirable genes to crops and reducing the time to breed specific varieties. The implication for the winning firm is that it gains knowledge in the process of the race that is useful and can be productively employed in further research (by it or others). That is, the more money it expends on the race today, the greater is the likelihood of winning *and* the greater will be the cost savings on future research or production. Moreover if we do not assume a winner-take-all situation the endogeneity of the prize value also has repercussions for the losing firm.[5] That is, the losing firm, which moves on to the next stage but uses the now inferior technology, will face lower profits owing to a decline in its market share. The decline in market share for the losing firm, and by extension profits, is a function then of the cost reduction implied by winning the prize (for the winning firm) and how much research was expended to achieve that prize (by the winning firm).[6]

To capture more accurately the microfoundations of an R&D race observed in genomics research, this chapter characterizes research as a two-stage process. The research effort in the first stage reduces the cost of applied research or production in the second stage. We use this framework to gain insights into a traditional theme in industrial organization research that of the relationship between market structure and innovation. Economists have been interested in this issue ever since Joseph Schumpeter's seminal work hypothesizing a positive correlation between market power and innovation. Schumpeter (1934) argued that a few firms were more likely efficiently to develop and employ more advanced technology than a competitive industry. Formal models of firms' innovation-seeking behaviour have evolved, that have either confirmed or refuted the so-called 'Schumpeterian trade-off'.[7] Similar to the mixed theoretical results, the findings of the

[5] The notion of 'winning' and 'losing' in the context of genome research may seem inappropriate as firms cannot obtain patents for simply sequencing a certain DNA strand nor are there any immediate commercial benefits from the knowledge of such sequences. Nevertheless, it has been observed that private firms are more reluctant to disclose their sequences (e.g. both Syngenta and Celera did not make their discoveries public through a commonly use public database). By effectively using trade secrecy to protect their sequences (especially large assembled sequences), the private firms hope to appropriate any return that may arise at later stages. This is in contrast to public efforts in genomic research, which have made the sequences available to all without any strings attached.

[6] For example, assume a strategic game between two firms in the output stage where the cost of production for firm i before the innovation is $c_i[q_i,\gamma]$ where q_i is the amount produced and γ a technology parameter. After the innovation race, the winning firm will maximize its profits $\pi_w = P[q_w,q_l]q_w - c[q_w,\gamma_w,x_w]$ choosing q_w (where x_w is the research expended in the racing stage by the winner) and the losing firm maximizes $\pi_l = P[q_w,q_l]q_l - c[q_l,\gamma_l = \gamma]$ choosing q_l (assume $\partial c_i/\partial\gamma_i, \partial c_w/\partial x_w < 0$ and $\gamma_w > \gamma_l$). Solving for the equilibrium properties it can be shown that profits of the winning firm will be greater than that of the losing firm $\pi_w^* = \pi_w^*[\gamma_l^*,\gamma_l^*,x_w^*] > \pi_l^* = \pi_l^*[\gamma_l^*,\gamma_l^*,x_w^*]$.

[7] For a review of this literature see Kamien and Schwartz (1982) or van Cayseele (1998). Among the many writers who subscribe to the Schumpeterian view are Kamien and Schwartz (1982). The claim has been challenged by Arrow (1962) and Dasgupta and Stiglitz (1980).

vast empirical literature on the Schumpeterian trade-off are mixed and ambiguous as no obvious relationship between industrial concentration and R&D performance emerges from the data.[8]

Specifically we ask what kind of market structure and circumstances promote R&D when the nature of R&D is as described for genomics. Does the public sector serve a useful purpose by performing genomics R&D when all evidence suggests that there are willing private firms that can do the same type of research more quickly and at lower cost? Do public sector efforts have a role to play in genomics R&D? We show that this question can be answered in the affirmative, if certain sufficient conditions regarding the prize value and the nature of the innovation process are met. For the current analysis we compare the currently observed market structure – that of a mixed oligopoly – with that of plausible alternative – a pure duopoly.[9] Note that it is not the intention of our analysis to comment on the desirability of a mixed market over the other two (which requires comparison to the first-best outcome), but to provide a set of sufficient conditions whereby one would *expect* the public sector to conduct more R&D relative to firms in the other regimes. It is in this sense that we suggest that the public sector has a role to play in conducting research of the kind observed in genomics R&D.

The Model

Consider a two-stage game. In the first stage, firms choose their R&D investment and in the second stage they compete further by conducting more applied R&D or competing in the product market.[10] The first stage is modelled as an innovation race where firms compete for the rights to an infinitely live patent. The innovation embodied in the patent allows firms to lower the cost of research in the second stage (or the cost of production, if the second stage is modelled as a product market). Through backward induction, the profits from the second stage determine the value of the first stage patent. The firm that innovates first is awarded the patent and gets the exclusive right to use the more productive technology forever. The losing firm, on the other hand, has to continue using the pre-innovation race technology in the second stage and hence accrues a lower profit than the winning firm and possibly even lower than its own profits before innovation.

The research effort employed in the R&D race not only determines the outcome of the race, but also results in the generation of knowledge that is valuable to the winning firm. This knowledge can be used in later stages to complement with the winning technology and lower costs in those stages even further. In this respect the value of the prize for the winner is endogenous and an increasing function of research expenditure in the R&D race. R&D effort thus has a two-pronged direct effect on the winning firm; allowing it to win the race *and* lowering the cost in later stages. The losing firm is also affected by the amount of research effort employed by the winning firm in the first stage (see footnote 6). Since we assume the strategic game in the second stage as well, an increased market share for the winning firm from the lowering of its cost will imply lower profits for the losing firm, *ceteris paribus*.

To fix these ideas, assume that two firms play the following two-stage game. In the first stage firm i independently takes action, denoted x_i, regarding the current research market. In a patent race set-up, x_i represents the flow cost of research where its probability of being successful at or prior to date t is $1-e^{-th[x_i]}$. The instantaneous conditional probability that firm i will be first to innovate at time t, given no success to date, is therefore $h[x_i]$. Firm i's expected benefits after a discovery are determined by both firms

[8] See Dosi (1988) for a review.

[9] The chapter is organized as follows. The next section introduces the modelling framework, the assumptions and our approach in comparing the two market structures. The following section discusses the comparative analysis and the chapter ends with concluding remarks.

[10] The specific characteristics of the second stage are not of concern here. We note only that the pay-off (or prize) from undertaking R&D in the first stage is function of the market structure in the second stage and the amount of research performed in the first stage.

actions (b_i,b_j) and are denoted by $W_{wi}(b_i[x_i], b_j[x_j])$ if the firm emerges as the winner and $W_{Li}(b_i[x_j],b_j[x_j]$ when it loses the race. In a Cournot set-up, b_i would represent output whereas in a Bertrand game it would be price. A strategy of firm i in this entire game can then be written as $s_i \equiv (x_i,b_i(\bullet))$ where $b_i(\bullet)$ is a function specifying firm i's post innovation action conditional on first stage actions, in particular on the amount of research done by the winning firm. Given (s_i,s_j), the pay-off to the private firm i is shown in Equation 1a and 1b (see bottom of page).

As we shall see, the value of the innovation will be different for the private and public firms. For the public firm the value of the innovation is the total welfare generated by it. For the private firm it is the value of the private benefits or profits. Next we proceed to characterize the equilibrium condition in R&D for the market structures of interest, namely pure duopoly, and mixed duopoly. We make progress by first characterizing the best response function for each firm in the three markets.

Pure duopoly equilibrium condition

In the pure duopoly case (two profit maximizing firms), firm 1 chooses x_1 and firm 2 chooses x_2 to maximize the pay-off function. Due to symmetry ($n = 2$, therefore, $x_1 = x_2 = x_D$), the best response function for each firm is defined by the condition $\frac{\partial V_D}{\partial x_D} = 0$, which is true, if and only if, $R_D[x_D] = 0$, where (see Equation 2, bottom of page).

Following Delbono and Denicolo (1993), the difference between the profit from winning and the firm's current profit $(W_{WD}[x_D] - W_D)$ measures the incentive to innovate in the absence of rivalry and is referred to as the 'profit incentive'. Further in a duopoly there is rivalry as each firm anticipates research by the other. In the presence of rivalry the incentive to invest in R&D is also reflected in the difference between the flow of profits should it win the race and should it not, $W_{WD}[x_D] - W_{LD}[x_D]$. Call this the 'rivalry incentive'. The presence of both pre-innovation profits and post-innovation profits for the loser induces firms to delay the expected date of innovation, that is higher pre-innovation and loser profits decrease the profit and rivalry incentives. The smaller the profit and rivalry incentive, the more time it will take for innovations to arrive.

Mixed duopoly equilibrium condition

In the mixed duopoly the private and public firms choose x_i but maximize different pay-offs.

$$V_i[W_{Wi}, W_{Li}, \{x_i, x_j\}] = \int_0^{\infty} e^{-(h[x_i]+h[x_j]+r)t}(h[x_i]W_{Wi}[x_i]/r + h[x_j]W_{Li}[x_j]/r + W_i - x_i)dt \qquad (1a)$$

$$= \frac{(1/r)(h[x_i]W_{Wi}[x_i] + h[x_j]W_{Li}[x_j]) + W_i - x_i}{h[x_i] + h[x_j] + r} \qquad (1b)$$

where

r is the discount rate

x_i is firm i's R&D expenditure is firm i's instantaneous probability of innovating or the hazard rate. The hazard rate, $h[x_i]$, is twice differentiable, strictly increasing and satisfies (i) $h[0] = 0 = \lim_{x \to \infty} h'[x]$, (ii) $h'[x_i] > 0$, 3) $h''[x_j] < 0$

$W_{Wi}[x_i]$ is the value of innovation accruing to firm i if it wins the race

$W_{Li}[x_j]$ is the value of innovation accruing to firm i if it loses the race where j is the winning firm

W_i is the pre-innovation benefits accruing to firm i

$$R_D[x_D] \equiv \begin{bmatrix} h'[x_D](W_{WD}[x_D] - W_D) + (1/r)h'[x_D]h[x_D](W_{WD}[x_D] - W_{LD}[x_D]) - \\ r - 2h[x_D] + x_D h'[x_D] + h[x_D]W'_{WD}[x_D](1 + (2/r)h[x_D]) \end{bmatrix} = 0 \qquad (2)$$

The private firm maximization problem remains unchanged from that of a firm in the pure duopoly case. That is, it maximizes Equation 1 (for $i = 2$), where we denote firm 1 as the private firm (P) and firm 2 as the public-sector firm (S). Since symmetry no longer holds (as the public firm's pay-off is different), the private firm's best response function is defined by the condition $\partial V_P/\partial x_P = 0$, which is true if and only if $R_P[x_P, x_S] = 0$ where (see Equation 3, bottom of page).

The best response function of the private firm in the mixed duopoly is similar to the one in pure duopoly. In both cases, there exists a profit incentive as well as a competitive threat, faced by the private firm. The only difference between the two is that the source of the competitive threat in the pure duopoly case is another, identically specified private firm, whereas in the mixed duopoly case it is the welfare-maximizing public firm.

The public firm in our model is a welfare maximizer to whom the value of the prize is not just the private profits it embodies but the social welfare as well. This means, first, that the 'W's' in the pay-off function are greater for the public firm than they will be for the private (more discussion on the assumption and relative values of the prize follows). Second, the public firm, as a social welfare maximizer, takes into account the flow cost of research incurred by all firms in the economy. In the duopoly case where one firm is public and the other private, the public firm's pay-off function is written as in Equation 4a and 4b (see bottom of page).

The maximization of Equation 4 yields the public firm's reaction curve which is defined by the condition $\partial V_S/\partial x_S = 0$, which is true, if and only if, $R_S[x_S, X_P] = 0$ (see Equation 5, bottom of page).

The reaction curve of the public firm in the mixed oligopoly also reveals that the public firm faces the 'profit incentive' and the 'rivalry threat' from the opposing firm. For the public firm the 'profit incentive' is a misnomer (since the value of the prize to it is total welfare, and not private profits as the name implies), although it is still the incentive to innovate in the absence of a rival. As with the earlier case, the smaller the rivalry and profit incentives the later is the date of innovation

In summary, Equations 2, 3 and 5 represent the best response functions for a duopolist in a pure duopoly (Equation 2), private firm in the mixed duopoly (Equation 3) and the public firm in a mixed duopoly (Equation 5). In all these cases we see that each firm faces a profit incentive and a competitive threat. The profit incentive is a function of how large the difference is between current profits and profits if the firm wins, and similarly the competitive threat is a function of how big the difference in profits is between winning and losing. Clearly, if the differences are small, then the firms will be conducting less research, which will delay the expected date of innovation. For the public firm in the mixed duopoly, the profit incentive and the competitive threat also matter (only that it is not profits that the public-sector firm is after, but welfare). But since it takes into account the total R&D cost, the effect of R&D cost on the public firm ($x_P + x_S$), relative to private firm (x_P), is to bring closer the expected date of innovation.

Lastly we note the presence of the term $\partial W_{Wi}/\partial x_i$ in each of the four best response functions. This term, which reflects the marginal change in the prize value due to a change in own research, is a direct consequence of our assumptions regarding the endogeneity of the value of

$$R_P[x_P, x_S] \equiv \begin{bmatrix} h'[x_P](W_{WP}[x_P] - W_P) + (1/r)h'[x_P]h[x_S](W_{WP}[x_P] - W_{LP}[x_S]) - r - \\ h[x_P] - h[x_S] + x_P h'[x_P] + h[x_P]W'_{WP}[x_P](1 + (1/r)(h[x_P] + h[x_S]) \end{bmatrix} = 0 \tag{3}$$

$$V_S = \int_0^\infty e^{-(h[x_P] + h[x_S] + r)t}(h[x_S]W_{WS}[x_S]/r + h[x_P]W_{LS}[x_P]/r + W_S - x_S - x_P)dt \tag{4a}$$

$$= \frac{(1/r)(h[x_S]W_{WS}[x_S] + h[x_P]W_{LS}[x_P]) + W_P - x_P - x_S}{h[x_S] + h[x_P] + r} \tag{4b}$$

$$R_S[x_S, x_P] \equiv \begin{bmatrix} h'[x_S](W_{WS}[x_S] - W_S) + (1/r)h'[x_P]h[x_S](W_{WS}[x_S] - W_{LS}[x_P]) - \\ r - h[x_P] - h[x_S] + h'[x_P](x_P + x_S) + \\ h[x_S]W'_{WS}[x_S](1 + (1/r)(h[x_P] + h[x_S])) \end{bmatrix} = 0 \tag{5}$$

the prize. If we assume that profits are concave with respect to own research then the endogenous nature of the prize value brings closer the expected date of innovation. This implies all firms carry out a greater amount of research relative to the case where the prize value is exogenous. To put it differently, when R&D complements the innovation at a later stage, firms have an incentive to increase their research effort. High amounts of research or the so-called over-investment problem (Dasgupta and Stiglitz, 1980) could therefore be explained by the presence of such a complementary effect.

Comparative Analysis

Having established the nature of the game and market structure we now turn our attention to the relative ranking of the equilibrium research effort by individual firms (x_D, x_P, and x_S) as well the industry ($2x_D$, and $x_P + x_S$). First, however, we need to make certain assumptions about the relative value of the prize in the two markets and for the different firms in these markets. Since the relative values of W will determine the equilibrium values of research effort for all the firms, we need to make our assumptions regarding them explicit.

Assumption 1: The winning firm profits (for the private firms) or welfare (for the public firm) is greater than current profits/welfare. That is $W_{Wi}[x_i] > W_i$. This assumption ensures that the 'profit incentive' to innovate is always positive.

Assumption 2: In the pure duopoly the profits in the winning state are greater than or equal to in the losing state. Moreover profits are increasing in R&D for the winning firm and decreasing for the losing firm. Assume that firm 1 emerges as the winner and it gets the rights to a superior cost-reducing technology. In the second stage, the two firms play a Cournot game. If we assume increasing costs of production in the second stage, then it can be shown that in equilibrium the winning firm will produce more than the losing firm. Consider, for example, a second stage product market where $P = a - bQ$ is the inverse demand function (for $Q = q_{WD} + q_{LD}$). Assume that before the race, the cost of production for both firms was $C[\gamma,q] = \gamma q^2/2$ where the parameter γ represents the technological opportunity due to the successful research, such that $\partial C/\partial\gamma < 0$. At the completion of the race, the losing firm will continue to produce at the pre-innovation cost, but the winner obtains a better technology such that it lowers its costs to $C[\gamma,q] = \bar{\gamma} q^2/2(1 + x_D)$ where $\bar{\gamma} < \gamma$. The $(1 + x_D)$ term reflects the fact that the winning firm also gains from its research effort of the first stage. The post-innovation maximization problems for the winning and losing firm are therefore, respectively

$$\max_{q_{WD}} W_{WD} = Pq_{WD} - C[\bar{\gamma}, q_{WD}, x_D^*] \tag{6}$$

$$\max_{q_{LD}} W_{LD} = Pq_{LD} - C[\gamma, q_{LD}] \tag{7}$$

The two firms acting simultaneously and non-cooperatively solve their maximization problem. From the first order condition for a maximum it can be easily shown (see Naseem, 2002) that the standard Cournot equilibrium when costs are asymmetric will prevail, that is:

$$q_{WD}^*[x_D^*] \geq q_{LD}^*[x_D^*],\ W_{WD}^*[x_D^*] \geq W_{LD}^*[x_D^*],\ \partial W_{WD}^*[x_D^*]/\partial x_D^* > 0,\ \partial W_{LD}^*[x_D^*]/\partial x_D^* < 0$$

Assumption 3: In the mixed duopoly, no public production or further research takes place in the second stage. Should the public firm win the race, it licenses its technology to the private firm. Alternatively, if the private firm wins the race in the mixed case, it does not face a rival in the second stage and assumes the role of a monopolist. On the other hand if the public firm wins the first stage race, and in the absence of further research or production by the public firm in the second stage, it licenses the technology to the private firm. While the private firm is still a monopolist even with the licence (by virtue of the fact that no rival exists), we assume that the terms of the licence are such that it is not allowed to produce at the profit maximizing level (where marginal cost (MC) would equal marginal revenue (MR)) but rather at a level where the welfare losses associated with monopoly production are minimized, though not necessarily eliminated. This is because the terms of the licence also have to be incentive compatible in the sense that the profits for the private firm from production using the new licensed technology are greater than or equal to the profits associated with the older technology.

Therefore it is reasonable to think that the welfare attained with a public firm innovating is higher than the welfare attained when the innovator is a private firm, simply because a public firm would license and contract its innovation to the private firm which would have to price its innovation in order to maximize social welfare taking into account consumer surplus. With $W^*_{WS}[x^*_S]$, as the total welfare generated by the innovation when the winning firm is the public firm, and $W^*_{LS}[x^*_P]$ as the total welfare generated when the winning firm is the private firm. Finally, $W^*_{WP}[x^*_P]$ is the value of the private benefits when the private firm innovates, and $W^*_{LP}[x^*_S]$ when it does not. Given these definitions, one has

$$W^*_{WS}[x^*_S] > W^*_{LS}[x^*_P] \geq W^*_{WP}[x^*_P] \geq W^*_{LP}[x^*_S]$$
$$\text{and } q^*_{LP}[x^*_S] \geq q^*_{WP}[x^*_P]$$

(see Naseem, 2002 for a more formal treatment).

The generalized functional form of the hazard rate and the prize value does not permit us to solve explicitly for equilibrium research effort in each case that could be compared across the three different market structures. However if we assume that the equilibrium research condition x^*_M, x^*_D, x^*_P, and x^*_S solves their respective best response function, we can derive a set of sufficient conditions to evaluate the relative magnitude of research among the firms. To illustrate this approach, assume that, in a two-firm strategic game, x^*_i and x^*_j solve the best response function $R_i[x_i, x_j]$ and $R_j[x_i, x_j]$ for firms i and j, respectively (that is $R_i[x^*_i, x^*_j] = R_j[x^*_i, x^*_j] = 0$). Assume also that there is an asymmetric relationship between the two firms such that, a priori, we are unable to determine the relative levels of research i.e. $x^*_i \gtreqless x^*_j$. In such cases it is possible to derive a set of sufficient conditions that will satisfy $x^*_i > x^*_j$, $x^*_i = x^*_j$ and $x^*_i < x^*_j$. We do so by taking the difference of the best response function of one firm evaluated at the other firm's optimal research levels (i.e. $R_j[x^*_i, x^*_i]$) and the best response of the other firm evaluated at its optimal level (i.e. $R_i[x^*_i, x^*_j] = 0$). If we assume that the underlying value function for both firms is concave with a relative maximum at x^*_i and x^*_j, and that the second order condition is satisfied $(\partial^2 V_i[x_i]/\partial x_i^2 < 0)$, then we can claim that the conditions under which $R_j[x^*_i, x^*_i] - R_i[x^*_i, x^*_j] \gtreqless 0$ imply $x^*_i \gtreqless x^*_j$.

We apply this strategy to compare and derive the sufficient conditions to evaluate the equilibrium research effort between firms in a duopoly and firms in the mixed duopoly.

Mixed duopoly vs. pure duopoly

We seek to derive the sufficient conditions under which $x^*_D \gtreqless x^*_P$ and $x^*_D \gtreqless x^*_S$. If the sufficient conditions for $x^*_D > x^*_P$ are not inconsistent for those that would satisfy $x^*_D > x^*_S$, then we can claim to have found the sufficient condition for $2x^*_D > x^*_S + x^*_P$. Similarly if the sufficient conditions for $x^*_D < x^*_P$ and $x^*_D < x^*_S$ are not inconsistent, then those conditions would imply that the aggregate research effort in the mixed market is greater than that in the duopoly market (i.e. $2x^*_D < x^*_S + x^*_P$).

We first examine the sufficient conditions that would satisfy $x^*_P > x^*_D$ implied by $R_P[x^*_D, x^*_S] - R^*_D[x^*_D] > 0$, where (see Equation 8, bottom of page).

Sufficient conditions for $R_P[x^*_D, x^*_S] - R^*_D[x^*_D] > 0$ are that each term in Equation 8 be non-negative. For the first term in Equation 8 to be non-negative would require that the equilibrium research effort by a firm in the duopoly be greater or equal than that of public firm $(x^*_D \geq x^*_S)$. The second term will be non-negative if we assume that profits for firms in the duopoly are equal, win or lose (i.e. $W_{LD}[x^*_D] = W_{WD}[x^*_D]$). The third term is always non-negative and follows from assumption 2 and the earlier sufficient condition $x^*_D \geq x^*_S$ (hence, $W_{WP}[x^*_D] \geq W_{LP}[x^*_S]$). For the fourth and fifth terms to be

$$\begin{aligned} R_P[x^*_D, x^*_S] - R^*_D[x^*_D] = h[x^*_D] - h[x^*_S] + \\ (1/r)h'[x^*_D]h[x^*_D](W_{LD}[x^*_D] - W_{WD}[x^*_D]) + \\ (1/r)h'[x^*_D]h[x^*_S](W_{WP}[x^*_D] - W_{LP}[x^*_S]) + \\ h'[x^*_D](W_D - W_{WD}[x^*_D]) + (W_{WP}[x^*_D] - W_P) + \\ h[x^*_D](W'_{WP}[x^*_D] - W'_{WD}[x^*_D]) + \\ (1/r)h[x^*_D](W'_{WP}[x^*_D](h[x^*_D] + h[x^*_S]) - W'_{WD}[x^*_D](2h[x^*_D])) \end{aligned} \quad (8)$$

non-negative implies that the profits from winning in the mixed duopoly case increases faster than in the pure duopoly case, such that $W'_{WP}[x_D^*] \geq W'_{WD}[x_D^*]$ and $|W_D - W_{WD}[x_D^*]| \leq |W_{WP}[x_D^*] - W_P|$. The last term will be non-negative if and only if $W'_{WP}[x_D^*](h[x_D^*] + h[x_S^*]) \geq W'_{WP}[x_D^*](2h[x_D^*])$.

A key sufficient condition in establishing that $x_D^* < x_P^*$ has been the assumption that $x_D^* \geq x_S^*$. Are there sufficient conditions that would satisfy $x_D^* \geq x_S^*$ consistent with those for $x_D^* \geq x_P^*$? We examine this next. For $x_D^* \geq x_S^*$ to hold implies that $R_S[x_P^*, x_D^*] - R_D^*[x_D^*] \leq 0$ also needs to hold, where (see Equation 9, bottom of page).

For $R_S[x_P^*, x_D^*] - R_D^*[x_D^*] \leq 0$, a sufficient condition is that each term in the equation be less than or equal to zero. The first term will be less than or equal to zero by assumption 2, $W_{LD}[x_D^*] \leq W_{WD}[x_D^*]$. However to be consistent with the sufficient conditions that satisfy $x_D^* < x_P^*$ requires the stronger condition $W_{LD}[x_D^*] = W_{WD}[x_D^*]$, which is maintained here as well. The sign on the second term is ambiguous, as assumption 2 no longer holds due to asymmetric research effort. That is since $x_D^* < x_P^*$, the term $[W_{WS}[x_D^*] - W_{LS}[x_P^*]]$ cannot be signed without further assumptions on the value of the welfare in the losing and winning states. Therefore we require the stronger condition that the public's welfare in the winning state (evaluated at the research effort of the duopolist) be equal to or less than the losing state (evaluated at the research effort of the private firm in the mixed duopoly). The third and fourth terms, taken together imply that relative to the pre-innovation profits, the gains from innovation are greater to the winning duopolist than they are for the public firm should it win. That is $W'_{WS}[x_D^*] \leq W'_{WD}[x_D^*]$ and $|W_D - W_{WD}[x_D^*]| \geq |W_{WS}[x_D^*] - W_S|$. The fifth term will be less than or equal to zero if and only if $W'_{WS}[x_D^*](h[x_D^*] + h[x_P^*]) < W'_{WD}[x_D^*](2h[x_D^*]) + 1$. Note that this sufficient condition is consistent with our assumptions that $x_P^* < x_D^*$ and $W'_{WS}[x_D^*] > W'_{WD}[x_D^*]$ which implies $W'_{WS}[x_D^*](h[x_D^*] + h[x_P^*]) > W'_{WD}[x_D^*](2h[x_D^*])$. The last term will be less than or equal to zero if and only if $x_P^* h'[x_D^*] < h[x_P^*]$.

To derive the sufficient conditions that would guarantee the reverse relationship, i.e. $x_P^* < x_D^* < x_S^*$, simply implies that we reverse the above condition. For the sake of brevity we present those results and a summary of the sufficient conditions in Table 9.1. The conditions summarized in table three cannot, however, be used to make statements on the aggregate research relationship between the mixed and pure duopoly. To do so requires that we derive conditions for $x_D^* \gtrless x_S^*, x_P^*$, which implies $2x_D \gtrless x_S + x_P$. Table 9.2 summarizes these sufficient conditions and follows from Equation 9. That is to derive the sufficient conditions that would simultaneously satisfy $x_D^* > x_S^*$ and $x_D^* > x_P^*$ requires that both Equations 8 and 9 be non-negative. These sufficient are summarized in the first column of Table 9.2. The second column lists those sufficient conditions that would reverse this relationship, such that they satisfy $x_D^* < x_S^*$ and $x_D^* < x_P^*$.

Discussion

The sufficient conditions that have been derived, which allow us to rank the research effort in the two markets, are primarily a function of three properties. First is how the profits/welfare are distributed, post-innovation, in the pure and mixed duopoly. Second, how the profits/welfare are increasing, for the winning firm, in *x*. Lastly how the hazard rate function changes in *x* (or the curvature properties of the hazard rate). In all the comparison we note that our assumptions about these three properties allow us to derive the sufficient conditions and hence a particular ranking. What do the

$$\begin{aligned} R_S[x_P^*, x_D^*] - R_D^*[x_D^*] = {} & (1/r)h'[x_D^*]h[x_D^*](W_{LD}[x_D^*] - W_{WD}[x_D^*]) + \\ & (1/r)h'[x_D^*]h[x_P^*](W_{WS}[x_D^*] - W_{LS}[x_P^*]) + \\ & h'[x_D^*]((W_D - W_{WD}[x_D^*]) + (W_{WS}[x_D^*] - W_S)) + \\ & h[x_D^*](W'_{WS}[x_D^*] - W'_{WD}[x_D^*]) + \\ & (1/r)h[x_D^*](1 + W'_{WS}[x_D^*](h[x_D^*] + h[x_P^*]) - W'_{WD}[x_D^*](2h[x_D^*])) \\ & x_P^* h'[x_D^*] - h[x_P^*] \end{aligned} \quad (9)$$

Table 9.1. Sufficient conditions for ranking pure duopoly and mixed duopoly research effort.

$x_P^* > x_D^* > x_S^*$	$x_P^* < x_D^* < x_S^*$
1. Post-innovation profits in the duopoly are equal across states ($W_{LD}[x_D^*]=W_{WD}[x_D^*]$) **2.** Post-innovation welfare for the public firm in the mixed duopoly is equal across states ($W_{LS}[x_D^*]=W_{WS}[x_P^*]$) **3.** Private firm's profits in the mixed duopoly are always greater in the winning state than in the losing state **4.** Relative to the pre-innovation profits the winning firm in a pure duopoly gains less than the winning private firm in the mixed duopoly, such that $W'_{WP}[x_D^*]>W'_{WD}[x_D^*]$ and $\lvert W_D - W_{WD}[x_D^*]\rvert < \lvert W_{WP}[x_D^*]-W_P\rvert$ **5.** Relative to the pre-innovation profits the winning firm in a pure duopoly gains more than the welfare gains of the winning public firm in the mixed duopoly, such that $W'_{WS}[x_D^*]<W'_{WD}[x_D^*]$ and $\lvert W_D - W_{WD}[x_D^*]\rvert < \lvert W_{WS}[x_D^*]-W_S\rvert$ **6.** $W'_{WP}[x_D^*](h[x_D^*]+h[x_S^*])>W'_{WD}[x_D^*](2h[x_D^*])$ **7.** $W'_{WS}[x_D^*](h[x_D^*]+h[x_P^*])<W'_{WD}[x_D^*](2h[x_D^*])+1$ **8.** $x_P^* h'[x_D^*]<h[x_P^*]$	**1.** Post-innovation profits in the duopoly are equal across states ($W_{LD}[x_D^*]=W_{WD}[x_D^*]$) **2.** Post innovation profits for the private firm in the mixed duopoly are equal across states ($W_{LP}[x_D^*]=W_{WP}[x_S^*]$) **3.** Public welfare is always greater in the winning state than in the losing state **4.** Relative to the pre-innovation profits the winning firm in a pure duopoly gains more than the winning private firm in the mixed duopoly, such that $W'_{WP}[x_D^*]<W'_{WD}[x_D^*]$ and $\lvert W_D - W_{WD}[x_D^*]\rvert < \lvert W_{WP}[x_D^*]-W_P\rvert$ **5.** Relative to the pre-innovation profits the winning firm in a pure duopoly gains less than the welfare gains of the winning public firm in the mixed duopoly, such that $W'_{WS}[x_D^*]>W'_{WD}[x_D^*]$ and $\lvert W_D - W_{WD}[x_D^*]\rvert > \lvert W_{WS}[x_D^*]-W_S\rvert$ **6.** $W'_{WP}[x_D^*](h[x_D^*]+h[x_S^*])<W'_{WD}[x_D^*](2h[x_D^*])$ **7.** $W'_{WS}[x_D^*](h[x_D^*]+h[x_P^*])>W'_{WD}[x_D^*](2h[x_D^*])$

Table 9.2. Sufficient conditions for ranking pure duopoly and mixed duopoly research effort.

$x_D > x_S, x_P$ (implies $2x_D < x_S + x_P$)	$x_D < x_S, x_P$ (implies $2x_D < x_S + x_P$)
1. The profits from winning are strickly greater than that from losing in the duopoly case ($W_{LD}[x_D^*]<W_{WD}[x_D^*]$) **2.** The profits/welfare from winning or losing in a mixed duopoly are equal. $W_{LS}[x_P^*]=W_{WS}[x_D^*]$ and $W_{LP}[x_S^*]=W_{WP}[x_D^*]$ **3.** The change in profits (welfare) for the private (public) firm in the mixed duopoly is less than that for the duopolist. That is $W'_{WS}[x_D^*]<W'_{WD}[x_D^*]$ and $W'_{WP}[x_D^*]<W'_{WD}[x_D^*]$ **4.** $\begin{bmatrix} h[x_D^*](1+W'_{WP}[x_D^*](h[x_D^*]+h[x_S^*])- \\ W'_{WD}[x_D^*](2h[x_D^*])) \end{bmatrix} < h[x_S^*]$ **5.** $\begin{bmatrix} 1+W'_{WP}[x_D^*](h[x_D^*]+h[x_P^*])- \\ W'_{WD}[x_D^*](2h[x_D^*]) \end{bmatrix} < \frac{h[x_P^*]-x_P^* h'[x_D^*]}{h[x_P^*]}$	**1.** The profits from winning or losing in the duopoly case are equal ($W_{LD}[x_D^*]=W_{WD}[x_D^*]$) **2.** The profits/welfare from winning are strickly greater than losing in a mixed duopoly. $W_{LS}[x_P^*]<W_{WS}[x_D^*]$ and $W_{LP}[x_S^*]<W_{WP}[x_D^*]$ **3.** The change in profits (welfare) for the private (public) firm in the mixed duopoly is greater than that for the duopolist. That is $W'_{WS}[x_D^*]>W'_{WD}[x_D^*]$ and $W'_{WP}[x_D^*]<W'_{WD}[x_D^*]$ **4.** $\begin{bmatrix} h[x_D^*](1+W'_{WP}[x_D^*](h[x_D^*]+h[x_S^*])- \\ W'_{WD}[x_D^*](2h[x_D^*])) \end{bmatrix} > h[x_S^*]$

sufficient conditions mean and what are the implications?

Our interest in the comparison of mixed and pure duopoly was to explore the conditions under which a mixed duopoly would perform more research than a pure duopoly (or vice versa; refer to Table 9.2). For the research intensity of the mixed market to be greater than that of the pure duopoly it is observed that four conditions need to be satisfied. The first condition, that profits be distributed equally across both the winning and losing states in the pure duopoly, will be satisfied if one assumes that the losing firm easily imitates the innovation. This would arise if property rights protection is weak, and results in an inability of the winning firm to appropriate the returns to the innovation.[11]

The second and third conditions relate to the profits/welfare function in the winning state are strictly greater than the losing state. The second condition is satisfied if we assume that the first-stage R&D enters into the firm's second stage profit function as an autonomous cost reduction. For example, if we assume that the cost of production, post-innovation, for the winning firm is $C[\underline{\gamma}, q_{\mathrm{WD}}, x_{\mathrm{D}}^{*}] = (\gamma)q_{\mathrm{WD}} - x_{\mathrm{D}}^{*}$, and for the losing firm $C[\gamma, q_{\mathrm{LD}}] = (\underline{\gamma})q_{\mathrm{LD}}$ then it can be shown that $W_{\mathrm{L}i}[x_j^*] < W_{\mathrm{W}i}[x_g^*]$ for i = S, P, D and $\forall x_g, x_j > 0$ (see Naseem, 2002).

The third condition relates to the curvature properties of the profit (for the private firm) and welfare (for the public) functions. That is if the profit function of the winning private firm (in the mixed duopoly) and the welfare function of the public firm are 'more' concave in x than the profit function of the winning duopolist, then this condition is satisfied.

The last sufficient condition for the mixed duopoly to perform more R&D than that in the pure duopoly can be restated as such:

$$1 + W'_{\mathrm{WP}}[x_{\mathrm{D}}^{*}](h[x_{\mathrm{D}}^{*}] + h[x_{\mathrm{S}}^{*}]) - W'_{\mathrm{WD}}[x_{\mathrm{D}}^{*}](2h[x_{\mathrm{D}}^{*}]) > \frac{h[x_{\mathrm{S}}^{*}]}{h[x_{\mathrm{D}}^{*}]} > \text{for } x_{S}^{*} > x_{D}^{*} > 0$$

A case where this condition is satisfied is when we assume that the hazard rate is bounded such that $h[x] \to 1$ as $x \to \infty$ (e.g. a logarithmic function) and that equilibrium R&D is 'high' for both the firm in the private duopoly and the public firm (i.e. $\frac{h[x_{\mathrm{S}}^{*}]}{h[x_{\mathrm{D}}^{*}]}$ approaches 1 from above). Since $W'_{\mathrm{WP}}[x_{\mathrm{D}}^{*}](h[x_{\mathrm{D}}^{*}] + h[x_{\mathrm{S}}^{*}]) > W'_{\mathrm{WD}}[x_{\mathrm{D}}^{*}](2h[x_{\mathrm{D}}^{*}])$, it follows that the left hand side will be greater than $\frac{h[x_{\mathrm{S}}^{*}]}{h[x_{\mathrm{D}}^{*}]}$, if, $W'_{\mathrm{WP}}[x_{\mathrm{D}}^{*}] >> W'_{\mathrm{WD}}[x_{\mathrm{D}}^{*}]$. An interpretation of this sufficient condition is that the innovation is drastic for the winning private firm in the mixed duopoly and non-drastic for the winning duopolist (in the pure duopoly).

In summary, the key conditions for the mixed duopoly to perform more R&D than the private duopoly are heuristically interpretable as conditions that profits from winning exceed profits from losing, and that appropriability is problematic. Limited appropriability suggests an underinvestment in research, so that it is exactly in the underinvestment case that the public intervention increases research. Thus, the public sector investment is expected to increase social welfare.

We note that an analogous interpretation holds for the opposite case. The sufficient conditions for the mixed oligopoly to perform less research are conditions that can be heuristically interpreted to be associated with excess competition and overinvestment. Thus, in this case as well, public sector intervention is expected to improve social welfare – in this case, by reducing the research overinvestment.

Concluding Remarks

The analysis of this chapter was motivated by the observation that in genomics research, the amount of R&D expended to win an R&D race affects not only the probability of success but also downstream profits. Further we observe that the public research sector is engaged in fierce competition with private firms in a variety of genomics projects. Since it was not clear, a priori, the reasons for the public firm to undertake genomics research (especially in light of the fact that activities are similar to those of the private firms), we set out in this chapter to derive a set of sufficient conditions under which R&D across different plausible and observed markets in genomics research could be ranked. It was found that the sufficient conditions relate to

- the concavity properties of the profit/welfare function with respect to first stage research;

[11] A simple example illustrates this point. Consider that post-innovation, the winning duopolist is able to lower its cost to $C[\gamma, q_{\mathrm{WD}}, x_{\mathrm{D}}^{*}]$. If the losing firm is able to imitate the technology than it too will have the same technology and cost function, that would result in equilibrium conditions where both firms produce the same amount and make equal profits (i.e. with all firms having the same cost function, the symmetric result would hold).

- the distribution of profits across firms in the duopoly markets;
- the magnitude of the gains in the second (profit/welfare) from the innovation relative to other firms as well as pre-innovation profits/welfare; and
- the curvature properties of the hazard rate.

One interpretation of the concavity of profit/welfare is how useful first stage R&D (or the knowledge gained in the racing stage) complements the innovation that is eventually employed in applied research or production. For the firm for which the complementarity effect is the greatest, the incentive to conduct more research will also be larger.

When comparing with the duopoly market, how the profits are distributed across the two firms affects has implication for the sufficient conditions. Distribution of profits across firms can be affected by the nature of the innovation (whether drastic or non-drastic) and/or the ease to which the winning innovation can be imitated by the losing firm. For example, if the innovation is minor such that the distribution of profits of the two firms remains relatively unchanged then the incentives to innovate are less. Similarly if the innovation is easily copied, due to weak property rights perhaps, then the incentives remain weak. In such cases, we find that the mixed market undertakes more R&D then the pure duopoly, and that the usual argument for public sector research when appropriability is weak holds.

Interestingly, social optimum may be less if the returns for the winning and losing firms in the pure duopoly are different, as there may be excess R&D competition between the two private firms. We have derived the sufficient conditions for the mixed R&D to be lower than the pure duopoly, which would imply that if those conditions are met, then the mixed duopoly may be more socially desirable than the pure duopoly, as the R&D conducted in a mixed duopoly will be less than the R&D conducted in the pure duopoly.

References

Arrow, K.J. (1962) Economic welfare and the allocation of resources for invention. In: Nelson, R. (ed.) *The Rate and Direction of Inventive Activity: Economic and Social Factors*. Princeton University Press, Princeton, New Jersey.

Bennetzen, J. (2002) Opening the door to comparative plant biology. *Science* 296(5565), 60–63.

Dasgupta, A. and Stiglitz, J. (1980) Industrial structure and the nature of innovative activity. *Economic Journal* 90(358), 266–293.

Delbono, F. and Denicolo, V. (1993) Regulating innovative activity: the role of a public firm. *International Journal of Industrial Organization* 11(1), 35–48.

Dosi, G. (1988) Sources, procedures, and micro-economic effects of innovation. *Journal of Economic Literature* 26(3), 1120–1171.

Kamien, M. and Schwartz, N. (1982) *Market Structure and Innovation*. Cambridge University Press, New York.

Naseem, A. (2002) Essays in mixed oligopoly and agricultural R&D. PhD thesis. Department of Agricultural Economics, Michigan State University, East Lansing.

Reinganum, J.F. (1989) The timing of innovation: research, development, and diffusion. In: Schmalensee, R. and Willig, R.D. (eds) *Handbook of Industrial Organization*. Elsevier Science Publications, Amsterdam, p. 182–264.

Sabido, A.C. (1994) R&D investment in strategic settings: a survey of patent races. *Estudios Economicos* 9(1), 61–118.

Schumpeter, J.A. (1934) *The Theory of Economic Development*. Harvard University Press, Cambridge, Massachusetts.

van Cayseele, P.J.G. (1998) Market structure and innovation: a survey of the last twenty years. *De Economist* 146(3), 391–417.

10 The Case for Differentiated Appropriability in Intellectual Property Rights for Plant Varieties

Frank van Tongeren and Derek Eaton[1]
Agricultural Economics Research Institute, Wageningen University and Research Centre, The Hague, The Netherlands

Abstract

This chapter asks the question whether uniform international standards for intellectual property rights over plant varieties lead to improved research results in genetic improvement and improved welfare outcomes compared with a system that allows for a differential treatment between 'North' and 'South'. Studies of the effects of Plant Variety Protection legislation in various countries have found inconclusive evidence in terms of increased private-sector investments in breeding. With the increased investments involved in exploiting the tools of biotechnology for varietal improvement, including the development of genetically modified organisms, some industrialized countries have opted for patent protection for plants or plant varieties. This chapter adapts a recently published model of innovation in the seed industry by endogenizing the appropriability of the research benefits, allowing an optimum level of protection in two segmented markets to be determined by a social planner. The most important conclusion of our simulations with the model is the demonstration of the economic case for differentiated levels of Intellectual Property Rights protection between North and South. By allowing more lax standards in the South, more farmers could benefit from innovations, and they would be able to contribute their share to development costs. If developing countries were forced to adopt equally strict standards of protection, then adoption would be lower, and the farmers in this region, as well as the breeding firm, would suffer. These losses would be only partly recuperated by farmers in northern countries who may benefit from lower prices.

Introduction

This chapter asks the question whether uniform international standards for intellectual property rights over plant varieties leads to improved research results in genetic improvement and welfare outcomes compared with a system that allows for a differential treatment between 'North' and 'South'. Developing countries are required under the Agreement on Trade Related Aspects of Intellectual Property Rights (TRIPS Agreement) to introduce, as a minimum, some form of plant varietal protection but can choose the scope of protection they wish to offer. The policy question is to what extent the adoption of industrialized country Intellectual Property Right (IPR) standards will lead to increased provision of improved varieties by the international private sector and what the resulting distribution of the benefits is.

[1] Corresponding author.

The chapter adapts an existing model of innovation in the plant breeding sector to demonstrate that common standards are not optimal from a global welfare point of view. Global welfare, and particularly that of the 'South', can be improved by allowing for more relaxed appropriability in markets where buyers have a relatively lower willingness to pay. The case for this differentiation is based on a variant of third-degree price discrimination. The second section reviews briefly the policy issues relating to Plant Variety Protection (PVP) from an economic perspective, with particular attention to the TRIPS Agreement of the World Trade Organization (WTO). The third section then elaborates a variant of a model of innovation developed by Alston and Venner (2002) allowing for differentiation between markets with different purchasing power. We calibrate the model in the fourth section to run some hypothetical simulations involving uniform versus varying appropriability between the two markets. The fifth section discusses the results. In the conclusions, we offer some recommendations for future research.

PVP

IPR provisions for agricultural plant varieties were developed in industrialized countries during the 20th century to provide incentives for private sector plant breeders. For this purpose, PVP[2] was created, a specialized form of IPR that is more limited in scope than patents available, for example, for industrial innovations. In general, PVP allows breeders to restrict the commercial production and sale of an improved variety, provided that a number of conditions are satisfied.[3] In some instances, further restrictions may be granted concerning the use of the variety in breeding programmes or the use of progeny (second generation seed).

Industrialized countries developed their PVP legislation at different stages but most are now members of the International Union for the Protection of New Varieties of Plants (UPOV). UPOV is a convention that defines specific provisions of PVP legislation which signatories agree to incorporate in their legislation. Successive versions of the UPOV convention have increased the scope and duration of protection offered with the 1991 Convention being the most recent. In some industrialized countries, stricter patent protection for plants (or their components) has also been made available as a result of rapid developments in agricultural biotechnology.

In general, developing countries have considered IPR on agricultural plant varieties as a policy that serves only the interests of private sector breeders in industrialized countries, at the expense of the agricultural sector in their own countries. Many developing countries have only nascent commercial breeders who rely heavily on breeding in either the public sector or foreign breeders. Some higher income developing countries began to implement PVP legislation in the late 1980s and 1990s (e.g. Argentina, Mexico) but the most important force behind the current elaboration of PVP legislation in the developing world has been the WTO-TRIPS Agreement. Article 27(3)b of the TRIPS Agreement obliges member states of the WTO to provide for the protection of plant varieties either by patents or by an 'effective *sui generis* system or by any combination thereof'. This requirement resulted essentially from pressure from industrialized countries. The concession that developing countries were able to obtain was the latitude left in the specific provisions for such a system, embodied in the term '*sui generis* system'. Indeed, the Agreement does not specify at all the scope of protection that must be offered. Some countries have seen it to be in their interest to draft relatively lax PVP legislation, with only the most limited of restrictions, while others have become signatories to UPOV (1991).

The position of industrialized countries is relatively clear: promoting the commercial interests of their breeding companies. By embedding such an Agreement under the WTO, instead of other bodies such as the World Intellectual Property Organization (WIPO), industrialized

[2] Plant varietal protection (PVP) is also referred to by the legal concept of 'plant breeders' rights' (PBR).

[3] These conditions usually comprise the genetic characteristics of Distinctness, Uniformity and Stability, thus DUS (see Leskien and Flitner, 1997; Ghijsen, 1998).

countries were able to force compliance through the dispute settlement mechanism of the WTO with its potential threat of trade sanctions. On the other hand, the TRIPS Agreement is viewed by others as a 'mistake' for many developing countries (Srinivasan, 2002), including the agricultural plant variety sector. Many developing country governments do not perceive the potential incentive effect for innovation or access to foreign varieties in their domestic breeding sector resulting from PVP as compensating for the outflow of royalties to foreign breeding companies.

Economic studies of the impact of PVP on breeding efforts and productivity are relatively scarce and almost all examine the effects of the US PVP Act of 1970 (Lesser, 1997).[4] The results find only limited evidence of an incentive effect for PVP. Butler and Marion (1985) found evidence of increased R&D investments by breeders for a few specific crops but it is difficult to ascertain to what extent this is due to other factors (such as increasing market demand for crops in question; see also the follow-up study by Butler 1996). Perrin *et al.* (1983) concluded that the US PVP Act did have a positive effect on research efforts on some crops, including soybeans and some cereals, even taking into account demand factors. But they found only weak evidence of improved productivity in the case of soybean varieties. This analysis is still so limited though that it cannot be seen as representative. The general view of economists seems to be that the effect on breeding efforts is positive (Fuglie *et al.*, 1996; Lesser, 1997). But critics highlight that conclusive proof is still lacking (e.g. Rangnekar, 2000).

Alston and Venner (2002) recently undertook a relatively thorough analysis of the effects of the US PVP Act on wheat breeding in the USA. Their theoretical model illustrates how increased appropriability leads to increased breeding efforts and seed quality. In testing this hypothesis econometrically, Alston and Venner find no evidence that the 1970 Act had any measurable effect on private sector breeding efforts or commercial yields in the wheat sector. The weak empirical link between PVP and innovation, although still based on limited evidence, has led to the hypothesis that the protection offered is too limited to have the intended incentive effect (Srinivasan and Thirtle, 2002). The major limitation of PVP is usually considered to be the competition provided by farm-saved seed and also the inability to exclude other breeders from using a protected variety in their own research efforts. Even if the farmers' exemption is precluded by legislation, serious enforcement difficulties remain. This has been interpreted as a rationale for allowing stricter IPR protection on plant varieties, such as patents (or the stricter version of UPOV), and it has also been seen as a rationale for permitting or even encouraging the development of GURTs (Srinivasan and Thirtle 2002).[5]

The following section develops a theoretical model to examine the effects of PVP on access to foreign varieties in addition to that on innovation itself. A recent study of international plant variety transferability found limited evidence of this effect (Srinivasan *et al.*, 2002). On the other hand, a review of the Canadian 1990 Plant Breeders' Rights Act highlights the potential role of PVP in inducing foreign seed suppliers to market their varieties internationally (Canadian Food Inspection Agency, 2001).

Two-country Model of Property Rights Protection for Genetic Improvement

The starting point for our model is Alston and Venner (2002). In their set-up a monopolistic innovator develops a new variety and grants seed propagators the right to market seed of that variety to farmers in exchange for royalty payments. In this chapter, we extend the Alston and Venner model in two respects. First, we allow for market segmentation. The two markets may differ in their willingness-to-pay for improved seed varieties. As a consequence,

[4] There is one published review of experience in developing countries. Jaffe and van Wijk (1995), who carried out a study based on interviews of breeding firms in four Latin American countries, found little link between PVP and breeding firms' investment decisions.

[5] The argument is based partly on the relatively higher investments in hybrid crops which also have a built-in biological benefit appropriation (see Fuglie *et al.*, 1996 for a review of these figures).

different market prices (royalties) will result in the two markets, and welfare effects from sharpened IPR enforcement will differ across the two markets. Second, we extend the Alston and Venner model by allowing the optimal degree of appropriability to be determined endogenously. A social planner sets IPR standards for both markets such that global welfare is maximized. In this decision, the social planner weighs the positive effects of tighter IPR protection that result in increased R&D investments and higher seed quality against the losses to farmers that stem from higher prices, and hence lower adoption rates.

The model is set up as a three-stage simultaneous move game: in the first stage the social planner determines the optimal IPR regime where appropriability is represented by the fraction of seed sales on which royalties are paid to the innovator. In the second stage, the monopolistic, multinational innovator determines his optimal level of genetic improvement (R&D), given the IPR regime. In the third stage, the seed propagators determine their optimal supply to farmers in a non-cooperative, oligopolistic setting, given the quality provided by the innovator. The equilibrium of the model is found by solving the third stage first and working backwards.

Demand for seed in both markets is determined from the (inverse) demand relations:

$$r_i = r_i(Q_i, z) \text{ with } Q_i = \sum_j q_{i,j} \qquad (1)$$

Where $i = 1, 2$ indexes the two markets; $j = 1, 2 \ldots n_i$ is the index for the propagators in both markets, and z measures seed quality. The demand functions are continuous, convex and downward sloping in quantities, Q. $\frac{\partial r_i}{\partial Q_i} < 0; \frac{\partial^2 r_i}{\partial Q_i^2} \geq 0$, and seed demand reacts positively to increased quality z, but possibly at a diminishing rate: $\frac{\partial r_i}{\partial z} > 0; \frac{\partial^2 r_i}{\partial z^2} \leq 0$. Negative prices are ruled out, $r_i(Q_i,z) \geq 0$ and there is a maximum demand, at a given quality $\bar{z}$, and at royalty rate equal to zero $\bar{Q} = r_i^{-1}(0, \bar{z})$. At each quality $\bar{z}$, aggregate willingness to pay in market 1, North, is at least as high as willingness to pay in the South:

$$\int_0^{\bar{Q}_1} r_1(Q_1, \bar{z}) dQ_1 \geq \int_0^{\bar{Q}_2} r_2(Q_2, \bar{z}) dQ_2$$

The third stage of the game: seed propagators determine royalty rates and seed supply

In each region, seed propagators independently maximize their profits, given the seed quality, z, provided by the innovator. They pay a fraction α_i of royalties to the innovator. Marginal reproduction costs are zero (see Alston and Venner, 2002), but they incur a fixed cost F_i to set up operations. Each firm j in market i sets its own $q_{i,j}$ given the choice of its competitors. We allow for free entry, so that the symmetric Cournot equilibrium in each market is characterized by the set of first-order conditions for profit maximization and the zero profit condition for each firm, and where the identical firms produce $q_{i,j} = q_i$ for all $j = 1 \ldots n_i$. From these conditions we can derive the optimal seed supply for each firm and the optimal royalty rate per unit (see also Alston and Venner, 2002)

$$q_i^* = q_i(F_i, \alpha_i, r_i^*) = \left(\frac{-F_i}{(1-\alpha_i)} \frac{1}{\partial r_i / \partial Q_i} \right)^{1/2} \qquad (2)$$

$$r_i^* = r_i(F_i, \alpha_i, q_i^*) = \left(\frac{-F_i}{(1-\alpha_i)} \frac{\partial r_i}{\partial Q_i} \right)^{1/2} \qquad (3)$$

And the equilibrium number of firms in market i is given by:

$$n_i = \frac{Q_i(z, r_i^*)}{q_i^*(F_i, \alpha_i, r_i^*)} \qquad (4)$$

Consider the response of seed propagators to tightened appropriation, i.e. increasing α. This works like an ad valorem tax, which is partially passed on to farmers. The royalty rate increases (see Equation 3) and this leads to a drop in demand. This in turn lowers the number of firms, as fixed costs can only be covered by increased sales per firm while free entry and exit leads to profits being zero. A new equilibrium is reached with a lower adoption rate, a higher royalty per unit, a smaller number of suppliers, but higher seed sales per firm.

The second stage of the game: the innovator determines supply of seed quality (R&D)

The innovator takes as given the decisions of the seed propagators and the resulting equilibrium royalty rates r_i^* in each of the two markets. The innovator chooses seed quality z so as to maximize his profits:

$$W(z) = \sum_i \alpha_i r_i^* Q_i(z, r_i^*) - c(z) \quad i = 1,2$$

Where the cost function $c(z)$ is a reduced form of an R&D cost function. The R&D process is assumed to exhibit decreasing returns to scale,[6] so that the cost function has positive first and second derivatives (i.e. is increasing in z and marginal costs are increasing).

The first order conditions for profit maximization require that:

$$\sum_{i=1,2} \alpha_i \cdot \frac{\partial Q_i(z, r_i^*)}{\partial z} \cdot r_i^* - \frac{\partial c(z)}{\partial z} = 0 \qquad (5)$$

and where r_i^* is determined from Equation 3. Solving Equation 5 results in the optimal seed quality:

$$z^* = z(\alpha_1, \alpha_2, r_1^*, r_2^*) \qquad (6)$$

It can be shown that:

- Quality improves if appropriability is tightened, that is, the optimal z^* is increasing if either of the α_i rises.
- The royalty fee increases if appropriability is tightened.
- Diffusion of the improved quality will be lower, if the demand-enhancing effect of the quality improvement does not dominate the price effect.

The first stage of the game: the social planner determines optimal appropriation levels

Finally, the social planner's problem is to choose optimal levels of α in both markets. He maximizes Net Social Welfare (NSW), which is the sum of the innovator's profits, the seed propagators' profits and farmers' ('consumer') surplus in both markets. Since the seed market displays zero profits by assumption, NSW is given by:

$$NSW(\alpha_1, \alpha_2) = \sum_i \alpha_i, r_i^* Q_i(z^*, r_i^*) - c(z) + \int_0^{Q_1(r_1^*, z^*)} r_1(Q_1, z^*) dQ_1 + \int_0^{Q_2(r_2^*, z^*)} r_2(Q_2, z^*) dQ_2 \qquad (7)$$

In this formulation, the welfare improvement accruing to farmers is approximated with the concept of the consumer surplus which uses the area under the demand curve as a money measure of utility at given price–quantity combinations.[7] Equation 7 is maximized subject to Equations 1, 3 and 6. This highly non-linear equation is maximized with numerical methods.[8,9]

Simulations of Differentiated Standards

For the numerical illustration, we shall assume the (inverse) demand function to be linear and a quadratic R&D cost function:

$$r_i(Q_i, z) = a_i - b_i Q_i + \gamma_z$$

$$c(z) = Az^2 \qquad A > 0$$

Using the relevant expressions, this leads to an optimal quality z^* as:

$$z(\alpha_1^*, \alpha_1^*) = \frac{\gamma}{2} \left[\frac{\left(\frac{F_1 b_1}{1-\alpha_1^*}\right)^{1/2} b_2 \alpha_1^* + \left(\frac{F_2 b_2}{1-\alpha_2^*}\right)^{1/2} b_1 \alpha_2^*}{A b_1 b_2} \right]$$

We present results for six simulation scenarios in Table 10.1 (numbered 1–6 in the columns). The benchmark solution, column 1, assumes both countries to be equal in their demand parameters (see table for values). Since the two regions are identical in all respects, the

[6] Decreasing returns in knowledge production has been a popular assumption at least since Nordhaus' (1969) seminal contribution.

[7] A more complete analysis would derive the factor demand for seed from a fully specified production model of the farm allowing for feedback effects from the product market.

[8] Numerical simulations are performed with Mathcad 2000 Professional. The Mathcad code can be made available on request.

[9] Existence conditions for a maximum have been derived, but are omitted here. For the existence of a maximum it is important that the z-effect in the demand function does not dominate the r-effect.

Table 10.1. Simulation of differentiated standards.

		Benchmark; both regions equal demand parameters (1)	Region 2 has lower income (lower a_2) (2)		Region 2's fixed cost F_2 decreases (3)		Region 2's fixed cost F_2 decreases and $\alpha_2 = 0$ (4)	R&D subsidy: A decreases from 5 to 2.5 (5)		R&D subsidy and $\alpha_2 = 0$ (6)
			Unconstrained	Constrained: $\alpha_1 = \alpha_2$	Unconstrained	Constrained: $\alpha_1 = \alpha_2$		Unconstrained	Constrained: $\alpha_1 = \alpha_2$	
Appropriability, North	α_1	0.51	0.46	0.30	0.47	0.43	0.46	0.59	0.58	0.56
Appropriability, South	α_2	0.51	0.02	0.30	0.34	0.43	0.00	0.56	0.58	0.00
Global welfare	NSW	98.3	49.6	46.3	56.0	55.7	54.6	66.6	66.6	60.4
Grower surplus, North	CS_1	36.4	34.4	39.0	35.5	36.9	34.3	40.0	40.8	35.1
Grower surplus, South	CS_2	36.4	3.8	0.0	7.1	6.0	8.9	8.9	8.4	11.6
Innovator benefits	W	25.5	11.4	7.3	13.4	12.8	11.3	17.7	17.4	13.7
Royalty revenues	$\alpha^* r^* Q$	27.6	11.8	7.9	14.3	13.7	11.7	22.3	22.2	15.1
R&D expenditures	$c(z)$	2.1	0.4	0.6	0.9	0.9	0.4	9.2[a]	8.1[a]	2.9[a]
Royalty rate North	r_1	6.4	6.1	5.3	6.1	5.9	6.1	7.0	6.9	6.8
Royalty rate South	r_2	6.4	4.5	5.3	3.9	4.2	3.2	4.8	4.9	3.2
Seed quality	z	0.6	0.3	0.3	0.5	0.4	0.3	1.4	1.4	0.8
Number of seed firms, North	n_1	0.7	0.7	0.9	0.7	0.8	0.7	0.6	0.6	0.6
Number of seed firms, South	n_2	0.7	0.2	0.0	0.4	0.3	0.7	0.3	0.3	0.8
Seed quantity sold, North	Q_1	4.3	4.2	5.0	4.3	4.5	4.2	4.4	4.5	4.0
Seed quantity sold, South	Q_2	4.3	0.8	0.0	1.5	1.2	2.1	1.6	1.5	2.6
Constant term seed demand, North	a_1	10.0	10.0	10.0	10.0	10.0	10.0	10.0	10.0	10.0
Slope parameter seed demand, North	b_1	1.0	1.0	1.0	1.0	1.0	1.0	1.0	1.0	1.0
Constant term seed demand, South	a_2	10.0	5.0	5.0	5.0	5.0	5.0	5.0	5.0	5.0
Slope parameter seed demand, South	b_2	1.0	1.0	1.0	1.0	1.0	1.0	1.0	1.0	1.0
Fixed cost seed firms, North	F_1	20.0	20.0	20.0	20.0	20.0	20.0	20.0	20.0	20.0
Fixed cost seed firms, South	F_2	20.0	20.0	20.0	10.0	10.0	10.0	10.0	10.0	10.0

Notes: $\gamma = 1$; $A = 5$ in all simulations, except columns 5 and 6.
[a]Private plus public research expenditures.

benchmark case is a symmetric equilibrium and optimal IPR protection is found to be equal in both regions: $\alpha_1^* = \alpha_2^* = 0.51$. The innovator receives 27.6 monetary units of royalty revenues, of which he spends 2.1 monetary units on quality improvements. The optimal seed quality index equals 0.6. In both markets 4.3 units of seed are sold, which generates a surplus to farmers that equals 36.4 monetary units.

Next, column 2 in the table reports the case where region 2 ('South') has a lower income, represented by a lower demand shift parameter a_2, than region 1 ('North'). All other parameters remain the same. The optimal α's decrease in both markets, as does seed quality, z. At any royalty rate, demand shifts downward, relative to the benchmark. This forces the seed-producing industry in the South to contract output, and to reduce the royalty rate. This in turn reduces the incentives for the innovator to invest in R&D at any given set of α's. R&D costs decrease faster than the innovator's revenues as he decides to lower z. This allows the social planner to set lower α in both regions, since seed buyer's surplus (consumer surplus) increases if α is decreased. The fall in the optimal α_2 is considerably larger than the fall in α_1. This follows from the fact that there is more surplus to be captured from the richer North. With reduced incomes, seed propagators in the South find it difficult to sell the improved quality. Only if they are allowed to keep a larger portion of the seed revenues, can a small number of seed companies continue business and cover their fixed costs.

If the α's were constrained to be equal to each other in the two regions, a sharp drop in adoption in the South would occur. With equally tight appropriation systems, and no differences in seed propagators' fixed costs, the royalties per unit in both countries have to be equal to each other (see Equation 5). The optimal level of appropriation implies a higher α in the South. This leads to a further contraction of the propagating industry in the South, as seed demand collapses in the face of rising royalty rates. The royalty fees to be paid to the innovator are simply too high to generate enough sales in the South, and adoption in the South drops to zero. The logic is completely analogous to a third-degree price discrimination argument. In segmented markets with different demand characteristics such as different income and price elasticities, welfare is generally optimized with the supplier charging different prices in each market. A uniform high price[10] would mean that the innovator forgoes sales in the price-elastic market segment, while the price-inelastic, segment is evidently willing to pay a more.

Simulation (column 3) adds some realism to the setting by allowing fixed costs for seed propagators to be lower than that of their colleagues in the North.[11] The lower fixed costs in the South leads to higher optimal α's compared with the unconstrained simulation (column 2). More firms enter the South market and adoption is improved. This allows the social planner to set a higher level of appropriation in the South. The surplus thus extracted generates more revenues to the innovator such that seed quality can be further improved. However, if the IPR regime is constrained to be globally uniform, the adoption in the South drops, as well as global welfare (second column under (column 3)). Comparison of the unconstrained and constrained case shows clearly that surplus is transferred from farmers in the South to farmers in the North. Farmers in the North benefit twice: first from a lower royalty rate they have to pay, as a larger share of revenues comes from the South, and second by improved seed quality. While farmers' welfare in the South drops by −1.1, welfare of farmers in the North increases by +1.4, and innovator's profits drop by −0.6.[12] It is thus not in the innovator's own interest to push for globally uniform standards, nor is it in the global interest. Some 0.3 monetary units are 'wasted' in the process of making IPR standards uniform. By allowing for a lower standard in the South, more farmers could benefit from innovations, and they would be able to contribute their share to development costs.

In simulation (column 4), we relax IPR protection in the South completely; propagators would not pay royalties to the innovator, but would charge a market-based fee to farmers.

[10] The effective price that the innovator receives in each of his markets equals $\alpha_i \cdot r_i$.

[11] Lower F can also be seen as representing reduced excludability in the South, following the interpretation of Alston and Venner (2002).

[12] The difference of 0.1 is an error due to rounding.

This can also be interpreted as there being no enforceability in the South, while seed propagators can still produce the innovator's seed (implying that excludability is certainly not complete, as reflected in lower F). In this case, both total R&D expenditures and seed quality would drop. With lower seed prices in the South, adoption would improve, and more propagators would find it profitable to remain in business. This setting leads to a re-distribution of benefits from the innovator and from farmers in the North to farmers in the South. While the farmers in the North would not witness a change in the price they are facing for seed (their royalty rate remains almost constant) they have to accept lower seed quality. Because no royalties are skimmed off the Southern seed market, the total royalty revenues drop and the innovator can perform less R&D.

One way to align public benefits from innovation with the private innovator's benefits is to subsidize the R&D process. Would a subsidy on seed improvements be well spent? We simulate an R&D subsidy as a downward shift in private R&D costs by reducing the constant term in the R&D cost function to half its original value. This corresponds to a 'matching funds' subsidy: the innovator now faces marginal cost of $A \cdot z$, instead of $2A \cdot z$. For each unit of quality generated, the subsidy covers half the additional development cost. The decrease in private marginal R&D costs provides a considerable boost to quality improvements, as the cost function exhibits decreasing returns to scale (see column 5 in Table 10.1). Global welfare is clearly improved.[13] Among the farmers, the biggest relative gains are achieved in the South, where farmers' welfare increases with 25%, against a 13% welfare increase in the North. The optimal levels of IPR protection, α's, are higher than in the no-subsidy case. This is a consequence of the fact that the seed quality improvement is so large that the demand for improved varieties is boosted sufficiently to allow for the setting of tighter IPR standards in both regions.

Under a subsidy, constraining IPR regimes to be equal in both regions seems again not to be a good idea, although the effects on all variables are very small. The main loser would be the innovators and the farmers in the South, who would face a slightly higher royalty rate. An interesting alternative is to provide the R&D subsidy and link this to the requirement to make the innovations freely available to propagators in the South. Such solutions have recently been proposed, for example with respect to the diffusion of 'Golden Rice', a variety of rice genetically engineered for a higher beta-carotene content.[14] Simulation (column 6) presents an extension of simulation (column 5) in which appropriability in the South, α_2, is set to be zero. Overall R&D investments drop and seed quality is lower compared with the unrestricted case. This is a consequence of the fact that less royalty revenue is generated through the market mechanism. The main beneficiaries of such a solution are clearly the farmers in the South, both in terms of their welfare and seed adoption. Compared with the no-subsidy case, farmers in the North and the innovator do not appear to be affected very much, while seed quality is notably improved.

The subsidy clearly carries a positive welfare multiplier. The largest welfare multiplier in the South is achieved when the innovation is made freely available to propagators in the South. In this case, \$1 of subsidy generates \$1.8 additional welfare in the South. A similar picture emerges from the seed adoption: each dollar subsidy generates 0.7 units of additional seed sales in the South if the innovation is distributed freely.

Conclusions

The most important conclusion of our simulation with the model is the demonstration of the economic case for differentiated levels of IPR

[13] We did not correct global welfare for the subsidy (which varies with the amount of quality improvements). The subsidy is transferred to the innovator, who spends it on R&D. Hence, it is already accounted for in the calculations of innovator's benefits. Of course, the welfare picture is partial, since we do not model the financing of the subsidy, e.g. through taxes in the North and South. Adding such a general equilibrium view would certainly reduce the overall benefits of the subsidy.

[14] The analogy is not perfect as the main intention of the innovation is for diffusion in 'the South' without an immediate market in the North.

protection between North and South. This substantiates the point made by McCulloch *et al.* (2001):

> clearly the optimum level of protection is likely to vary by country, as Subramanian (1990) suggests. If activities infringing IPRs (such as commercial copying) in small developing countries do not affect world prices, then the only effect on their IPR regulation is to transfer benefits from such countries to inventor companies. In such cases it would be optimal for countries to have very lax or non-existent protection of IPRs.

In our case, though, one need not base the argument on a provision that world prices are not affected by partial enforceability or excludability, as suggested by McCulloch *et al.* We have demonstrated the argument for differentiated levels with an application of third-degree price discrimination in an extension of a model of partial appropriability of benefits in plant breeding.

Under a globally uniform system of IPR protection, there is limited possibility to discriminate among different markets, other than charging different royalty rates. This results in an inefficient outcome, with a lower global welfare level and lower benefits for the innovator, who could logically benefit from price discrimination. On top of that, the relative burden is highest for the South. This is developed in a model in which the South is being served by the same breeder/innovator as the North. Here we are considering PVP in developing countries more as an incentive for international breeders to export to these markets and thus to take them into account in their investment and marketing decisions.

The case for differing standards has to be viewed somewhat separately from arguments for increased tightening of PVP. The limited evidence available indicates that PVP protection may not be offering a clear incentive for many crops in industrialized countries. While there may be an economic case for tightening these further, we argue here that developing countries need not necessarily follow suit. This underlines the value of the approach taken in TRIPS which sets almost no specific provisions for the design of a *sui generis* system. Furthermore, this allows countries with different incomes to set their systems at the appropriate level. The appropriate scope of PVP protection increases with income.

The model also indicates that the commercial development of genetic use restriction technologies, which amount to a technological mechanism for maximum appropriability would almost certainly be negative for farmers in the South, as well as being welfare-reducing for the world as a whole. At the other end of the spectrum, a separate issue concerns whether any protection at all is warranted in the South. Our results do indicate though that although farmers in the South may benefit from free access to improved varieties from the North, the quality of such varieties could be considerably lower. This issue requires further research, including other specifications of the role of seed quality in the demand for an improved variety. Further research could also attempt to represent specific provisions of PVP legislation that are under discussion, such as the farmers' privilege and the breeders' exemption. It might also be possible to represent relevant differences between types of crops, such as self-pollinated, cross-fertilizing and hybrid varieties.

The simulation results indicate the potential value of a combination of policy measures to promote innovation in crop breeding. In this case, we have looked at a combination of PVP with a research subsidy (or contract). The results indicate the advantages of a performance-linked scheme, resulting from the specific cost function chosen. If such a subsidy is being granted in the interests of promoting diffusion in developing countries, then concessions on the appropriability of the resulting innovations could further improve farmers' welfare in that region. In general, our model suggests that governments should continue to remain active in agricultural research.

References

Alston, J.M. and Venner, R.J. (2002) The effects of the US Plant Protection Act on wheat genetic improvement. *Research Policy* 31, 527–542.

Butler, L.J. (1996) 'Plant breeders' rights in the U.S.: update of a 1983 Study'. In: Van Wijk, J. and Jaffe, W. (eds) *Proceedings of a Seminar on the Impact of Plant Breeders' Rights in Developing Countries* held at Santa Fe Bogota, Colombia, 7–8 March, University of Amsterdam, Amsterdam, pp. 17–33.

Butler, L.J. and Marion, B.W. (1985) *Impacts of Patent Protection in the US Seed Industry and Public Plant Breeding*. North Central Regional Research Bulletin No. 304, University of Wisconsin, Madison, Wisconsin.

Canadian Food Inspection Agency (2001) *10-Year Review of Canada's Plant Breeder' Rights Act*. Ottawa, Canada.

Fuglie, K., Ballenger, N., Day, K., Klotz, C., Ollinger, M., Reilly, J., Vasavada, U. and Yee, J. (1996) *Agricultural Research and Development: Public and Private Investments Under Alternative Markets and Institutions*. AER-735. Economic Research Service, United States Department of Agriculture, Washington, DC.

Ghijsen, H. (1998) Plant Variety Protection in a developing and demanding world. *Biotechnology and Development Monitor* 36, September/December, pp. 2–5.

Jaffe, W. and van Wijk, J. (1995) The Impact of Plant Breeders' Rights in Developing Countries. Inter-Am Institute for Cooperation on Agriculture, University of Amsterdam, The Netherlands.

Leskien, D. and Flitner, M. (1997) *Intellectual Property Rights and Plant Genetic Resources: Options for a* sui generis *System*. Issues in Genetic Resources No. 6, International Plant Genetic Resources Institute (IPGRI), Rome.

Lesser, W. (1997) Assessing the implications of intellectual property rights on plant and animal agriculture. *American Journal of Agricultural Economics* 79, 1584–1591.

McCulloch, N., Winters, L.A. and Cirea, X. (2001) *Trade Liberalization and Poverty: a Handbook*. Centre for Economic Policy Research, London.

Nordhaus, W.D. (1969) *Invention, Growth and Welfare*. MIT Press, Cambridge, Massachusetts.

Perrin, R.K., Hunnings, K.A. and Ihnen, L.A. (1983) *Some Effects of the US Plant Variety Protection Act of 1970*. Economics Research Report No. 46, Department of Economics and Business, NC State University, Raleigh, North Carolina.

Rangnekar, D. (2000) *Intellectual Property Rights and Agriculture: an Analysis of the Economic Impact of Plant Breeders' Rights*. ActionAid, London.

Srinivasan, T.N. (2002) *Developing Countries and the Multilateral Trade System after Doha*. Yale University Economic Growth Center Discussion Paper No. 842. New Haven, Connecticut.

Srinivasan, C.S. and Thirtle, C. (2002) Impact of terminator technologies in developing countries: a framework for economic analysis. In: Swanson, T.M. (ed.) *Biotechnology, Agriculture and the Developing World: the Distributional Implications of Technological Change*. Edward Elgar, Cheltenham, UK.

Srinivasan, C.S., Shankar, B. and Holloway, G. (2002) An empirical analysis of the effects of plant variety protection legislation on innovation and transferability. Paper presented at the *Conference of the European Association of Agricultural Economists*, Zaragoza, Spain, August 2002.

Subramanian, A. (1990) TRIPS and the paradigm of the GATT: a tropical temperate view. *World Economy* 13, 509–521.

11 Biotechnology and Developing Countries: the Struggle over Intellectual Property Rights and Implications for Biodiversity Conservation

Ottavio Janni
Instituto Sperimentale di Studi Socio-Economici sull'Innovazione e le Politiche della Ricerca, Milan, Italy

Introduction

The growing importance of biotechnology products in the global economy has engendered a fierce debate over intellectual property rights (IPR), which has been embodied by the different normative approaches to property rights to genetic material inherent in the Convention on Biological Diversity (CBD) and the Trade Related Aspects of Intellectual Property Rights (TRIPS) agreement under the World Trade Organization (WTO). Broadly speaking, these two different institutional regimes can be said to reflect the positions of developing (in the case of the CBD) and developed countries (in the case of TRIPS). The CBD calls for national sovereignty and equitable sharing of the genetic resources that are fundamental to the biotechnology industry, while the TRIPS agreement favours patent legislation in all technological fields, including biotechnology. One field in which this struggle has potentially important repercussions is that of biodiversity conservation.

This chapter will analyse the implications for biodiversity conservation that arise from the property rights regimes supported by TRIPS and the CBD, and the different ways in which property rights regimes are implemented throughout the world. Because the most biodiverse areas on the planet fall almost entirely within the developing world, it is important that developing countries benefit from the use of genetic and biotechnology resources if they are expected to conserve their biological patrimony, the primary source of much of the genetic material used by the biotechnology industry. The chapter will briefly summarize the close links between biotechnology and biodiversity conservation, especially as they apply to developing countries, before exploring in greater detail the implications that different property rights regimes have for biodiversity conservation, and the potential future evolution of this issue.

Biotechnology, Biodiversity Conservation and Developing Countries

Biodiversity loss and climate change are the two most visible environmental issues of global reach, and also the most pressing. Broadly speaking, biodiversity encompasses the diversity of life forms present on the planet. The importance and visibility of biodiversity conservation as a crucial international issue have been greatly increased since the signing of the CBD in 1992, and the CBD's definition best broaches

 The Regulation of Agricultural Biotechnology (eds R.E. Evenson and V. Santaniello)

the different views of biodiversity: 'biological diversity means the variability among living organisms from all sources, including *inter alia* terrestrial, marine and other aquatic ecosystems and the ecological complexes of which they are part; this includes diversity between species, within species, and of ecosystems'.

Developing countries hold both the richest areas for biodiversity and the most threatened ones, and as such biodiversity conservation is a particularly pressing issue for them. A recent survey (Mittermeier *et al.*, 1997) has identified 17 countries – the so-called 'megadiversity' countries – which alone account for over 80% of the planet's biodiversity, and which hold the highest levels of endemism and the most critically threatened ecosystems. The majority of these 17 countries are in the developing world, particularly in the Andean region, the Amazon basin, and in south Asia along with several other large, mostly developing countries.

Why are biotechnology and biodiversity so closely linked then? The short answer is that the biotechnology industry, whether in agriculture, medicine, or any other field, could not exist without the tremendous genetic variety found in nature. Biodiversity can either be a stock of information for the biotechnology industry, or it can be the primary stock itself. The biotechnology industry can either develop new products based on the observed characteristics (or phenotype) of a particular genetic resource, or it can directly use such biological or genetic resources for a particular purpose (Swanson and Goschl, 2000). For the purposes of this chapter, biotechnology can best be defined as 'any technique that uses living organisms (or parts of organisms) to make or modify products, to improve plants or animals, or to develop microorganisms for specific uses' (USOTA, 1991). In particular, the pharmaceutical, agricultural, environmental and genetics industries are those who are most active in biotechnology.

The biotechnology industry's boom is a relatively recent one, and as of now firms are overwhelmingly located in the developing world. The lack of resources devoted to scientific research and weak institutional regimes have made it difficult for biotechnology firms to establish themselves in developing countries. Although a fledgling biotechnology industry exists in east Asia, as well as Brazil, Mexico, Cuba, India, and China, in much of the developing world, especially in Africa, biotechnology research efforts are even further behind. It would be therefore expected for the biotechnology industry to reflect the views of developed countries in international forums dealing with trade and environment issues.

How can biotechnology then help to promote biodiversity conservation? The basic premise is that biodiversity contains hidden assets of potentially huge value to humanity, such as new or better food crops. The search for new commercial applications for plant and animal species therefore gives biodiversity a significant enough innovation option value that biotechnology companies would be willing to pay for its preservation, and as such conserving a patch of biodiversity-rich rainforest, for example, becomes more financially viable than converting it to farmland. The opportunity costs of biodiversity conservation would then be offset by the potential gains. Furthermore, as the private sector recognizes the economic value of biodiversity, landowners and local communities in biodiversity-rich areas will recognize the value and the potential benefits of their natural resources and will find it profitable to work towards their conservation (Simpson *et al.*, 1996; Barrett and Lybbert, 2000).

The way benefits from biotechnology are distributed is extremely important in determining whether they can serve as an incentive for biodiversity conservation. Existing property rights regimes are a key factor in influencing the distribution of benefits, and an understanding of the current international institutional regime dealing with property rights is therefore crucial. One must also keep in mind that different countries have different national property rights regimes, and these are also crucial in any analysis of the problem.

Normative Approaches to Property Rights: TRIPS vs. CBD

The international institutional framework within which the links between biotechnology, property rights and biodiversity conservation overlap is currently rather muddled. In fact, the struggle over property rights is one of the main differences between the institutional framework created by the CBD and the TRIPS regime of the WTO.

The reasons behind this conflict have to do with the different nature of the two agreements, and the interests they represent. The CBD's main purpose is to protect biological diversity and the sustainable use of its components, while TRIPS mainly exists as a framework to strengthen and harmonize IPR systems in all fields, especially as they relate to international trade. These two regimes reflect different sectors of the global community. Developed world interests are strongly reflected in TRIPS, with its emphasis on patents and property right, while the CBD's emphasis on the equitable sharing of the benefits from genetic resources reflects the views of developing countries which often harbour such resources.

Perhaps the single most important bone of contention between CBD and TRIPS is the question of what can be patented. Again, this rises in part out of the conflict between developed and developing nations. While developing nations host the bulk of the world's biodiversity, and therefore its genetic resources, developed countries have the technological and scientific know-how to elaborate and commercialize these resources. Each therefore claims ownership over these products. The biotechnology sector sustains significant research costs and would like to see compensation in order to make its efforts economically profitable, while the developing world also wants to profit from what it sees as its resources being marketed. At issue is whether biological products and processes are true inventions and whether they can be patented, as TRIPS argues.

The key provisions of both treaties with regards to property rights and genetic resources are worth looking at in some detail. Within the TRIPS agreement, perhaps the most important provision in this regard is Article 27.1, which states that national laws should provide patent protection to inventions, without discrimination as to the field of technology concerned, therefore including biotechnology. This is a major bone of contention between developed and developing countries, as the latter only hold a very small percentage of worldwide patents. Developing countries were particularly opposed to patenting plant and animal species as well as genetic materials, as such a system would not recognize the important contributions made by traditional and indigenous knowledge.

These concerns were addressed to some degree in TRIPS. Article 27.3(b) states that countries may exclude from patentability 'plants and animals and other micro-organisms, and essentially biological processes for the production of plants and animals other than non-biological and micro-biological processes'. Furthermore, Article 27.3(a) grants countries the right to exclude from patentability diagnostic, therapeutical, and surgical methods for the treatment of humans and animals, therefore addressing some of the moral concerns over IPR. It should be noted that this article is optional, and that countries that do want to patent plants or animals may do so. A review of this Article is currently underway in the TRIPS council.

The general outlook of TRIPS remains rooted in developed country interests, and some of its provisions seem to contradict Article 27.3. TRIPS establishes an IPR regime over plant varieties, either through patenting or through establishing a *sui generis* legal system, which could be achieved through joining the Union for the Protection of New Varieties of Plants (UPOV). TRIPS calls for the member countries that choose the *sui generis* option to do so by January 2000; least developed countries have until January 2005 to do so. A problem with this aspect of TRIPS is that it is not entirely clear what an effective system to protect plant varieties is, or to what extent this provision conflicts with the option to exclude plants from patentability as stated in Article 27.3.

The concerns of developed countries were addressed more specifically in the CBD, particularly with regards to ownership of genetic resources. While under TRIPS IPR override national sovereignty, the CBD regimes regards animal, plant and genetic material as part of a nation's patrimony. As such, under the CBD individual nations have the right to regulate access to their genetic resources, through prior informed content and on mutually agreed terms. The CBD is not retroactive and under its terms developed countries still have free access to the genetic material deposited in international gene banks.

The most important article of the CBD with regards to the debate on biotechnology and property rights is Article 15, which recognizes 'the sovereign rights of States over their natural resources' and states that 'the authority to determine access to genetic resources rests with the national governments and is subject to national legislation'. Furthermore, it calls for 'sharing in a fair and equitable way the results from research and development and the benefits arising from the commercial and other utilization of genetic resources with the Contracting Party providing such resources'. Article 19 addresses this issue even more explicitly, stressing once again the fair and equitable access to benefits from the use of genetic resources, and setting provisions for the safe use of 'any living modified organism resulting from biotechnology that may have adverse effect on the conservation and sustainable use of biological diversity'. Finally, Article 8(j), obligates member nations to work to preserve traditional and indigenous knowledge on biodiversity use, to involve such communities in wider applications of their knowledge, and to share any benefits arising from such knowledge both with the relevant indigenous communities and with the countries that host them.

A recent development took place within the framework of the Sixth Conference of the Parties in April 2002. The CBD adopted a set of global guidelines on genetic resources, known as the Bonn Guidelines on Access to Genetic Resources and Fair and Equitable Sharing of the Benefits Arising out of their Utilization. These guidelines address the growing concerns of developing countries that the economic gains from their genetic resources were being reaped by companies based in developed countries. The guidelines base themselves on a number of successful bioprospecting agreements, and encourage them as a way to help developing country governments set fair conditions for users seeking genetic resources, in return for such benefits as profits, royalties, scientific collaboration and training (CBD Press Release, 2002). Several such agreements have successfully managed to use benefit sharing as a way to conserve biodiversity (Day-Rubenstein and Frisvold 2001). It should be noted however that these guidelines are voluntary, and therefore unenforceable.

Implications for Biodiversity Conservation

The fact that the current institutional framework is so contradictory has major implications for biodiversity conservation policy. In particular, one major question arises: can the CBD be effectively implemented if the property rights regime on genetic material that it calls for is contradicted by TRIPS?

In light of this, the relative strength of the CBD *vis-à-vis* TRIPS is worth analysing. There are many factors that make the TRIPS regime much stronger that the CBD. These factors are both political and normative. On the political front, TRIPS represent the interests of the far more powerful developed countries. While it is true that the initial impetus behind the CBD was also driven by developed countries, the negotiating process gradually tilted in favour of developing countries. While CBD has gained acceptance in the developed world as well, the USA has refused to sign it. Furthermore, some newly industrialized countries, having recognized the growing importance of patents and property rights to their own economies, are beginning to side with developing country views in some cases (Rosendal, 1999).

Another factor that makes TRIPS de facto stronger that the CBD is the fact that TRIPS deals mainly with trade while the CBD deals with environmental issues. The economic ramifications of a trade regime are greater than those of an environmental regime, and have wider ranging impacts on overall government policies. While developing countries may not agree with all of TRIPS provisons, they are too dependent on access to markets and on good trade relations with developed countries to seriously challenge the TRIPS/WTO regime.

The main reasons why TRIPS is a stronger regime are not political but normative. There are a number of weak points in the CBD that undermine its effectiveness. Under the CBD, access to biological and genetic resources takes place under prior informed consent, which means that the country that provides the genetic resources must have legislation in place to regulate the appropriation of such material. If this legislation is lacking, then users of the genetic material in question may continue to exploit it free of charge (Rosendal, 1999). However, many developing countries have a poorly developed legal framework, which makes it difficult for them to approve and enforce such laws, and makes them vulnerable to abuse.

There is also a strong need to better define ways to protect traditional knowledge in the CBD. Traditional knowledge is a comparative advantage of developing countries which needs to be protected in order for those countries to compete better in the global market. However, while the CBD calls for the encouragement and protection of traditional knowledge, it does not provide an explicit framework to implement this. A working group of the CBD is addressing the issue, but there is still no consensus on whether existing mechanisms of intellectual property rights should be used to protect traditional knowledge, or whether a new, *sui generis* system needs to be developed (World Intellectual Property Organization, 2000).

The biggest advantages that TRIPS has over the CBD are its enforcement mechanisms and timetables. If members of the WTO refuse to abide by TRIPS provisions, they become liable for economic sanctions. This is a very powerful incentive for countries to sign up to TRIPS, and particularly developing countries which are especially vulnerable to sanctions. The CBD on the other hand cannot impose economic sanctions, and thus compliance is essentially voluntary. The Bonn Guidelines on Access to Genetic Resources are also voluntary. Additionally, the CBD has no timetables for compliance. In contrast, the WTO lays out explicit timetables for its members to comply with TRIPS, in accordance with each country's capability. It is this striking difference in enforcement mechanisms that makes TRIPS a regime to be complied with while the CBD, lacking teeth, becomes little more than a declaration of principles.

National Property Rights Regimes and their Significance

While the international institutional framework obviously plays a huge role in how property rights on genetic resources are established, national policies also need to be taken into account. In fact, the two go hand in hand, as one often influences the other. The CBD for example makes demands on national policies on property rights. In general, developed countries have been quick to draft legislation that complies with CBD provisions, but slow to approve it and enact it.

In the case of developing countries, the two biggest sources of conflict in national legislation are those between tenure and legal property rights, and between different property rights regimes at the national, legal, and local levels. Problems dealing with indigenous and traditional communities, and their rights, are also at the forefront. Furthermore, in many cases the conflict between tenure and legal property rights is exacerbated by the fact that a third conflict arises when the rights over genetic resources are taken into consideration, as the landowner does not necessarily have property rights over the genetic resources on his or her land.

There are significant differences between nations as to how property rights are applied. In the majority of cases, there is a distinction between the rights over land, and those over genetic material. This has been the case in a number of high profile genetic access agreements. The agreement between INBio and Merck Pharmaceutical is frequently cited as a pioneering example of such agreements. In this case, the land containing the genetic resources belongs to the landowner, but the genetic resources themselves are in the public domain. In the Philippines, both access to genetic resources agreements approved so far reflect national legislation: while traditional communities can lay claim to ancestral lands and have property rights over them, the genetic material contained therein remains the property of the national government.

Where the legal framework over property rights is not so clear, companies seeking to sign agreements on access to genetic resources have attempted to skirt the problem by asking for access to public lands only. This was the case in another agreement, that between Diversa and the US National Parks Service, giving the company the right to conduct research on heat-resistant microorganisms found with Yellowstone NP (Nunes and van der Bergh, 2001). In this case, ownership over the genetic material found on such public lands rested with the US government. A number of such agreements have taken place in developed countries as well: the National Cancer Institutes, the Missouri Botanical Garden, and Purdue University, have reached agreements the University of Yaoundé and the Cameroonian government to conduct bioprospecting efforts focusing on the plant *Ancistrocaldus korupensis*. Their efforts however take place entirely within Korup National Park, and the Cameroonian government retains clear property rights over genetic resources found on these public lands (Porzecanski *et al.*, 1999).

Another way to deal with this issue, especially when indigenous groups are concerned, is to sign agreements directly with them. In India, a deal between the Tropical Botanic and Research Institute, the Kani tribal people, and the drug company Aryavaidya Pharmacy Coimbtore Ltd to market a herbal tonic derived from the plant *Trichopus zeylandicus* has so far generated US$21,000 to be shared between the Kani community and the Research Institute. In this case, the Kani, who used *T. zeylandicus* in their traditional medicine, receive compensation for contributions made by their traditional knowledge. Such direct agreements are the best way to involve indigenous groups, as they are a way to circumvent sometimes uncertain property rights regimes when dealing with such groups, which often have little political input in national legislation.

When looking at IPR the picture remains muddled. This is particularly true with regards to indigenous knowledge. Many modern discoveries in biotechnology have been based on natural compounds whose properties were known to indigenous peoples, but this knowledge was obviously never patented by them. In many developing countries, there is no national legislation that protects indigenous knowledge, or that equates it with intellectual property. This is true in such important countries for biodiversity conservation as Colombia and Brazil, both of which have also been active in signing bioprospecting agreements with foreign firms.

An exception to this rule is the Philippines, which has strong legislative protection for indigenous rights. This is thanks to Executive Order 247, which was signed in 1995 specifically to regulate access to biodiversity. This piece of legislation, along with the Indigenous People's Rights Act of 1997, recognizes 'the rights of indigenous cultural communities/indigenous peoples and other Philippine communities to their traditional knowledge and practices when this information is directly and indirectly put to commercial use'. The legislation also recognizes that indigenous peoples have full ownership, control, and protection over their cultural and intellectual rights.

The picture with regards to national legislation on IPR is therefore mixed. However, as the issue continues to gain visibility, it is likely that more and more developing countries will follow the Philippines' lead and grant stronger property rights to indigenous peoples for their traditional knowledge.

Implications of Local Property Rights Regimes for Biodiversity Conservation

The different property rights regimes described above obviously have different implications for biodiversity conservation. Biodiversity conservation is hindered both in cases where property rights are unclear, and where local or indigenous communities do not hold real property rights over their lands, or IPR over their traditional knowledge.

Bioprospecting agreements can certainly take place in cases where national legislation over property rights are unclear. However, their effectiveness as incentives for biodiversity conservation diminishes in such cases. This is because in the absence of clearly defined property rights, stakeholders in bioprospecting agreements focus on public lands where the property rights issue is moot. The natural choice becomes national parks, where large tracts of habitat are preserved and biodiversity levels are also high. However, since national parks are already protected to begin with, bioprospecting agreements focusing on protected areas provide no incentive to preserve biodiversity in areas where it is unprotected (Porzecanski *et al.*, 1999). At best, the most successful and profitable ones can emphasize the economic viability of protected areas and encourage their creation elsewhere.

The main hindrance to biodiversity conservation efforts however comes from situations in which local communities do not hold property rights over the genetic resources contained in their lands. As we have seen, this is the most common scenario. However, in such cases, local inhabitants, who bear the brunt of the opportunity costs of habitat protection, cannot participate in negotiations over bioprospecting agreements, nor can they reap any of the financial benefits. In most major bioprospecting deals so far, local communities have been bypassed in genetic sharing agreements. This means that local communities do not have the necessary incentive to participate actively in the conservation of natural resources, and greatly weakens the argument that such agreements can help to support biodiversity conservation.

Conclusions, Recommendations and Outlook for the Future

Both international and national property rights regimes with regards to genetic resources ultimately determine who benefits from such resources. While the conflict within the international regime can best be interpreted as developed versus developing countries, conflicts within national property rights regimes typically pit local communities – in particular with regards to indigenous peoples – against central or local governments. Both these conflicts affect the way biotechnology and access to genetic resources agreements impact biodiversity conservation.

At the international level, the conflict between the TRIPS regime and CBD provisions needs to be resolved. Because of TRIPS' enforcement mechanisms, a more biodiversity-friendly property rights regime would see the incorporation of some of the CBD's principles into TRIPS. There is some evidence that this might be happening, based on the events at the Fourth WTO Ministerial Conference in November 2001 in Doha.

Two important developments took place there. The first was the Declaration on the TRIPS Agreement and Public Health. This declaration recognizes the gravity of the public health crisis in developing countries, and that while intellectual property is important in developing new medicines, prohibitively high prices for such medicines run counter to public interest and result in medicine not being available for all. The declaration reiterates that TRIPS is flexible on such matters, and that each member has the freedom to determine the grounds upon which compulsory licences are granted, the right to determine what constitutes a national emergency (explicitly recognizing that public health issues can constitute such an emergency), and leaves each member free to establish its own regime for the exhaustion of IPR without challenge. Furthermore, the declaration recognizes that members with insufficient capacities in the pharmaceutical sector could have difficulties in making use of compulsory licensing under TRIPS, and sets up a framework to deal with this issue by the end of 2002.

While the above may not seem directly related to biodiversity conservation or biotechnology, it is in fact highly relevant to the debate. One of the arguments that the CBD regime uses against excessive patenting of biological and genetic material is that such a policy runs counter to moral aspects, that such materials are of basic importance to human needs and as such no exclusive rights can be granted over them. The declaration on public health implicitly recognizes this argument, and while it limits it to the field of public health it opens the door for a wider application of this principle. The fact that members are allowed to determine what constitutes a national emergency also suggests that in the future such arguments could be applied to environmental emergencies such as biodiversity loss.

The Doha conference also explicitly addressed the conflict between TRIPS and the CBD. Ministeral Declaration 19 states that, as part of the review of the Article 27.3(b), the TRIPS council needs to examine 'the relationship between the TRIPS agreement and the CBD, the protection of traditional knowledge and folklore, and other relevant new developments raised by members pursuant to Article

71.1'. This is another important step that shows that biodiversity-related concerns are coming to the forefront in the debate over TRIPS. It should be noted that Article 27.3 of TRIPS has been one of the agreement's most hotly debated provisions, and a number of countries have put forward proposals to amend Article 27.3. These include suggestions that the article be rewritten to exclude from patents any organisms or genetic materials; define in detail what an effective plant protection systems is, extend protection to indigenous or traditional knowledge, and make explicit linkages with obligations for the conservation and use of biodiversity. Ministerial Declaration 19 is rather vague as to what should be done, but it does directly acknowledge some of these suggestions – most notably with regards to indigenous knowledge – and further incorporates biodiversity issues into the TRIPS decision-making process.

There seems to be a growing consensus that the discrepancies between TRIPS and the CBD with regards to property rights over genetic resources need to be addressed. However, no concrete steps have been taken so far, and until they are international property rights issues over genetic resources will remain muddled.

With regards to national legislation, the picture is mixed, as these vary widely. While many of the countries that are of critical strategic importance for biodiversity conservation, such the Andean ones, have quickly drafted legislation to comply with CBD standards, they still have major gaps in extending intellectual property rights to indigenous communities. In order for a biodiversity-friendly property rights regime to emerge, more countries will have to follow the example of the Philippines, which have extended intellectual property rights to the traditional knowledge of their indigenous communities. At a more basic level, property rights over genetic resources for local landowners would go a long way in encouraging biodiversity conservation.

Biotechnology and biodiversity conservation are undoubtedly closely linked. The nature of the property rights regimes over genetic resources will go a long way in determining to what extent biodiversity conservation can be helped by bioprospecting and access to genetic resources agreements.

References

Barrett, C.B. and Lybbert, T.J. (2000) Is bioprospecting a valuable strategy for conserving tropical ecosystems? *Ecological Economics* 34, 293–300.

CBD Press Release (2002) First ever global guidelines adopted on genetic resources. Biodiversity conference also acts on forests and invasive species. Available at http://www.biodiv.org/doc/meetings/cop/cop-06/other/cop-06-pr-end-en.pdf

Day-Rubenstein, K. and Frisvold, G. (2001) Genetic prospecting and biodiversity development agreements. *Land Use Policy* 18, 205–219.

Mittermeier, R., Gil, P.R. and Mittermeier, C.G. (1997) *Megadiversity: Earth's Biologically Wealthiest Nations*. Conservation International, Washington, DC.

Nunes, P.A.L.D. and van den Bergh, J.C.J.M. (2001) Economic valuation of biodiversity: sense or nonsense? *Ecological Economics* 39, 203–222.

Porzecanski, A.L., Sears, R., Grant, T., Putzel, L., Dávalos, L., Barnes, T., Cross, H., Raygorodetsky, G. and Simmons, B. ('EPS Workshop') (1999) Access to Genetic Resources: an Evaluation of the Development and Implementation of Recent Regulation and Access Agreements. Published by the School of International and Public Affairs at Columbia University. Available at http://www.columbia.edu/cu/sipa/FUNC/EPS/wkpaper-4. pdf

Rosendal, G.K. (1999) Biodiversity: between diverse international arenas. In: Bergesen, H.O., Parmann, G., Thommessen, Ø.B. and Reklev, S. (eds) *Yearbook of International Co-operation on Environment and Development*, The Fritjof Nansen Institute, Norway.

Simpson, D.R., Sedjo, R.A. and Reid, J.W. (1996) Valuing biodiversity for use in pharmaceutical research. *Journal of Political Economy* 104, 163–185.

Swanson, T. and Goschl, T. (2000) Property rights issues involving plant genetic resources: implications of ownership for economic efficiency. *Ecological Economics* 32, 75–92.

[USOTA] United States Congress Office of Technology Assessment (1991) *Biotechnology in a Global Economy*. US Congress Office of Technology Assessment, US Government Printing Office, Washington, DC.

World Intellectual Property Organization (2000) *Intellectual Property Rights and Genetic Resources: an Overview*. Available at http://www.wipo.int/globalissues/biotech/documents/word/ipgr002.doc

12 Intellectual Property Strategy in the Context of Inter-organizational Relations: the Case of International Agricultural Research

Eran Binenbaum[1] and Philip G. Pardey[2]
[1]*School of Economics, Adelaide University, Australia;* [2]*Department of Applied Economics, University of Minnesota, St Paul, Minnesota, USA*

In recent decades, intellectual property (IP) strategy has become a more prominent part of the management of many organizations. This chapter demonstrates the complexity of IP strategy. Using a taxonomy of inter-organizational relations, we discuss the IP strategies of the Consultative Group on International Agricultural Research (CGIAR, or CG for short). However, the core content of this chapter is methodological and has much wider applicability. Our ideas apply to non-profit and for-profit organizations as well as to other aspects of strategic management, such as funding, partnering, and technology positioning.

The CGIAR, formed in 1971, and its 16 research centres (CG Centres), played a pivotal role in the Green Revolution, thanks in part to a global network of transfers of data, genetic resources, and technologies. In the 1990s, proliferating IP claims caused CG policy makers to become increasingly concerned about the functioning of this network. The Centres' freedom to operate in research and development (R&D) could no longer be taken for granted. In response, the CG's Central Advisory Service on Intellectual Property (CAS-IP) was established in 1999, and several Centres began hiring IP specialists.

Three underlying themes run through this chapter: partnerships, technology positioning and segmentation. Partnerships can be defined as medium-term cooperative inter-organizational relations. Examples include repeated transfers of R&D inputs or outputs, repeated funding and joint R&D projects. Partnerships tend to reward cooperation and punish non-cooperation.

Technology positioning is a crucial determinant of each Centre's inter-organizational relations. The CG System and its Centres are (or ought to be) constantly reassessing their role in the global division of labour in agricultural R&D. Technologies may be either vertically or horizontally related. They are vertically related if one technology enables another, either physically or conceptually. Horizontal relations between technologies include the opposites of complementarity – implying synergistic effects of combining the technologies – and substitutability – implying that alternative technologies may be used with similar results. Some types of technology relations appear to have 'natural' counterparts in inter-organizational relationship types. For example, substitutability between two technologies may be associated with competition, although it may also be associated with collaboration or exchange. Suppose A is an established technology (say, sexual reproduction of hybrid crop varieties) while technology B (say, apomixis as applied to hybrids) is a new and potentially competing technology. Firms that control A may seek to purchase the rights to B, or may seek

 The Regulation of Agricultural Biotechnology (eds R.E. Evenson and V. Santaniello)

collaboration to secure access to B. This example, taken from a collaborative agreement of the International Wheat and Maize Improvement Center (CIMMYT), illustrates the complexity of the link between technology relations and inter-organizational relations.

Segmentation of markets and clients is important in partnerships between for-profits and non-profit organizations. Non-profit organizations want their client groups to benefit from R&D products on concessional terms, while for-profit organizations want to minimize those concessional terms with respect to their most lucrative markets. Various criteria may be used for segmenting markets (CGIAR, 1998; Falcon, 2000; Byerlee and Fischer, 2001; Nottenburg *et al.*, 2002).

Sketch of the Relevant Agents

The Centres are the primary decision-making units of the CG System. In addition, a number of units provide System-wide intellectual property support such as CAS-IP. The balance between Centre authority and System-wide policies is a recurring concern. While Centres have significant leeway in interpreting System-wide guidelines and in specific IP and funding issues, they cannot afford to deviate drastically from the guidelines and common practices in the System. For example, a Centre does not need permission from anybody for a patent application, but it would probably run into trouble with donors if it suddenly announced it would fund half its budget from licensing revenues.

Inter-Centre cooperation can be somewhat problematic. We do not doubt that decision makers at each Centre hold a common commitment to food security, poverty alleviation, and environmental sustainability goals. Each Centre has its own mission statement that supports the CGIAR's overall mission. But there is a certain degree of inter-Centre rivalry, not least because most Centres are funded by the same donors and funding for international agricultural R&D has stalled over the past decade. On balance collaborative forces are likely to outweigh competitive ones in inter-Centre relations. Centre rivalry can be seen as the mirror image of inter-firm collaboration. For-profit organizations that are active in the same industry are primarily competitors, but there are also significant collaborative or collusive tendencies which may interfere with competition. Conversely, inter-Centre relationships are primarily collaborative, but there might also be competitive tendencies that hamper collaboration. The degree of Centre rivalry is difficult to assess, as it is an informal phenomenon.

Several commentators (CGIAR, 1998; CGIAR, 2000b; Byerlee and Fischer, 2001) suggest that collaboration between Centres is in some sense suboptimal. If inter-Centre collaboration (or the search for information in preparation for such collaboration) yields positive net expected benefits for all parties involved in spite of its higher costs, then how could Centres forfeit such opportunities? Here is an example of the relevance of game theory. Inter-Centre versus intra-Centre collaboration could be modelled as simple Prisoner's Dilemma with a suboptimal outcome.[1] The premise of CGIAR (2000b) is that there is a large untapped potential for inter-Centre collaboration. The main logic here is combinatorics: the number of possible combinations of researchers increases exponentially with the size of the relevant pool of researchers. If all researchers in the CG System are part of a single pool, the probability of initiating a highly successful research project may be greatly increased. This reasoning is in line with Pinchot and Pellman's (1999) advocacy of 'intrapreneuring': to maximize possibilities for internal entrepreneurship[2] ('intrapreneurship') in large organizations, employees must be encouraged to reach across divisional boundaries. The Change Design and Management Team instituted by the CG in 2000 recommended enhancing incentives for researchers to reach across Centre boundaries through Global Challenge Programs (GCPs) (CGIAR, 2000b). The GCP approach was adopted at the CG's 2001 Midterm Meeting.

[1] In a Prisoner's Dilemma each player can either cooperate or not cooperate; it is characterized by pay-offs that give each player incentives not to cooperate, even though all would be better off if all cooperated.

[2] Entrepreneurship can be defined as innovative action that creates some structure that yields social benefits. Entrepreneurship thus defined need not be primarily motivated by material pay-offs.

Centres' IP issues relate to many agents, including donors, non-governmental organizations (NGOs), government research agencies, universities conducting agricultural research, international research organizations, other non-profit organizations with missions similar to or relevant to the CGs, life sciences multi-nationals, smaller firms connected to agriculture, as well as farmers' and consumers' organizations.

The donors are mostly government agencies and various private foundations. Groups critical of biotechnology influence the System's relations with the private sector, causing the Centres to be circumspect in their dealings with life sciences firms. Because of the wide variety of donors, beneficiaries, NGOs, and other stakeholders that shape CG structures and policies, and the consensus mode of decision making within the CG, it is difficult to reform the CG radically. Many decisions on contentious IP and funding issues reflect compromise rather than visionary or even properly informed leadership. Furthermore, stakeholders' insistence on full accountability and consultative processes entail the danger of bureaucratic inefficiency and rigidity. Thus, the designers of Global Challenge Programs must avoid the creation of another layer of bureaucracy.

Some relevant funds come from donor initiatives that do not directly conduct R&D, but rather allocate money to collaborative R&D of others. The longest-established of these is the Indo-Swiss Collaboration in Biotechnology (ISCB) initiated in 1974, which 'is not directly related to agricultural issues, but rather focuses on methodology development, capacity building and technology transfer' (Komen, 2000, p. 20). Another important initiative in this category is the Agricultural Biotechnology Support Project (ABSP), launched in 1991 and funded mostly by the US Agency for International Development (USAID).[3] With a single government agency as the dominant donor, such initiatives are more flexible than the CG to pioneer novel funding and IP arrangements.

National Agricultural Research Systems (NARS) in developing countries, consisting of government agencies, universities, and farmers' organizations, are the Centres' primary local partners and are involved in technology transfer to poor farmers. NARS differ widely in their agbiotech capacity (Skerritt, 2001; Trigo *et al.*, 2001).

Some national R&D institutes in the developed world target less developed country (LDC) agriculture. CIRAD (the Centre de Coopération International en Recherche Agronomique pour le Développement) of France is among the biggest agencies of this kind, with an annual budget of about US$200 million. It conducts research on a wide variety of crops in some 50 different countries (Komen, 2000, p. 20; Pardey and Beintema, 2001, footnote 14). In its developing-world orientation and the diversity of its research, CIRAD is comparable to the CG System. CIRAD conducts R&D on a range of major export crops, in contrast to the CG Centres which emphasize food crops grown for subsistence and local markets. Thus, CIRAD is in a different position *vis-à-vis* the private sector. Some developed-world university institutes are devoted to LDC agriculture, such as the Cornell International Institute for Food, Agriculture and Development (CIIFAD). Partnerships with developed-world universities are more common in the CG system than partnerships with multinationals. ISCB and ABSP are each managed by a donor-country university (Ives *et al.*, 1999; Jenny and Schaltegger, 1999).

Some international agricultural research centers are not part of the CG. Examples include the Asian Vegetable Research and Development Center (AVRDC), headquartered in Taiwan, and the International Center of Insect Physiology and Ecology (ICIPE) headquartered in Kenya.[4] The International Center for Genetic Engineering and Biotechnology (ICGEB) has many LDC members, conducts research at campuses in Trieste (Italy) and New Delhi (India), and has human health and plant biology components. ICGEB pursues a fairly aggressive patenting policy that yields substantial licensing revenues. Another relevant international organization is CAB International (CABI), of which 41 countries

---

[3] USAID solicited applications for a follow-on, 5-year, US$15 million ABSP II project in March 2002 that was awarded to a consortium led by Cornell University in late 2002.

[4] See Pardey *et al.* (1991) for a more comprehensive listing.

are members. CABI Bioscience division integrates four former international biological institutes and operates worldwide from six centres, four of which are located in LDCs. CABI is highly active in partnerships.

Private firms play a direct and gradually growing, but still comparatively limited, role in transferring agricultural technologies to LDCs.[5] The potential here is mostly untapped. It is important to differentiate between multinationals and small-to-medium-sized firms (SMFs), some of which conduct cutting-edge research. Multinationals are a dominant force in ag-biotech in terms of research expenditures. However, SMFs play important roles in specific fields. Given their dominance and their strategic possession of important technologies and market development capabilities, multinationals may be more conspicuous as prospective partners. Moreover, multinationals have vastly more direct involvement in LDCs than do SMFs based in the North. Centres thus might overlook SMFs that possess useful IP and might be valuable partners. Several LDCs have SMFs with substantial R&D capacity.

Relevant private non-profits include *inter alia* the International Service for the Acquisition of Agri-biotech Applications (ISAAA) and the Center for the Application of Molecular Biology to International Agriculture (CAMBIA). ISAAA acts as an intermediary in transferring proprietary technologies to LDCs and also seeks to strengthen links between Southern and Northern (often private) research. CAMBIA pursues a market-segmenting strategy. It designs, develops and distributes new research methods and enabling biotechnologies on a preferential basis to LDCs, while recouping royalties from licensing these same technologies to Northern researchers.

The for-profit and non-profit funding potential for medical biotech and genomics is much greater than that of agricultural biotech. Private-sector non-profit organizations with more general biotech- or genomics-related missions may become important players in the CG's realm. The Institute for Genomic Research (TIGR) is a prime example. Established with grant money from the for-profit private sector, TIGR is supported by an increasing number of private grants and contracts as well as several large US federal grants. TIGR is involved in genomics partnerships with CG Centres. For example, it is working with the International Livestock Research Institute (ILRI) to sequence the genomes of selected cattle disease agents. Private-sector non-profit organizations such as ISAAA, CAMBIA and TIGR have much more manoeuvring space in their dealings with for-profit organizations than the CG.

Relation Types

Catalysis

Catalysis involves the creation or transformation of other organizations or of relations among them. Catalysis is an alternative to Centre activities of any kind. The CG and its Centres may not be suitable for certain activities because of their specific institutional constraints. Catalysis is important to the CG System's strategies for at least two fundamental reasons: (i) the System's uniquely central (i.e. connective) status as a set of nodes in the global network of partnerships in agricultural R&D; and (ii) the modest size of the System's budget – 1.5% of public agricultural R&D worldwide (or just 1% of total spending) (Pardey and Beintema, 2001). Its comparative smallness requires the System to leverage other organizations' budgets and activities; its connectedness makes the System highly suitable for this. Catalysis is closely linked to technology positioning; catalysis may be an alternative to Centre R&D. For effective catalysis and technology positioning, CG decision makers need to constantly track others' activities in agricultural R&D.

The formation of a strong local private agricultural sector and viable related industries is an integral part of any successful strategy for poverty alleviation in LDCs. One could thus argue that local private-sector development ought to be a significant component of the

[5] Pardey and Beintema (2001, p. 10) estimate that in the mid-1990s just 6% of the world's private agricultural R&D spending occurred in developing countries.

CG System's strategies. It should immediately be added that where public NARS are strong, this is primarily their responsibility. But where they are weak – as in many African countries – the Centres certainly may have a directly catalytic role to play. In addition, where the International Service for National Agricultural Research (ISNAR), a CG Centre headquartered in The Hague, and other Centres support the development of public NARS, they could catalyse the local public sector's catalytic role.

Traditional agricultural R&D has been a feature of farmers everywhere for millennia. The knowledge and experimental capacity of local farmers has become more recognized in the 1980s and 1990s. As a result, many Centres now have projects in place, often under the guise of 'participatory plant breeding' programmes, that substitute a two-way model of agricultural R&D for the traditional scientist led approach. Similarly, where Centres assist in the birth or growth of locally based (seed) firms, they should be cognizant of local (perhaps latent) R&D capabilities. Thus, Centres may in this way not only directly promote economic development, but also enhance their own R&D network. One further and related argument for Centre involvement in local private sector development is the potential for providing a modicum of counter-balance to the marketing power of multinational firms and their local alliances.

The International Institute for Tropical Agriculture (IITA) has assumed this catalytic role. By undertaking R&D that would have been prohibitively expensive for start-up firms, it 'played an important role in getting the private seed sector started in West and Central Africa' (IITA, 1997). In addition, it provided training and services to small-scale food processing companies (IITA, 2000, p. 52).

Where Centres take on this role, targeted IP strategies may be of assistance or otherwise relevant. Consider a few examples:

- As part of a segmentation strategy in a partnership between a Centre and a multinational, permission to use a proprietary technology owned by a multinational may be transferred to local firms. The corporate partner may perceive this to be in its interest.
- If problems with national claims to genetic resources accumulate, perhaps countries can still be persuaded to send materials to Centres for conservation and R&D if local firms can obtain certain IPR related to the materials.
- In a variety of partnerships, one or more local private- or public-sector partners could hold IP in coordination with one or more Centres.

There exist partnerships that have as their primary purpose catalysis of partnerships, like collaborative R&D. Examples of such meta-partnerships include:

- The Global Forum on Agricultural Research (GFAR), founded in 1996, involves seven stakeholder groups: donors, private sector, NGOs, Advanced Research Institutes (ARIs), farmers' organizations, IARCs, and NARS. It does not itself conduct research but aims to mobilize these groups in their efforts to alleviate poverty, increase food security, and promote sustainable use of natural resources.
- Many information networks involve subsets of these stakeholders, regions, areas of R&D, or crops. Examples include the Asian Rice Biotechnology Network (ARBN), in which the International Rice Research Institute (IRRI), a CG Center, plays a prominent role; the Global Program for *Musa* Improvement (PRO*MUSA*); and the Global Initiative on Late Blight (GILB). The latter two have an explicit meta-partnership function (Frison *et al.*, 1997).
- PRO*MUSA* was initiated in 1996 by the International Network for the Improvement of Banana and Plantain (INIBAP, part of the International Plant Genetic Resources Institute (IPGRI), a CG Centre) and the World Bank. Its working groups – consisting of scientists from a variety of organizations – focus on different themes in the genetic improvement of *Musa*. Each of these groups functions as an information exchange network and fosters collaborative projects.
- GILB was established in 1996. The International Potato Center (CIP), a CG Center, is its convener. GILB is intended to coordinate and enhance R&D efforts to fight late blight disease of potato, which results in annual worldwide losses of about US$3

> billion. Partners in its activities include ARIs, NGOs, NARS, farmers, the for-profit sector, and CIP scientists. 'Bilateral as well as multilateral partnerships among these participants' are 'encouraged and supported' by GILB (Frison *et al.*, 1997, p. 23).

Meta-partnerships can contribute to a favourable atmosphere for sharing information and IP and may foster innovative arrangements, such as clearinghouse mechanisms for IP.

Transfers

Transfers of proprietary research inputs to Centres

Proprietary research inputs come in bundles composed of four elements: (i) codifiable information; (ii) materials embodying the information; (iii) IPR – rights to use and benefit from the information; and (iv) human capital, especially tacit knowledge. Transfers of proprietary technology, be they gifts or exchanges, are complex affairs, and these four elements need to be distinguished carefully. Licences and material transfer agreements (MTAs) may be used for transferring IP and materials, respectively.

Many of the problems relating to Centres' use of research inputs owned by others can be summarized in the question: What must be provided in return? The quid pro quo may be money, or perhaps a restriction beyond the input's immediate use. For example, a research licence restricts use rights to research only. Such restrictive clauses hamper technology transfer. It may be better for the Centre to pay royalties for a licence rather than obtain it free if less restrictive conditions can be obtained (Nottenburg *et al.*, 2002).

According to a 1998 survey, Centres frequently obtained permission through MTAs, licences, or sublicences to use proprietary inputs. However, almost as frequently, permission to use proprietary inputs was absent or unknown (Cohen *et al.*, 1999). The primary risk of unauthorized use is not legal action against Centres, but rather IP owners' reluctance to share their properties with Centres in the future (CGIAR, 1998, p. 6). The outright purchase of IP by Centres is much less common than the use of proprietary inputs. We know of only one such case, the purchase of the rights to a *Bt* gene by a public-sector consortium led by IRRI (Byerlee and Fischer, 2001, p. 13). In apparent contrast to the CG Centres, Latin American NARS do frequently purchase IP rather than license it (Cohen *et al.*, 1999). A variant of this approach would be to contract with the private or public supplier, perhaps through competitive bidding, to *develop* a specific tool, while retaining ownership of the product (Byerlee and Fischer, 2001, p. 10).

Motivations for a firm to 'donate' proprietary inputs include: (i) limited commercial viability of many applications that benefit the poor; (ii) public relations; (iii) connections with a network of non-profit organizations; (iv) the crossing and testing of crops in different environments, generating valuable data for subsequent crop improvement; (v) an opportunity to demonstrate the benefits of the technology; (vi) encouragement of governments to put in place regulations on safety and intellectual property; (vii) strengthening market presence; and (viii) philanthropic motives (CGIAR, 1998, pp. 9–10).

When will companies be *un*willing to license? Companies' considerations here revolve around control of technology:

> [C]ompanies are unwilling to license if it leads to their losing control over the licensed technology. Lack of control may cause technical problems: for example, in the case of Bt maize it could result in the companies being unable to ensure a suitable management regime that would minimize the build-up of insect resistance. It will also cause major commercial problems if the licensed technology is used to compete with the licensor in profitable markets
> (CGIAR, 1998, p. 10)

Because licences and MTAs may contain restrictions, it makes sense for Centres to negotiate terms rather than simply accept a gift. Awareness of the jurisdictional extent of the IP is a prerequisite for such negotiations (Binenbaum *et al.*, 2003).

MTAs and licences may sometimes be used in the same R&D trajectory at different points in time. The MTA may, in effect, be an incomplete contract, to be followed up later by licensing negotiations (Byerlee and Fischer, 2001, p. 10). As is typical of incomplete contracting, the party making the greatest investments before renegotiation has a bargaining disadvantage. This is known as the 'hold-up problem'.

Problems with proprietary inputs can arise in relations with non-profit as well as for-profit organizations. For example, countries may be tempted to stake out claims to 'their' genetic resources. The Convention on Biological Diversity, which entered into force in 1993, provided a framework for such claims. CG Centres signed an agreement with the Food and Agriculture Organization (FAO) in 1994 whereby most of the materials (so called 'designated material') in the Centres' genebanks are held 'in trust' as a common property resource. Under this in-trust agreement, Centres may not seek IPR over designated materials, and are required to ensure that subsequent recipients will not do so either. These provisions aim to reassure nations that materials provided to the CG will not be appropriated by anyone, thus providing them with incentives to keep sharing genetic resources. However, these incentives may not be especially strong, in part because NARS are not party to the agreement. To provide a firmer basis for continued germplasm exchange, many countries were involved in 23 years of negotiations dubbed 'the International Undertaking on Plant Genetic Resources for Agriculture'. In November 2001, a draft International Treaty on Plant Genetic Resources was adopted by 116 nations. Japan and the USA abstained from voting on the treaty although the USA did sign the treaty in late 2002. The treaty, which comes in to force when ratified by 40 countries, 'establishes a multilateral system of access and benefit sharing for 64 crops and plants [including maize, wheat and rice but excluding soybeans, tomatoes, groundnuts and tropical grasses] that are fundamental to food security (FAO, 2002)'. The intent is to ensure the pool of genetic resources encompassed by the treaty will be freely available to plant breeders in countries that adopt the treaty, in exchange for royalties if the seeds are used to develop commercial varieties. Determining these royalties implies keeping track of breeding pedigrees, an issue yet to be resolved in the context of this treaty. Problems might also arise with technologies developed by non-profit organizations. In the USA, the 1980 Bayh-Dole Act mandated that the US Government cede ownership of intellectual property, emanating from government-sponsored research, to the recipient institution. As a result, some universities now hold significant IP portfolios. Negotiating use rights for publicly held intellectual property can be more problematic than for IP held by private firms: public agencies like universities may be hamstrung by regulations or bureaucracies, or royalty sharing arrangements with faculty (Nottenburg *et al.*, 2002).

The party transferring the proprietary input is likely to possess superior information about its cost and potential, and may exploit this information advantage in negotiations. Hence, it is valuable for the Centres to have information about cost, commercial value, and jurisdictional validity of proprietary inputs. It may be cost-effective for such information to be provided through System-wide services such as CAS or the informatics systems related to intellectual property being developed at CAMBIA.

Exchanging materials for data

Contractual provisions often mandate data-for-materials exchange. A standard MTA reads: 'Recipients are requested to furnish [Centre] with data and information collected during evaluations of the material' (SGRP, 2000, p. 12). The International Network for the Genetic Evaluation of Rice (INGER), established in 1977, is an example of a system of transfers of germplasm and information unencumbered by IP. Four Centres and scientists from NARS involved in rice breeding participate in INGER. The requirement – included in MTAs – that recipients supply INGER with relevant varietal performance data on the material distributed for evaluation via the network is crucial in this arrangement, which can be seen as a repeated game. Failure to collect and report data would save costs in the short run, but would eventually lead to exclusion from the network. On the other hand, interactions in a network like INGER may often be based on a vague sense of *quid pro quo* and on a culture of information sharing and cooperating for a common purpose, rather than on explicit costs–benefit calculations. An increase in 'territorial' behaviour in recent years has reduced the scope of INGER.

Funding issues

A fundamental issue in funding is the degree of influence exerted by the funding entity on the recipient's research. For the past decade, the CG's budget has failed to grow while the demands on the System in terms of its commodity coverage, research problem orientation, and accountability, have continued to expand. At the same time, an increasing proportion of the budget consists of restricted (programme and project) funding. In addition to concerns about undue influence, this has also raised the transaction costs of funding (CGIAR, 2000b, pp. 17–18). In the late 1990s, there was a growing awareness in the System that non-traditional sources of funding may have been neglected. This awareness led to Future Harvest, a joint initiative of the Centres, intended to deal with the problems of stagnant overall funding and declining unrestricted support. Future Harvest promotes the CG and its Centres to the public and others, targeting fundraising activities to foundations and companies not traditionally aligned with agriculture or the CGIAR (CGIAR, 2000c).

Another hitherto untapped source of funding is competitive grants:

> [I]t is possible that new sources of finance derived from Ministries of Science could also be accessed by the CGIAR if the funds were internally allocated on a competitive basis. At the moment, the CGIAR is not able to receive these funds because it does not have an internal competitive allocation mechanism
> (CGIAR, 2000b, p. 18)

The Global Challenge Programs are designed to be such an internal competitive allocation. The blueprints for Global Challenge Programs address the inter-Centre competition issue by stipulating that they 'require cooperative research, going beyond individual Center mandates' mechanism (CGIAR, 2001, p. 10).

Some of the Centres' clients reach beyond subsistence or low-income agriculture and may be willing to contribute to Centre R&D. A variation of this scenario is the catalysis of R&D consortia operating outside of the CG as an alternative to in-house R&D. At least two such consortia (FLAR and CLAYUCA, discussed below) demonstrate the potential for engaging agro-industries based in developing countries in research programmes similar to those at CG Centres. A sound IP arrangement may be critical to the success of such consortia.

Privately sponsored public R&D may combine elements of giving with elements of purchasing or collaboration. A prominent issue in sponsored research involves the disposition of rights to any IP arising from the research. For example, an agreement whereby Novartis (now Syngenta) funds plant biotech research at the University of California, Berkeley, grants Novartis the first right to negotiate for IPR to a fraction of the research results equivalent to the share of the budget provided by Novartis. Such provisions strengthen sponsor incentives but tend to spark controversy among the recipient's stakeholders, as has happened in this case. Centres are confronted with similar, potentially controversial, funding opportunities. Such difficulties are not confined to arrangements with the private sector. Increasingly government agencies that sponsor CG research seek a say over resultant IPR. A wide variety of such contractual clauses exist, some stating that IPR should not be sought on results of R&D funded by the donor; others that some might be expected, perhaps with defensive objectives; and still others that IPR be ceded to the donor. Some Centres have paid insufficient attention to such clauses, although recent IP audits have heightened awareness of these matters.

Technology transfer to the developing world

The CG System's technology transfer mechanisms function reasonably well – at least as far as the System's own research products are concerned. Still, the System's technology transfer role could be enhanced by an improved technology information system, including invention disclosures. The transfer of technology developed by others is an even greater challenge. ISAAA and ABSP are examples of other non-profit initiatives that may be more specialized than the CG System in the transfer of technology developed by others. Both benefit from links to US universities, to the for-profit sector (especially ISAAA), and to USAID (in the case of ABSP). CIIFAD (like ISAAA, hosted by Cornell) is an example of a university institute that is active in technology transfer to developing countries. A particularly interesting case concerns transgenic virus-resistant papaya.[6]

The more advanced NARS have the capacity for an office of technology transfer. The Brazilian Agricultural Research Corporation (EMBRAPA), which accounts for about half the country's agricultural R&D spending, already possesses such an office, called the Intellectual Property Secretariat (Maredia *et al.*, 2000). A natural role for the Intermediary Biotechnology Service of ISNAR (IBS) and CAS is to aid an international network of technology transfer offices.

Multinationals' presence in LDCs is mostly due to purchases of and partnerships with local firms rather than direct investment. Monsanto has reportedly purchased 16 local seed companies in Brazil, and most large Indian seed companies have formed alliances with global life sciences companies. Only a few LDC companies have a capacity in biotechnology research, and in nearly all cases, this research is carried out as part of an alliance with one of the global companies (Byerlee and Fischer, 2001).

In the traditional CG model of technology transfer, involving public entities in developing countries, there was no need for commercial development of Centre R&D outputs. This has begun to change. In certain kinds of technology, the for-profit sector often possesses superior development, production and distribution capacity. Examples include vaccines and biopesticides (see the LUBILOSA project below). In crop breeding, traditionally the CG's most important R&D activity, technology transfer via developing-country firms, including a downstream R&D role for those firms, is becoming more common in the Centres. For example, a partnership between Centro Internacional de Agricultura Tropical (CIAT) and Papalotla, a Mexican seed firm, involves the development and commercialization of new hybrid grass varieties of the genus *Brachiaria* for cattle farming. Multiplication of the grass seeds and other downstream activities require investments that are too large for most NARS. Thus, CIAT considered its best option to develop, multiply, and distribute the new grass varieties widely would be to partner with a firm such as Papalotla with established distribution channels. In this arrangement, Papalotla funds R&D through advance payments on future royalties; registers CIAT as the owner of the new grass varieties in relevant countries; and may sub-license the grasses to local firms. Without this IP arrangement – apparently novel in the CG System – this partnership might not have been possible. Moreover, because of Papalotla's exclusive rights, the firm can count on long-term relationships with its farmer customers. This makes it more feasible and attractive for the firm to play a role that includes farmer education, extension, and feedback from farmers (E. Binenbaum, P.G. Pardey and B.D. Wright, 2002, unpublished). Technology transfer in cases like LUBILOSA (see below) and Papalotla is similar to the prevailing pattern (after the Bayh-Dole Act) of university–industry technology transfer in the USA: the non-profit IP owner licenses to a for-profit partner. Here, IP serves as an incentive tool for private-sector development and commercialization of the technology (Mowery *et al.*, 2001; Parker *et al.*,

[6] Starting in 1986, Dennis Gonsalves, a New York State Experiment Station researcher, worked with the University of Hawaii and Upjohn Company to develop transgenic papaya for Hawaii. The new papaya was protected by IP and, in 1998, commercialized, allowing the Hawaiian papaya industry to recover from a papaya ringspot virus infestation. The licensing agreement used for this commercialization restricted the planting of the papaya developed in the programme up to that point to Hawaii only. However, the agreement did not preclude the development, use, or licensing of new transgenic papaya varieties with ringspot virus resistance. Gonsalves 'wanted to start a technology transfer program through which virus-resistant transgenic papaya could be produced for developing countries. CIIFAD was instrumental in getting the technology transfer program developed by giving me a small grant in 1992 to initiate work with personnel from [developing] countries'. Gonsalves worked with the Cornell Research Foundation to resolve IP issues to 'ensure that transgenic papaya developed at Cornell can be used in many countries, and that the benefits of this development can legally reach poor and middle-income farmers as well as more commercial farmers around the world' (Gonsalves, 1999). Relevant IPR are partly in the hands of multinationals (Lesser, 2000, Figure 1). Notably, a project that followed a conventional US domestic-technology-transfer pattern evolved into a programme of technology transfer to developing countries.

2001). To maintain an effective system of technology transfer, the CG System must make market segmentation an integral part of IP arrangements with technology suppliers. Such IP arrangements may be triangular – e.g. a research licence for a Centre plus a licence for NARS distribution of research products (Byerlee and Fischer, 2001, p. 18).

R&D outputs of interest to for-profits

R&D for the poor, a core activity of the CG System, may yield – as by-products – outputs of interest to the for-profit sector. Such outputs include varieties and genetic traits with food and non-food (e.g. fibres, energy) applications; methods for genetic conservation or transformation; methods and genes related to propagation and reproduction (e.g. apomixis); vaccines and (bio-)pesticides; and mechanical equipment. R&D outputs are a subset of valuable Centre assets. Well-organized direct and indirect information about the costs and benefits of these to private-sector players would be highly useful in order to notice opportunities for funding and collaboration, and anticipate strategic interactions. Given the present situation of ad-hoc information provision, a marginal investment in an information system would probably yield a substantial pay-off.

Collaboration

Collaborative R&D with a small number of partners

One strategy to deal with private IP ownership is to conduct collaborative R&D with private partners. This may give them a long-term stake in the relationship, thus encouraging them to share information, materials, and IP. There is much scope for R&D partnerships between Centres and life sciences firms – several of which have been initiated – because they have complementary strengths. Centres provide connections to a global R&D network, to testing sites, to local expertise. Collaborative R&D relations integrate elements of other relation types, including transfers of inputs and financial resources, and cannot properly be discussed without consideration of those categories. Conversely, however, the other categories are conceivable without collaborative R&D. The following three examples of collaborative R&D highlight the importance of IP.

1. Two ABSP projects. An early ABSP collaboration failed owing to insufficient initial specification of IPR: 'This was the case . . . with an ABSP collaboration between CRIFC in Indonesia and ICI Seeds [now AstraZeneca]. Ultimately, an agreement with ICI Seeds . . . for transfer of the Bt genes or maize transformation technology to CRIFC . . . could not be reached due to the lack of patent protection . . . Based on that early experience, ABSP and USAID have taken steps to address IPR concerns up front, and use the resolution of IPR issues as criterion for establishment of such public–private . . . collaborations' (Lewis, 2000, pp. 198–199). In contrast, a similar but later project was successful, partly because of initial specification of IPR segmentation: 'One of the most successful ABSP projects is a joint venture between Pioneer Hi-Bred International Inc., a large private multinational company [now acquired by DuPont], and the Applied Genetic Engineering Research Institute (AGERI), an Egyptian public research institute, to jointly develop Bt maize. In the collaboration, the Egyptian public system gains access to expertise to develop the local strain of Bt (the innovation) and to train its staff. In turn, the private company has access to the new Bt strain for use in markets outside of Egypt' (Byerlee and Fischer, 2001, p. 10).

2. LUBILOSA. The project called Biological Control of Locusts and Grasshoppers (French: LUBILOSA) was initiated by the International Institute of Biological Control – now CABI Bioscience. The idea was to develop a biopesticide for locust control, an attractive proposition for the donor community which spent hundreds of millions of dollars fighting a locust outbreak (1986–1989), using ecologically damaging pesticides. LUBILOSA began in 1989 as collaboration of CABI Bioscience, IITA, and an agency from Niger. LUBILOSA patented a biopesticide, Green Muscle™, in the mid-1990s and licensed it to two commercial partners, BCP in South Africa and NPP in France. These companies had the advantage of advanced technology and capacity for mass production. Independently acquiring such capacity would have required a major capital outlay. The relation with the companies includes a measure of R&D collaboration – as often occurs in technology transfer. For LUBILOSA to be able to patent Green Muscle™ the donors – development agencies – had to agree to transfer IPR to LUBILOSA. When selecting partners, LUBILOSA differentiates between SMFs and multinationals. Multinationals have the advantages of selling power, market penetration, expertise, and financial and logistic resources, but the disadvantages of inertia, time taken to establish markets, time taken to gain approval for partnerships, insistence on exclusivity, and lack of interest in small markets. SMFs have the advantages of knowledge of niche markets, flexibility and speed of action, a hunger for new products and opportunities, and a willingness to accept non-exclusivity agreements. Among these aspects, flexibility/speed and non-exclusivity are particularly important to LUBILOSA. It thus prefers SMFs and started a series of non-exclusive agreements with them, beginning with BCP and NPP (Dent, 2000).

3. CIMMYT Apomixis Partnership. In 1998 and 1999, CIMMYT entered into a number of R&D agreements with firms. Among these is an agreement with the French non-profit Institute de Recherche pour le Développement (IRD) and the firms Novartis, Limagrain, and Pioneer Hi-Bred, for the duration of 5 years. CIMMYT's objective is the 'development, evaluation, and distribution of apomictic hybrid maize to subsistence farmers'. CIMMYT receives 'access to scientific expertise and proprietary technologies; a paid-up, royalty free, worldwide, non-exclusive license (with the right to sub-license to non-profit institutions) to provide research products to subsistence farmers', as well as 'financial support for CIMMYT scientists involved'. In return, CIMMYT provides 'staff and laboratory resources; access to CIMMYT's and IRD's apomixis technology; [and] a paid-up, royalty-free, worldwide, co-exclusive [among the partners] license for research products'. Jointly, funds from the agreements amount to 3% of CIMMYT's budget (CIMMYT, 2000, p. 70).

Private-sector activity in crop breeding and seed sales has traditionally been based on the fact that hybrid seeds cannot be profitably replanted and thus must be bought annually. If apomixis – a form of asexual reproduction – is engineered into hybrid crops, the seeds only need to be bought once. Apomixis may therefore be of great strategic importance to seed companies. Anticipating this, CIMMYT obtained a maize apomixis patent. Note that apomixis is applicable to many crop species. A number of apomixis-related patent applications have been filed by universities, United States Department of Agriculture (USDA), CAMBIA, and private firms (Bicknell and Bicknell, 1999). The CIMMYT apomixis agreement illustrates the use of segmentation strategies, essential to enable both sides to pursue their objectives.

Based on these cases, we can identify several success factors in collaborative R&D between Centres and for-profit organizations. First, the CG System and for-profit firms tend to have complementary assets. If Global Challenge Programs function as envisioned, they will make Centres more attractive partners by exploiting untapped synergies between them. Second, IP provisions must be included in the initial agreement and should minimize ambiguity. Third, segmentation strategies are essential. Ideally, provisions should allow for preferential access both for 'subsistence' farmers in developing countries and for developing country farmers producing for domestic markets. Finally, in addition to contractual provisions, mutual trust, and a spirit of partnership extending beyond formal obligations are both crucial success factors in collaborative R&D (David Hoisington, personal communication; Mubashir, 1999). These elements, which are largely endogenous to the collaboration's design, highlight the value of personal interaction in research collaboration.

R&D consortia

Consortium can be defined as a partnership involving a number of member organizations. This is vague – as the number is unspecified – but useful analytically. Transaction costs in partner-specific arrangements rise rapidly with the number of partners. Consequently, consortia need to be based on more formulaic and less partner-specific arrangements. Consortia are a solution to the incentive problem of provision of public goods (defined as goods with the technical properties of non-rivalry and non-excludability) whose benefits are concentrated among the partners. R&D consortia can only function with clear IP arrangements, and may include IP clearinghouse mechanisms. A consortium may become institutionalized, i.e. an organization in its own right. In fact, both the CGIAR and its Centres can be viewed as consortia. There is a fluid line between R&D consortia and information exchange networks. When information exchange networks are institutionalized and assigned permanent and full-time staff, they become more like R&D consortia. The CG and its Centres are in a unique position to play a catalytic role in consortia. For example, the IPGRI developed 'the concept of regional, thematic and crop genetic resources networks. Today, over 150 countries participate in 50 such networks' (IPGRI, 1999, p. 7). As suggested by this quote, consortia and information exchange networks may be organized around (i) commodities, (ii) zones or regions, or (iii) technology themes. Examples include:

1. Commodity-based Consortia. The Latin American Fund for Irrigated Rice Research (Spanish acronym FLAR) is an example of a consortium involving farmers' organizations and CG Centres. In 2000, organizations from 11 countries, as well as CIAT and IRRI, contributed a total of US$625,000 to FLAR. The FLAR arrangement was inspired by the Australian Research and Development Corporations (RDCs). The RDCs are industry-specific and funded by check-offs collected from producers and supplemented by government matching grants (Alston *et al.*, 1999). FLAR was established in 1995 following a budget shortfall at CIAT and a shift in its priorities away from irrigated rice research. Thus, FLAR is an attempt to continue a valuable breeding programme through a different institutional setup – a public–private hybrid construction. In FLAR, each member country is represented by exactly one organization. Member contributions to FLAR are a function of domestic rice output. Transfers of FLAR products from members to multinationals and other third parties are subject to central approval. Member organizations are responsible for domestic protection of FLAR varieties. To provide incentives for members to share breeding materials, members are entitled to royalties to domestically registered FLAR varieties. Thus, FLAR's IP rules provide incentives for member support and the sharing of materials, in part by reassuring member organizations that leakage of FLAR products to third parties will be avoided (E. Binenbaum, P.G. Pardey and L.R. Sanint, 2002, unpublished). A recent external review of CIAT praises the CIAT/FLAR arrangement as 'the future for commodity improvement programs throughout the world' (CGIAR, 2000a, p. 33). A similar public–private consortium for cassava, the Latin American and Caribbean Consortium to Support Cassava Research and Development (Spanish acronym CLAYUCA) was established in 1999. Cassava and rice are both CIAT mandate crops, and CLAYUCA is also hosted by CIAT.

2. Biotechnology Consortia. Public-sector organizations active in biotechnology may often benefit from joining forces in consortia. Because of similarities in technological needs, such consortia are often regional. Examples include the Asian Rice Biotechnology Network, the Asian Maize Biotechnology Network and the Latin American Biotechnology Network. In order to handle IP optimally, such consortia need legal status and a central unit with IP expertise (Byerlee and Fischer, 2001, p. 17). R&D consortia and networks may help non-profit organizations to access IP. For example, 'Partly because of the Cassava Biotechnology Network, started by CIAT, some 50 laboratories – some third of them in developing countries – now use molecular markers, gene cloning, genetic transformation and disease diagnostics in cassava research' (CGIAR, 2000a, p. 28).

3. Genomics Consortia. Genomic data greatly help breeding. Consortia are the natural arrangement for genomics research. The leading agricultural genomics consortium is the International Rice Genome Sequencing Project (IRGSP), which is an international public-sector effort. A number of firms, including Monsanto, Syngenta, and Myriad Genetics are also active in rice genomics. All of these have indicated a willingness to share some information for non-profit purposes. Fischer *et al.* (2000) propose private-sector involvement in IRGSP. Part of this proposal is to use an MTA for transfer of genetic resources that, among other provisions: (i) allows patenting of resulting discoveries; (ii) requires licensing of such patents 'at a reasonable royalty for application in commercial markets of the developing world, and at zero royalty for application in non-commercial subsistence farming'; and (iii) requires free licensing of such patents for research purposes. Recently, a new banana genomics consortium was announced involving INIBAP, TIGR, and over 20 other institutes (Gewolb, 2001).

Competitive and adversarial relations

A widely accepted principle is that Centres should avoid activities for which market incentives are sufficient. Thus, on the face of it, competitive or otherwise adversarial relations with the for-profit sector would appear to be an irrelevant topic. None the less, such relations may occasionally be in accord with the CG mission.

The orphan crop argument

According to the orphan crop argument, Centres should focus on crops that are important to the poor but lack commercial appeal. Obviously this has IP implications: without commercial interests, IP trouble is less likely. The orphan crop argument can be qualified in a number of ways, all of which have to do with competition with private-sector products. First, while non-hybrid varieties may be of inferior quality to hybrid varieties of the same crop, they may nevertheless compete with hybrids, forcing firms to lower the price of hybrid seed. Thus, sophisticated multinational technology may become affordable to farmers (Barton and Berger, 2001). Second, the public sector may have a role in developing products that lack negative externalities – for example, environmental effects of pesticide use associated with pesticide-resistant varieties. Third, genomics and new applications of orphan crops could render them of interest to for-profit organizations, making them valuable Centre assets in relations with for-profit organizations. Fourth, the best opportunities for poor farmers may reside in commercial (not subsistence) activities that are new to them. But such commercial activities are of interest to private research firms as well.

Infringing others' rights

IPR infringement by Centres is most likely to occur inadvertently. Infringement is often far from obvious. Many patent claims are unresolved. The main danger of infringement may be a reduced willingness by IP owners to cooperate with Centres, rather than anti-Centre litigation which is unattractive from a public relations perspective. Infringement can only be prevented through investment in IP expertise and IP information systems.

'Inventing around'

Centres wishing to avoid infringement can opt for the strategy of inventing around the IP. This is especially relevant when, among a cluster of complementary IP items, some are much more difficult to obtain than others. Such 'gaps' in a 'toolkit' are prime targets to be designed around. For example, 'public research organizations in China and India appear to be have developed their own transformation protocols for Bt cotton although it is likely that they have used, legally or otherwise, some proprietary tools as well' (Byerlee and Fischer, 2001, p. 8, citing Pray). The product of inventing around may well be valuable – since it is a substitute for the original input, which was worth obtaining IP protection for. The new product is thus a potential competitor to the original input, and may serve as a 'bargaining chip' in various relations. This adds to the attractiveness of inventing around.

Stimulating local industry

Stimulation of local industries – e.g. seed or agbiotech firms – that compete with multinationals, thus reducing the latter's market power, may benefit farmers and consumers. The Centres have a catalytic role in this regard.

Contesting IPR

IP strategy is not complete without serious consideration – necessarily supported by costly expertise and information systems – of the option of contesting other people's IP: 'It can happen that a patent is granted which ought not to have been . . . Where it is clear that this has happened, it is wrong to collude in the patentee's unjust claims' (CGIAR, 1998, p. 12). CIAT may be the only Centre that has initiated litigation so far. It is contesting one patent, and has considered contesting another, on the grounds of prior art. Both concern varieties held in trust by CIAT, which considers itself partly responsible for keeping in-trust varieties in the public domain and wishes to set a precedent to discourage biopiracy. Once an IP information system is in place in the CG System, there may be more cases of Centre-initiated litigation. It may often be possible to find allies to pay for part of the litigation costs.

Relational Decomposition of IP Strategy

All of the relation types discussed in the previous section should be taken into account in the design of the Centres' IP strategy. This is true even when we confine our attention to a single decision problem. For example, suppose that a Centre needs a proprietary input for its R&D programme. What are the Centre's options? Byerlee and Fischer (2001) and Wright (2000) identify options. What follows is a somewhat expanded list of options, each of which (with one exception) involves one or several relation types.

1. The Centre may negotiate with the input's owner for a licence.
2. The Centre may unilaterally access the technology.
3. The Centre may contest the IPR in court or at the IP-granting agency.
4. The Centre may attempt to invent around the technology. This may result in a valuable asset in exchange or collaborative relations. Note that even when any of the potentially adversarial moves (the second, third or fourth options) are not carried out, they may still play a role as implied or explicit threats (Lesser, 2000, pp. 612–613).
5. The Centre may initiate collaborative research with the input's owner; use rights to the input may be included in the partnership.
6. The Centre may initiate a consortium and participate in it. The consortium may include other parties that would like to access the proprietary input, and may either focus on this particular input or have some broader theme.
7. The Centre may entirely abandon the R&D programme if the input is both critical and inaccessible.
8. The Centre may abandon the R&D programme, but catalyse other – non-profit or for-profit – organizations better able to deal with the input problem to undertake the R&D programme instead.
9. Specific funding opportunities might be available in combination with some of the aforementioned options. For example, perhaps the home government of the input's owner might help subsidize use of the input.
10. The need for proprietary inputs renders it more attractive for the Centre to seek IP on its own R&D products. Such a strategy would enhance many of the aforementioned options.

The method of relational decomposition is relevant for many choice problems in IP strategy, such as the question whether or not to seek an IPR product (E. Binenbaum, 2002, unpublished). The various IP choice problems are closely linked to each other; they are best integrated in a single strategic vision. For example, seeking IP protection is linked to accessing other people's IP. Game theory is helpful in solving IP choice problems. In the example given above, each option initiates a game. Lesser's (2000) sketch of adversarial and cooperative approaches to IP problems in agbiotech clearly has clear connections to game theory.

Thinking through IP strategies would be greatly aided by an appropriate information system that furnishes readily accessible information on relevant agents, their IP, and their inter-organizational relations. For example, when a Centre selects a partner for an R&D collaboration, it might already have multinational A in mind. Without a proper information system, the Centre might go ahead and accept A's terms for a partnership. With a proper information system:

- the Centre could check whether other partners could be selected instead of or in addition to A;
- it could examine the IP holdings of different partners;
- it could access reports on other partnerships of those prospective partners, drawing cautionary and innovative lessons from such information;
- it could negotiate a better deal perhaps increasing its R&D budget while enhancing its freedom to operate;

- it could strengthen its position by inclusion of other partners from the public sector, thus being able to insist more strongly on the interests of the poor;
- it would know its own potential bargaining chips as well as bargaining chips of other public institutions that might join the partnership, and so on.

It is only when managers are fully aware of the issues and the integrative approach discussed in this chapter, that they will be able to appreciate the value of an information system that offers these advantages.

Acknowledgements

The authors received valuable insights from conversations with Dave Hoisington, Bonwoo Koo, Zaida Lentini and Luis Sanint. This study was conducted as part of the research on intellectual property rights led by the International Food Policy Research Institute with financial support from the Swedish International Development Agency and the European Union, Additional support was provided by the Giannini Foundation of Agricultural Economics, and the Department of Agricultural and Resource Economics and the Graduate Division of the University of California, Berkeley.

References

Alston, J.M., Harris, M.S., Mullen, J.D. and Pardey, P.G. (1999) Agricultural R&D policy in Australia. In: Alston, J.M., Pardey, P.G. and Smith, V.H. (eds) *Paying for Agricultural Productivity*. Johns Hopkins University Press, Baltimore, Maryland, pp. 118–171.

Barton, J.H. and Berger, P. (2001) Patenting agriculture. *Issues in Science and Technology Online*. http://www.nap.edu/issues/17.4/p_barton.htm

Bicknell, R.A. and Bicknell, K.B. (1999) Who will benefit from apomixis? *Biotechnology and Development Monitor* 37, 12–21.

Binenbaum, E., Nottenburg, C., Pardey, P.G., Wright, B.D. and Zambrano, P. (2003) South North trade, intellectual property jurisdictions, and freedom to operate in agricultural research on staple crops. *Economic Development and Cultural Change* 51(2), 309–335.

Byerlee, D. and Fischer, K. (2001) Accessing modern science: policy and institutional options for agricultural biotechnology in developing countries. *IP Strategy Today* 1. http://www.biodevelopments.org/ip/

CGIAR (1998) *Report of the CGIAR Panel on Proprietary Science and Technology*. Technical Advisory Committee of the CGIAR, Rome.

CGIAR (2000a) *Fifth External Programme and Management Review of CIAT*. CGIAR Secretariat, Washington, DC.

CGIAR (2000b) *Report of the Synthesis Group*. CGIAR Secretariat, Washington, DC.

CGIAR (2000c) *Toward a Longer-Range Financing Strategy for the CGIAR*. CGIAR Secretariat, Washington, DC.

CGIAR (2001) *Charting the CGIAR's Future – Change Design and Management; Designing and Managing Change in the CGIAR*. Report to the Mid-Term Meeting 2001. CGIAR Secretariat.

CIMMYT (2000) *CIMMYT Annual Report 1999–2000; Science and Sustenance*. CIMMYT, Mexico, DF.

Cohen, J.I., Falconi, C., Komen, J., Salazar, S. and Blakeney, M. (1999) Managing proprietary science and institutional inventories for biotechnology. In: Cohen, J.I. (ed.) *Managing Agricultural Biotechnology: Addressing Research Needs and Policy Implications*. CAB International, Wallingford, UK, pp. 249–260.

Dent, D. (2000) *Intellectual Property, Partnerships, and the LUBILOSA Story*. CAB International, Wallingford, UK.

FAO (2002) United States Signs International Treaty on Plant Genetic Resources for Food and Agriculture. FAO, Rome. Accessed December 19. http://www.fao.org/english/newsroom/news/2002/10660-en.html

Fischer, K.S., Barton, J.H., Khush, G.S., Leung, H. and Cantrell, R. (2000) Collaborations in Rice. *Science* 290, 279–280.

Frison, E.A., Collins, W. and Sharrock, S. (1997) *Global Programs: a New Vision in Agricultural Research*. CGIAR Secretariat, Washington, DC.

Gewolb, J. (2001) Genome research: DNA sequencers to go bananas? *Science* 293, 585–586.

Gonsalves, D. (1999) Technology transfer program shares transgenic papaya. In: *Cornell International Institute for Food, Agriculture, and Development Annual Report 1998–1999*. CIIFAD, Ithaca, New York, p. 97.

IITA (1997) *International Institute for Tropical Agriculture 1967–1997: in the Vanguard of Sub-Saharan Sustainable Development*. IITA, Ibadan, Nigeria.

IITA (2000) *International Institute for Tropical Agriculture Annual Report 1999*. IITA, Ibadan, Nigeria.

IPGRI (1999) *Diversity for Development: the New Strategy for the International Plant Genetic Resources Institute*. IPGRI, Rome.

Ives, C.L., Maredia, K.M. and Erbisch, F.H. (1999) International collaboration: intellectual property management and partner-country perspectives. In: Cohen, J.I. (ed.) *Managing Agricultural Biotechnology: Addressing Research Needs and Policy Implications*. CAB International, Wallingford, UK, pp. 261–271.

Jenny, K. and Schaltegger, E. (1999) Indo-Swiss collaboration in biotechnology: lessons learned and future strategies. In: Cohen, J.I. (ed.) *Managing Agricultural Biotechnology: Addressing Research Needs and Policy Implications*. CAB International, Wallingford, UK, pp. 140–151.

Komen, J. (2000) International initiatives in agri-food biotechnology. In: Tzotzos, G.T. and Skryabin, K.G. (eds) *Biotechnology in the Developing World and in Countries in Economic Transition*. CAB International, Wallingford, UK, pp. 15–32.

Lesser, W.H. (2000) '"Holding up" the public agbiotech research sector over component technologies'. In: *Transitions in Agbiotech: Economics of Strategy and Policy; Proceedings of NE-165 Conference, June 24–25, Washington D.C.* Food Marketing Policy Center, Department of Agricultural and Resource Economics, University of Connecticut, Storrs, Connecticut and Department of Resource Economics, Amherst, Massachusetts.

Lewis, J. (2000) Leveraging partnerships between the public and the private sector: experience of USAID's Agricultural Biotechnology Programs. In: Persley, G.J. and Lantin, M.M. (eds) *Agricultural Biotechnology and the Poor: Proceedings of an International Conference, Washington D.C. October 21–22, 1999*. CGIAR, Washington, DC, pp. 196–199.

Maredia, K.M., Erbisch, F.D. and Sampaio, M.J. (2000) Technology transfer offices for developing countries. *Biotechnology and Development Monitor* 43, 15–18.

Mowery, D.C., Nelson, R.R., Sampat, B.N. and Ziedonis, A.A. (2001) The growth of patenting and licensing by U.S. universities: an assessment of the effects of the Bayh-Dole Act of 1980. *Research Policy* 30, 99–119.

Mubashir, E. (1999) Introducing transgenic crops in India: a joint venture approach. In: Cohen, J.I. (ed.) *Managing Agricultural Biotechnology: Addressing Research Needs and Policy Implications*. CAB International, Wallingford, UK, pp. 279–285.

Nottenburg, C., Pardey, P.G. and Wright, B.D. (2002) Accessing other people's technology for non-profit research. *Australian Journal of Agricultural and Resource Economics* 48, 389–416.

Pardey, P.G. and Beintema, N.M. (2001) *Slow Magic: Agricultural R&D a Century After Mendel*. IFPRI Food Policy Report. International Food Policy Research Institute, Washington, DC.

Pardey, P.G., Roseboom, J. and Anderson, J.R. (1991) *Agricultural Research Policy: International Quantitative Perspectives*. Cambridge University Press, Cambridge, UK.

Parker, D., Castillo, F. and Zilberman, D. (2001) Public–private sector linkages in research and development: the case of U.S. agriculture. *American Journal of Agricultural Economics* 83, 736–741.

Pinchot, G. and Pellman, R. (1999) *Intrapreneuring in Action: a Handbook for Business Innovation*. Berrett-Koehler, San Francisco, California.

SGRP (2000) *Annual Report 1999 of the CGIAR System-Wide Genetic Resources Programme*. International Plant Genetic Resources Institute, Rome, Italy.

Skerritt, J. (2001) Biotechnology policies for Asia: current activities and future options. In: Pardey, P.G. (ed.) *The Future of Food: Biotechnology Markets and Policies in an International Setting*. International Food Policy Research Institute, Washington, DC.

Trigo, E., Traxler, G., Pray, C. and Echeverria, R. (2001) Agricultural Biotechnology in Latin America and the Caribbean'. In: Pardey, P.G. (ed.) *The Future of Food: Biotechnology Markets and Policies in an International Setting*. International Food Policy Research Institute, Washington, DC.

Wright, B.D. (2000) International crop breeding in a world of proprietary technology. In: Santaniello, V., Evenson, R.E., Zilberman, D. and Carlson, G.A. (eds) *Agriculture and Intellectual Property Rights; Economic, Institutional and Implementation Issues in Biotechnology*. CAB International, Wallingford, UK, pp. 127–138.

13 R&D Incentives for GM Seeds: Restricted Monopoly, Non-market Effects, and Regulation

Robert D. Weaver
Department of Agricultural Economics, Pennsylvania State University, Pennsylvania, USA

Abstract

Conditions that motivate use of patent-based incentives for R&D are derived. These require what is defined as 'universal dominance' of a technology and provide a basis for monopoly pricing of access to the innovation. When alternative technologies exist or when non-market outputs are jointly produced, the universal dominance condition may not be met, resulting in a restricted monopoly where pricing of innovation is restricted to a range defined by incentives associated with competing technologies. Under these conditions, the return to the innovation is substantially reduced, rendered stochastic, and conditional on pricing of competitive technologies. Alternative pricing approaches are considered including uniform pricing, and contract-based pricing. The implications of regulatory standards and labelling are considered.

Introduction

The traditional model for financing R&D returns is grounded on a series of conditions that characterize the process, the economic characteristics of the innovation, and the demand for the innovation. Where these conditions do not exist it is important to reconsider the utility of this traditional model as well as to consider alternatives. This chapter reconsiders the implications of the patent-based R&D incentive model for innovations, related to genetic modifications, in the agricultural and food industries. The traditional model focuses on innovations that result in changes in private goods. Typically, these changes include increased productivity in quantity or enhanced quality attributes that are observable. In both cases, the effect of the innovation is appropriable and exhaustively consumed by consumption of the private good quantity. Where innovation impacts quality and the private good can be quality-differentiated, the innovation-related attributes can be priced by markets. The implicit price of the innovation-related quality attribute provides a return to the innovation. This chapter clarifies that only in the case where the innovation provides what is defined as a 'universally dominant' technology will a patent grant establish an opportunity for the innovator to price the innovation at a monopoly level. The chapter notes that the condition of universal dominance follows from economic determinants that render the innovation an optimal choice across all producers. Based on this universal dominance, the return to the innovation is appropriable by the innovator and its monopoly pricing provides a basis for financing the R&D.

 The Regulation of Agricultural Biotechnology (eds R.E. Evenson and V. Santaniello)

This traditional model may fail in the case of genetically modified (GM) seed innovations targeted for agricultural and food industries. This failure, as well as the feasibility of alternative solutions, follows from the unique characteristics of GM seed innovations. The potential of GM innovations to offer improved productivity and performance in agricultural and food systems has been widely cited. However, whether this potential is harvested will depend upon the existence of incentives that can finance the necessary R&D.

This chapter first reconsiders the universal dominance of GM seeds. Consistent with the heterogeneity of the net benefits associated with many GM-related innovations, the concept of *local dominance* of a technology is introduced. Based on a two-step, sequential decision model, the acceptance or adoption of these technologies will depend on an assessment that compares the value of the innovation technology against that of other alternative technologies. This assessment depends on perceived and actual profitability, fixed costs of transition, uncertainty of outcomes, as well as on preferences that may be held with respect to alternative technologies. The chapter describes these sequential decisions using a general conceptualization of relative willingness-to-pay that varies across technologies, crops and farm characteristics. Heterogeneity of relative willingness-to-pay for an innovation implies that the monopoly grant of a patent may offer only a *restricted monopoly* to set a price for the innovation (seed) within a limited range that is discontinuous in, though dependent on prices and other determinants of the value of alternative crops and technologies. In the absence of universal dominance, heterogeneity across producers implies that patent-based uniform pricing of a GM-related innovation leaves substantial benefits with producers, preventing full appropriation by the innovator. Vertical integration and contracting are shown to provide effective mechanisms for increased appropriation by the innovator. Next, the implications of labelling or segregation of the crop output of the GM-related innovation are considered to establish implications for the optimal seed price for a restricted monopolist.

A second feature of GM-related innovations is that they are known to result in potentially substantial changes that are not private good associated. These non-market quality effects include environmental performance data or knowledge, local or more generalized impacts (e.g. improved pest conditions, pest resistance, or environmental effects), and benefits that accrue at different stages of the supply chain (e.g. product homogeneity). In this case, such products may be quasi-public in nature implying they will not be priced in markets. The returns to the innovation will not be appropriable by either producer or innovator through a patent mechanism. The potential of alternative pricing mechanisms for valuation of non-market benefits is considered and a role for regulation is motivated.

Characteristics of GM-related Innovations

GM crop varieties that offer pesticide-resistance (e.g. herbicide tolerant varieties) or pest control activity (e.g. *Bt* maize or cotton) constitute a complex of changes in traditional technology. These changes go beyond single factor augmentation of a private good technology. Instead, GM-related innovations have been cited as offering a series of changes in production practices, allowing substantial changes in input mixes, changes in crop output yield, changes in risk exposures, as well as other non-market effects. Changes in practice include shifts to no-till planting, pre-emergence herbicide use, change in the type of active ingredients used, and perhaps substantial changes in environmental impacts of crop practices such as reduction in the average number of active ingredients used per hectare while increasing the average weight applied per hectare. Further, by providing a change in the local plant growth environment as well as the plant and plant product characteristics, GM-related innovations offer a means of augmenting the quality flows associated with the crop. Importantly, GM-related innovations offer changes in quality attributes that can be viewed as quantity-related private good outputs, e.g. changes in grain shell hardness, crop product uniformity, crop product attributes (sugar, oil, protein, or other functional content). However, GM-related innovations may also offer changes in non-market quality flows from the production process, e.g. environmental or ecological.

Consider the evidence for GM soybeans. Carpenter and Gianessi (1999) reviewed shifts in practices citing the role of GM soybeans as a natural extension of an evolution toward increased use of post-emergence herbicides, simplification of weed control programmes, and improved effectiveness of active ingredient applications; see e.g. Pike *et al.* (1991). Importantly, this shift in practice had substantial implications for tillage practices that had focused on field preparation, and post-emergence tillage. Given post-emergence herbicides, adoption of conservation tillage was facilitated; leading to over 50% adoption by 1998, see Kapusta and Krausz (1993). This shift was further extended by introduction of herbicide tolerant soybeans that allow post-emergence, broad-spectrum herbicide application at nearly any stage of plant growth. Second, improved post-emergence herbicides have led to reduction in row spacing and cultivation, improved weed control due to canopy closure, increased land area yield, and reduction of crop damage associated with herbicides (e.g. stunting, delayed canopy closure), see Padgette *et al.* (1996) and increased effectiveness of weed kill (Rawlinson and Martin, 1998, unpublished).

Following introduction of GM soybeans, the mix of both pre- and post-emergence herbicides has changed dramatically, see USDA/NASS (1991–1999). From an economic perspective, Roundup Ready® soybeans have offered significant cost reductions for many producers (Rawlinson and Martin, 1998, unpublished) due in part to rapid dissipation of Roundup. However, price reductions in conventional materials provide a means of defending market share and reducing adoption of Roundup Ready® soybeans. Yield impact of GM alternatives has also been debated since yields depend on weed control as well as the extent of adaptation of conventional varieties. In sum, these studies have found a complex set of effects of GM soybeans on production practices, input use, and costs.

A key characteristic of GM innovations is that their effects apparently vary across farm conditions. Bullock and Nitsi (2000) provide a review of the physical science evidence and found that the potential of GMs varied with extent of pest exposure and the type of pest control practices used. In the case of *Bt* innovations, benefits are clearly conditional on the extent and nature of insect exposure. Similarly, herbicide-tolerant seed GM innovations will pay off differentially depending on field conditions, planting history, and climatic conditions as well as on current machinery, managerial expertise, and local market conditions. Within this context, GM innovations can be distinguished from simple crop yield enhancing innovations for which the pay-off is universal.

External implications of these GM innovations have been evaluated and debated. For GM soybeans, reduced tillage has been credited with increased crop residue, reduced fuel, labour, and machine time, reduced wind and water erosion, see ASA (2001). Carpenter (2001) and Fernandez-Cornejo and McBride (2000) note that herbicide resistant soybeans require less herbicide application. Concern for possible negative effects has in the past led to calls for segregation of products by technology-of-origin.

To examine the incentives for R&D that support GM-related innovations, this chapter focuses on the pricing of GM seeds. Suppose the innovator is the supplier of the GM seed. Private returns to a seed-based innovation would follow directly from profits from the sale of the seed:

$$\pi_s^i \equiv \omega_c^i \sum_{j=1}^{J} s_c^i - c_s^i(s_c^i)$$

where J is the target market population of producers of the cth crop using the ith technology and a quantity of seed s_c^i sold at a uniform price ω_c^i, produced by the supplier at cost $c_c^i(s_c^i)$. Seed sales follow from farm level demand derived from operation of the ith production technology for the cth crop yielding output y_{jc}^i per land area, on the jth farm. Suppose that while a common technology is available across farms, productivity of each available technology is conditioned by farm specific quasi-fixed and fixed input flows represented by a vector, $\boldsymbol{\theta}_j$. Suppose crop output is also conditioned by a stochastic event (shock), ε_j, generated by a density function $g(\varepsilon_j|0,1)$, and by a vector of inputs, $\mathbf{x}_c^i$, that are associated with the ith technology. A technology specific production function reflects unique technological attributes such as planting flexibility, or management intensity: $y_{jc}^i = y_c^i(\mathbf{x}_c^i, \boldsymbol{\theta}_j, \varepsilon_j)$ and producer profit per land area for crop c produced with technology i as: $\pi^i_{jc} \equiv p^i_c y^i_{jc} - r^i_c \boldsymbol{x}^i_{jc} - \omega^i_c \delta^i_{jc}$ where p^i_c is the output price that is allowed to be technology differentiated, $\boldsymbol{r}^i_c$ is the input price vector, δ^i_{jc} is a farm-specific effective seeding rate per land area, and ω^i_c is the price paid for ith seed type for crop c. This simple characterization is sufficient for our purposes, though in reality, productivity is conditioned by exogenous damage processes, see e.g. Carpentier and Weaver (1997). Importantly, profits will be uncertain and dependent on the outcome of the random event.

From this notation, it is clear that farm level selection of technology to operate and demand for associated seed will be depend on farm-specific characteristics including the seeding rate, quasi-fixed and fixed input flows, and risk preferences. Further, it is useful to define seed demand to result from a sequence of decisions. Suppose selection of the optimal technology to operate for each crop that might be grown is made to maximize expected utility of profits, $EU(\pi^i_{jc} \mid \gamma_j)$ where γ^i is the operator's risk aversion. The result is an expected value function $V^i_{jc}(\boldsymbol{\rho}^i_{jc}) \equiv V^i_{jc}(p^i_c, \boldsymbol{r}^i_c, \omega^i_c, \delta^i_{jc}, \boldsymbol{\theta}_j, \gamma_j)$ where $\boldsymbol{\rho}^i_{jc}$ is introduced as a farm specific vector of determinants or 'incentives'. For each alternative technology i', for crop c, the producer's *relative willingness-to-pay* (RWTP) for technology i versus technology i' is simply $\omega^{ii'}_{jc} \equiv V^i_{jc}(\boldsymbol{\rho}^i_{jc}) - V^{i'}_{jc}(\boldsymbol{\rho}^{i'}_{jc})$. It follows that RWTP is farm specific and technology i can be defined as *locally dominant* when $\omega^{ii'}_{jc} > 0$ for all other technologies, i'. Given the role of farm specific determinants in the incentives vectors $[\boldsymbol{\rho}^i_{jc}\ \boldsymbol{\rho}^{i'}_{jc}]$, it follows that technologies such as GM seeds may not be locally dominant across all farms.

Consider the case of GM seeds. For GM innovations that introduce pest action, e.g. *Bt* maize, the expected value of alternative technologies will depend on the operator's view of the probability of pest presence and intensity, i.e. the stochastic event in our notation. Where pest exposure and damage potential is unlikely, the producer's RWTP for GM versus conventional technologies for maize may be negative. For other producers, where exposure and damage potential is high, this RWTP might be positive. Similar sensitivity to local determinants would exist for herbicide resistance GM-innovations. Depending on local 'incentives', the technology will be dominant on some farms and not on others. The conclusion can be drawn that for GM-related technologies, heterogeneity across farm specific 'incentives' may result in an equilibrium condition where these technologies are not *universally dominant*, i.e. they are not optimal for all producers. It follows that 'adoption' of this type of technology could never be expected to fully 'saturate' a population of producers. It is clear from the definition of RWTP that local dominance will depend on the 'incentives' for both the GM technologies, as well as those for alternative technologies. This suggests that competing technology suppliers may be able to influence the extent and distribution of local dominance across farms by strategically influencing elements of the 'incentives' vector associated with their competing technology. Pricing as well as innovation strategies by competing technology suppliers would, in this case, become an important factor in the final demand for a GM-related technology.

In contrast to this story for GM-related technologies, the traditional rationale for patenting as a means of creating incentives for R&D views the innovation as resulting in a technology that is universally dominant over a wide range of incentives for existing technologies. That is, the innovation will be universally adopted, if not instantly, over a discrete time period as barriers to its 'diffusion' such as market imperfections and bottlenecks in distribution are eliminated. Further, this traditional rationale supposes that the universal dominance of the innovation is robust over a wide range of 'incentives' for alternative technologies, i.e. $\omega^{ii'}_{jc} > 0$ over a wide range of $\boldsymbol{\rho}^{i'}_{jc}$. Under such conditions, alternative technology suppliers are not able to price their 'old' technologies strategically, i.e. control $\boldsymbol{\rho}^{i'}_{jc}$, to deter adoption of a new technology.

This result holds even more strongly when the choice of crop is considered. That is, suppose the optimal technologies for each crop have been selected, the farm must next choose which crops to produce. Again, this can be viewed as following from a consideration of relative willingness-to-pay, in this case for crop c relative to crop c' where each is produced by its respective optimal technology, (noted as the ith technology). That is, crop c will be selected if $\omega^i_{jcc'} \equiv V^i_{jc}(\boldsymbol{\rho}^i_{jc}) - V^i_{jc'}(\boldsymbol{\rho}^i_{jc'}) > 0$ for all alternative crops, c'. It follows that in order for a crop-based technology to be universally dominant, as supposed by patent-based theories of R&D incentives, a technology i for crop c would be locally dominant for a given 'incentive' vector $\boldsymbol{\rho}_{jc}^{i}$ across a wide range of values for incentive vectors ($\boldsymbol{\rho}_{jc}^{i'}$ and $\boldsymbol{\rho}_{jc'}^{i}$) associated with both alternative technologies and alternative crops. This is highlighted by the result that the demand for seed is defined as: $s_{jc}^{i} = s_{jc}^{i}(\boldsymbol{\rho}_{jc}^{i}) \equiv a_{jc}^{i}(p_{c}^{i}, \mathbf{r}_{c}^{i}, \omega_{c}^{i}, \delta_{jc}^{i}, \boldsymbol{\theta}_{j}, \gamma_{j})\delta_{jc}^{i}$ conditional on $\omega_{jc}^{ii'} > 0$ for all other technologies i' and $\omega_{jcc'}^{i} > 0$ for all other crops. Thus, while demand for the seed is continuously related to the 'incentive' vector $\boldsymbol{\rho}_{jc}^{i}$, that relationship is conditional on the incentive vectors for all other crops ($\boldsymbol{\rho}_{jc'}^{i}$) and those for all other technologies, $\boldsymbol{\rho}_{jc}^{i'}$. Only when the crop and technology are universally dominant is the seed demand independent of incentives associated with competing technologies and crops, a condition that supports monopoly power for the innovator to price the seed for technology i for crop c. When such universal dominance is not supported, monopolist pricing of GM-related innovations will not be feasible and incentives for GM-related R&D will be reduced by competition across competing technologies and crops.

Framework for GM Innovation Pricing

Restricted monopoly pricing

The implications of this theory of the value of innovation are best viewed graphically. In Fig. 13.1, relative willingness-to-pay functions for alternative crops $\omega_{jcc'}^{i}$, and for alternative technologies for crop c, $\omega_{jc}^{ii'}$, are plotted for given incentive vectors ($\boldsymbol{\rho}_{jc}^{i}$, $\boldsymbol{\rho}_{jc}^{i'}$, $\boldsymbol{\rho}_{jc'}^{i}$), against the seed price for the ith technology and cth crop, ω_{c}^{i}. The slope of this curve with respect to ω_{c}^{i}. reflects the demand for the ith technology and cth crop. By definition, it follows that for given incentive vectors, the locally dominant technology for crop c is defined in Fig. 13.1 as that drawn by the dashed-line curve if $\omega_{jc}^{ii'} < \omega_{jc}^{ii''}$ for all $i'' \neq i'$. In this sense, the RWTP curve $\omega_{jc}^{ii'}$ represents a frontier defining RWTP for technology i relative to the next best technology i'' that envelops all other RWTP functions for technologies, i''. By analogy, the RWTP of a locally dominant crop for each technology can be defined as the frontier $\omega_{jcc'}^{i}$ shown by the solid curve. Suppose that over a wide range of values for the incentive vectors, $\omega_{jc}^{ii'}$ remained positive. In this case, the innovator could behave as an *unrestricted monopolist*, setting the seed price, ω_{c}^{i}, to achieve maximum profits.

Figure 13.1 clarifies that only a *restricted monopoly* is associated with technologies that are not universally dominant. That is, for the given local incentive vectors ($\boldsymbol{\rho}_{jc}^{i}$, $\boldsymbol{\rho}_{jc}^{i'}$, $\boldsymbol{\rho}_{jc'}^{i}$), it is clear that for any seed price $\omega_{c}^{i} > \underline{\omega}_{jc}^{ii'}$, the ith technology will not be selected for use. Above this price, the i' technology for crop c would be selected as locally dominant on the jth farm. At even higher seed prices, $\omega_{c}^{i} > \underline{\omega}_{jcc'}^{i}$, even the cth crop is no longer preferred relative to c' when each is produced with its respective optimal technology. In the presence of economically feasible alternative technologies and crops, this clarifies that a patent grant transfers to the grantee what is at best is a restricted right to price the innovation. In Fig. 13.1, $\underline{\omega}_{jc}^{ii''}$ defines the upper limit of monopoly pricing power for the jth farm, a threshold price. When the incentive vectors are exogenous and not controlled by individual agents in the economy, a patent will imply the innovator holds a right to *restricted monopoly pricing*. As is clear from Fig. 13.1, the range of feasible monopoly pricing varies across farms as the elements of that farm's incentive vector varies, i.e.

$$\underline{\omega}_{jc}^{ii'} = g(\boldsymbol{\rho}_{jc}^{i} \mid \boldsymbol{\rho}_{jc}^{i}) \text{ where}$$
$$\boldsymbol{\rho}_{jc}^{i} \equiv (p_{c}^{i}, \mathbf{r}_{c}^{i}, \omega_{c}^{i}, \delta_{jc}^{i}, \boldsymbol{\theta}_{j}, \gamma_{j})$$

An important result is that when the seed supplier/innovator is a profit maximizer, the optimal seed price will be $\underline{\omega}_{jc}^{ii'}$. When $\underline{\omega}_{jc}^{ii'}$ is insensitive to farm specific elements of this 'incentives' vector, a common seed price will exist implying that optimal restricted monopoly price will be uniform across farms. The premium established for the innovation over the next best technology is defined as $\tau_{c}^{ii'} = \underline{\omega}_{jc}^{ii'} - \omega_{c}^{i'}$ and is positive whenever $\omega_{c}^{i} < \underline{\omega}_{jc}^{ii'}$. When $\underline{\omega}_{jc}^{ii'}$ is sensitive to the farm specific elements of the incentives vector, the optimal restricted monopoly price and its premium over competing technology seed prices will vary across farms.

When $\underline{\omega}_{jc}^{ii'}$ is sensitive to farm specific elements of this 'incentives' vector, the level set for a uniform seed price will determine the extent of 'market penetration' and explain the often observed 'incomplete' adoption of innovations,

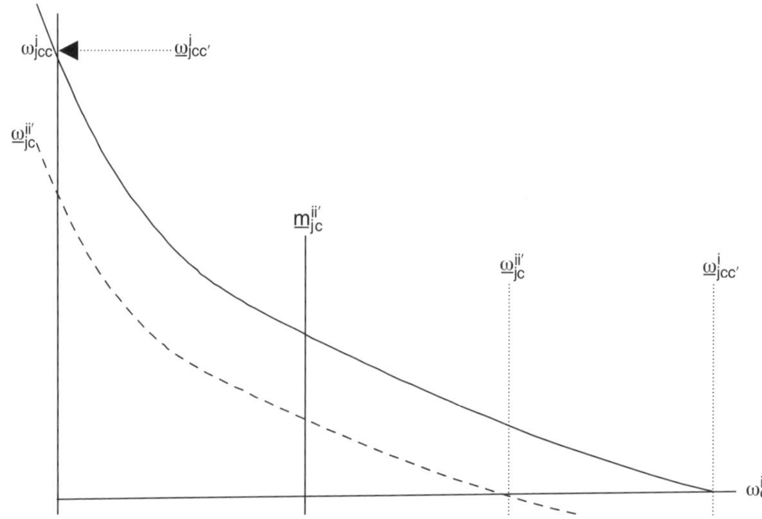

Fig. 13.1. Restricted monopoly pricing of innovation.

see Fig. 13.1. However, contrary to theories of diffusion based on constraints and market imperfections, incomplete adoption is rational and optimal even when information is costless and instantaneous. At any uniform seed price level, some farms will find the innovation to be dominant and buy the associated seed, others will find the uniform price to be greater than their threshold price, $\underline{\omega}^{ii'}_{jc}$. For a population of producers, $j \in J$, the maximum seed price for crop c produced by technology i that will ensure complete penetration (complete adoption) by each producer in the set J can be defined as the minimum value of $\underline{\omega}^{ii'}_{jc}$ across all producers in the set J. Calling that minimum for any J, $\underline{m}^{ii'}_{jc}$, many producers would be willing to pay more and would earn a net benefit equal to $\omega^{ii'}_{jc}(\underline{m}^{ii'}_{jc})$, see Fig. 13.1. Thus, under uniform pricing, the innovator transfers benefits to producers that would be willing to pay more. Further, these producers are able to appropriate benefits of the innovation that would be claimed by the innovator under unrestricted monopoly pricing. The extent of this appropriation will depend on the heterogeneity across producer incentive vectors and the sensitivity of producer RWTP functions to seed prices. Where heterogeneity is substantial, the loss of returns to the innovation would provide a strong incentive for the innovator to find an alternative mechanism for distribution of the seed.

Figure 13.1 also serves to illustrate the implications of changes in the elements of the incentive vectors. For example, if the crop, seed, or input prices change, this upper limit seed price $\underline{\omega}^{ii'}_{jc}$ will change. Consider a decrease in the crop price p^i_c. This would shift the relative willingness-to-pay functions in Fig. 13.1 towards the origin reducing the restricted monopoly price $\underline{\omega}^{ii'}_{jc}$. However, a similar shift would also result from an increase in the crop price $p^i_{c'}$ or $p^{i'}_c$. By the same argument, changes in any of the input prices $(\boldsymbol{r}^i_c, \boldsymbol{r}^i_{c'}, \boldsymbol{r}^{i'}_c)$ would result in shifts in the relative willingness-to-pay functions. Because the technology and crop choice are not continuous in these incentives, such changes will change these choices only when they change the sign of a relative willingness-to-pay function. As another example, given that the effective seeding rate, δ^i_{jc}, reflects the efficacy of the ith technology, as efficacy diminishes over time, e.g. because of pest resistance, the relative willingness-to-pay curve $\omega^i_{jcc'}$ would shift to the southwest implying that the maximum seed price, $\underline{\omega}^i_{jcc'}$, would be reduced.

Restricted strategic pricing

When the incentive vectors that determine the maximum seed prices $\underline{\omega}^{ii'}_{jc}$ are exogenous (not controllable) by other agents in the economy, the innovator can use restricted monopoly pricing. However, when suppliers of elements of competing technologies can react strategically to the innovator's pricing policy, the maximum seed prices $\underline{\omega}_{jc}^{ii'}$ will be conditional on prices set by competing technologies. Consider the case where other seed or input suppliers have market power that allows them to set the prices of their products. In this case, $\boldsymbol{\rho}_{jc}^{i}$ and $\boldsymbol{\rho}_{jc}^{i'}$ become endogenous to competitor policy, set by agents with competing crops, technologies, or inputs. Depending on the relative market power across agents, a multiple agent game would be defined. In such a game, players would be setting their own product prices subject to actual and anticipated restrictions imposed by other player's prices as represented by the maximum product prices defined in much the same way as the maximum seed price in Fig. 13.1.

Alternatives to monopoly-based uniform pricing

When uniform pricing leaves substantial benefits appropriated by producers using the innovation, integration offers a means for the innovator to internalize such benefits. Consider the case where the targeted market of J producers can be partitioned into two groups. Suppose the producers fall into one of two subsets: L, composed of low productivity, inefficient producers; and H, composed of high productivity, efficient producers. When setting the restricted monopoly price at $\omega_c^i = \overline{\omega}_{Lc}^{ii'}$ would result in substantial loss of appropriable benefits from the high productivity group, acquisition of that group would provide a means of internalizing these benefits.

Another mechanism is contracting. In this case, the extent of the producer appropriation is manageable through the use of a non-linear seed pricing rule that sets producer specific prices that transfer benefits that would be lost under uniform pricing back to the innovator. This approach could also be directed toward specific types of sharing of the benefits of innovation across the innovator and producers. This possibility is feasible through the use of contracts between the innovator/seed supplier and farms using the innovation. Through a contract, the distribution of the benefits of the innovation can be established depending on the bargaining power of the innovator relative to the firms using the innovation. Intuitively, a contracting approach establishes a price schedule such that each user firm pays a different price. By setting the price equal to $\underline{\omega}_{jc}^{ii'}$ the innovator appropriates the benefits of the innovation.

Contract-based seed pricing

Two cases are of interest with respect to pricing through a contract mechanism. Ignoring risk aversion for the moment, the cases are differentiated by the nature of information available to the innovator. In the first case, suppose the innovator has full knowledge (i.e. information) with respect to each user firm's characteristics (type) as represented by the quasi-fixed factor $\boldsymbol{\theta}_j$. In the second case, there is asymmetry in this information such that the innovator/seed firm only has knowledge of the distribution $f(\boldsymbol{\theta}_j)$. Suppose that the target market of J producers is predetermined.

In the first case, under full information, suppose the patent grant provides bargaining power to the innovator such that a set of user specific seed prices can be determined. This problem can be equivalently stated as a series of bilateral problems where the innovator chooses $\omega_{cj}^i \ \forall j \in J$ such that:

$$\omega_{cj}^i = \arg\max \pi_s^i \equiv \omega_{cj}^i s_c^i(p_c^i, \mathbf{r}_c^i, \omega_{cj}^i, \delta_{jc}^i, \boldsymbol{\theta}_j)\delta_c^i - c_s^i(s_c^i)$$

s.t.

$$(\mathrm{IR}_1)\ V_c^i(p_c^i, \mathbf{r}_c^i, \omega_c^i, \delta_{jc}^i, \boldsymbol{\theta}_j) \geq V_c^{i'}(p_c^{i'}, \mathbf{r}_c^{i'}, \omega_c^{i'}, \delta_{jc}^{i'}, \boldsymbol{\theta}_j)$$

$$(\mathrm{IR}_2)\ V_c^i(p_c^i, \mathbf{r}_c^i, \omega_c^i, \delta_{jc}^i, \boldsymbol{\theta}_j) \geq V_{c'}^i(p_{c'}^i, \mathbf{r}_{c'}^i, \omega_{c'}^i, \delta_{jc'}^i, \boldsymbol{\theta}_j)$$

The constraints (individual rationality, IR) define the conditions that must be met if technology i and if crop c are to be locally dominant given the existing values for the 'incentive' vectors. Where these incentive vectors are exogenous to other agents in the economy, the optimal seed price under contracting will be conditional on prices associated with all technologies and crops. With minor (though notationally complex) extension, the case of asymmetric information can be considered. In that problem, the innovator would choose a schedule of seed prices that varies with producer type, i.e. $\omega_c^i(\boldsymbol{\theta}_j)$ $\forall\ j \in J$ while optimizing the integral of profits taken over the distribution of producer type. The innovator's choice of the price schedule in this case is constrained by individual rationality constraints that ensure the price schedule results in the targeted producers adopting the seed technology, as well as incentive compatibility constraints that ensure the producers reveal their true type. Given the limited scope of this chapter, no further consideration is given to this case.

Regulation and restricted monopoly seed pricing

A key issue of concern with respect to GM-related innovations is that these innovations result in crop outputs that have quality attributes that are non-market. That is, where quality attributes are easily observable or verifiable at low cost, where such attributes are valued by consumers or users of the products or where they imply differential costs to their producers, quantity flows of the GM crop outputs could be quality attribute differentiated, e.g. from non-GM crop output, and markets would price these attributes. However, suppose there exist quality attributes that are not easily observable or *ex post* verifiable. In such a case, markets would not differentiate the quality of quantity flows of the crop output and would not price the attribute. Where the attributes are directly related to the technology of origin, one approach would be to establish traceability, perhaps through labelling and identity preservation. In the current notation, the crop product is labelled by technology of origin, i.e. the quantity flow from crop c and technology i is noted y_c^i with a quality specific price, p_c^i. Within the notation presented, it is clear that the premium or discount for the GM-related crop output will affect the dominance of the crop and technology, and therefore, the optimal restricted monopoly or contract-based price of seed. Define the non-GM-related crop output price as $p_c^{i'}$ and define the 'technology differential' $\upsilon_c^{ii'} = p_c^i - p_c^{i'}$. From Fig. 13.1, the extent of this differential will shift the relative willingness-to-pay functions to the left for positive differentials and to the right for negative differentials. When labelling is mandated by a regulation, arbitrage of land across the alternative technologies for crop c would result in $\upsilon_c^{ii'} = (p_c^i - p_c^{i'}) = AC_c^{ii'} + l_c$ where $AC_c^{ii'}$ is the differential in the average cost of producing the crop and l_c is the average cost of labelling and segregating the crop output.

In a more general sense, GM-related innovations have been cited as changing the external effects of crop production, e.g. the environmental effects. In the case of herbicide tolerance crops, quantity of chemical application may be reduced, tillage practice may shift toward no-till, and fuel and machinery use is reduced. Define the external, quasi-public goods flows associated with crop production by a vector, q_c^i. Where these quality flows are not priced in markets, an innovation that alters these flows will not result in a change in producer willingness-to-pay for the innovation. While an increase in this quality flow may be of value to society or subsets of agents, if it is not priced in the market, the producer's choice of dominant technologies and crops will not be conditioned by it. It follows that the producer's value function and seed demand associated with the selected technology and crop will be independent of this external output flow. Thus, the innovator is left with no appropriable return to an innovation's effect on this quality flow and related changes in the social welfare. When this quality flow is quasi-public in nature, though measurable, a regulatory standard could be imposed. In this case, the regulation would constitute a constraint, i.e. $q_{jc}^i \geq q_c$, imposed on the producer's choices and affecting the producer's value function. It follows that the regulation would result in producers considering the contribution of alternative technologies to achieving the regulatory constraint.

Conclusions

This chapter presents an evaluation of market incentives for R&D through patent-based approaches. The context of GM seeds involves an innovation that is not universally dominant across producers. Instead, it is locally dominant conditional on incentives for other technologies (e.g. traditional varieties and chemical regimes) as well as local field conditions. The implications of this reality are considered within framework of sequential choice of dominant technologies followed by choice of crop. The chapter demonstrates that patent-based approaches provide what is labelled as a 'restricted monopoly' where only limited monopoly pricing power exists. The range of pricing power is defined and shown to depend on the incentives and other determinants of value of alternative techniques and crops. The use of a uniform price for seeds is shown to result in appropriation of benefits of the innovation by farmers. Non-linear pricing based on contracting and vertical integration of highly productive farmers are shown to provide mechanisms that allow for increased appropriation by the innovator. Within this context, local dominance of an innovation is shown to be conditional on the incentives associated with alternative technologies and crops implying a motivation for the innovator to integrate elements of alternative technologies to allow optimal pricing of both the innovation and the alternative technology incentives (e.g. traditional seed and chemicals). While GM seed innovations offer proprietary benefits that are consistent with traditional patenting, they also offer a series of non-market effects the appropriation of which are not feasible under traditional patenting: data, environmental effects, production characteristics (improved weed, pest control, and resistance effects), reduced chemical use and tillage and associated environmental benefits. Regulatory standards are shown to incorporate these types of quality flows in the choice of technologies by producers providing a mechanism for valuation of the contribution of innovations to this type of non-market quality flow.

References

[ASA] American Soybean Association (2001) *ASA Study Confirms Environmental Benefits of Biotech Soybean*.

Bullock, D.S. and Nitsi, E.I. (2001) Roundup Ready soybean technology and farm production costs: measuring the incentive to adopt. *American Behavioral Scientist* 44, 1283–1301.

Carpenter, J.E. (2001) *Case Studies in Benefits and Risks of Agricultural Biotechnology: Roundup Ready Soybeans and Bt Field Corn*. National Center for Food and Agricultural Policy, Washington, DC.

Carpenter, J. and Gianessi, L. (1999) Herbicide tolerant soybeans: why growers are adopting Roundup Ready varieties. *AgBioForum* 2(2), 65–72.

Carpentier, A. and Weaver, R.D (1997) Damage control productivity: why econometrics matters. *American Journal of Agricultural Economics* 78, 47–61.

Fernandez-Cornejo, J. and McBride, W.D. (2000) *Genetically Engineered Crops for Pest Management in U.S. Agriculture: Farm-Level Effects*. Agricultural Economics Report No. 786. Economic Research Service, USDA.

Kapusta, G. and Krausz, R.F. (1993) Weed control and yield are equal in conventional, reduced-, and no-tillage soybean (*Glycine max*) after 11 years. *Weed Technology* 7(2), 443–451.

Padgette, S.R., Re, D.B., Barry, G.F., Eichholtz, D.E., Delannay, X., Fuchs, R.L., Kishore, G.M. and Fraley, R.T. (1996) New weed control opportunities: Development of soybeans with a Roundup Ready gene. In: Duke, S.O. (ed.) *Herbicide-Resistant Crops*. CRC Press, Boca Raton, Florida, pp. 53–84.

Pike, D.R., McGlamery, M.D. and Knake E.L. (1991). A case study of herbicide use. *Weed Technology* 5(3), 639–646.

[USDA] United States Department of Agriculture/ [NASS] National Agricultural Statistics Service (1991–1999) *Agricultural Chemical Usage: Field Crops Summary*. USDA/NASS, Washington, DC.

14 Agricultural Biotech R&D Structure: Cyclical or Not?

James F. Oehmke,[1] Christopher A. Wolf,[1] Kellie C. Raper[1] and Anwar Naseem[2]

[1]*Department of Agricultural Economics, Michigan State University, East Lansing, Michigan, USA;* [2]*Department of Agriculture, Food and Resource Economics, Rutgers State University of New Jersey, New Brunswick, New Jersey, USA*

Abstract

Since 1995, the most noticed feature of the agricultural biotech industry structure is the high level of concentration, spurred by mergers and acquisitions (M&A) activity. This empirical observation has spurred various conceptual discussions of why the industry is evolving the way it is and what is causing the M&A activity. Following a cyclical model developed by Oehmke, Wolf and Raper, we bring to bear empirical evidence on these questions. Results indicate that the cyclical model is consistent with the data on industry evolution as characterized by firm numbers, concentration, and the level of inventive activity. The model is inaccurate with respect to predictions about M&A activity.

Introduction

Since 1995, the most noticed feature of the agricultural biotech industry structure is the high level of concentration, spurred by mergers and acquisitions (M&A) activity. In an industry with only four firms generating meaningful revenue from transgenic plant varieties, there were at least 26 major acquisitions or alliances in 1999 alone, some involving multiple firms (James, 2000).

The seminal literature on innovation and industry evolution (Scherer, 1967; Kamien and Schwartz, 1976) modelled innovations as if they were exogenous, as if they were available on a regular basis to the highest bidder. Vickers (1986) generated persistent asymmetries in a duopoly market structure using this type of framework, but has no underlying explanation of the level of innovative activity. Loury (1979), Lee and Wilde (1980) and Dasgupta and Stiglitz (1980) introduced stochastic innovation probabilities, as functions of the level of R&D, but focused on a single innovation. Reinganum (1985) extended this aspect to sequences of stochastic R&D races to formalize the idea that the current market leader could be displaced by an innovating firm, which then became the market leader. Although the market structure in Reinganum is stable (the leader may change over time, but there is a single leader at any instant), this work paved the way for models of cyclical market structure. For example, the sequential R&D race literature evolved into quality improvement (ladders) models (Segerstrom *et al.*, 1990; Grossman and Helpman, 1991; Aghion and Howitt, 1992), which form the basis of Segerstrom's (1991) model of innovative and imitative R&D races and resulting industry monopoly–duopoly–monopoly cycles. Other models of cyclic behaviour include those driven by waves of major or revolutionary technical innovation followed by minor innovation (Cheng and Dinopoulos, 1992; Iosso, 1993). Folster and Trofimov (1997) generate cycles as out-of-steady-state behaviour in a model with multiple equilibria.

The literature on the evolution of the agricultural biotechnology industry is more focused on M&A activity than overall industry evolution, with a variety of explanations. For example, explanations of agricultural biotechnology M&A activity usually focus on explaining the recent flurry of M&A activity. Popular explanations include M&As as means of avoiding high transactions costs of strategic alliances between firms, and or combining complementary assets in a more efficient manner. The standard implication is that concentration will increase over time and firm numbers will decline (Artuso, 1999; Graff *et al.*, 1999; Johnson and Melkonian, 1999; Rausser *et al.*, 1999). Graphically, these models suggest that the biotechnology industry evolution, as represented by the number of firms, would follow a unimodal pattern over time, which is consistent with the dominant pattern of industry evolution (Dosi, 1988). In contrast to this approach, a separate branch of the literature views the recent flurry as part of a recurrent pattern of behaviour (Kalaitzandonakes and Hayenga, 1999; Peña *et al.*, 1999). Oehmke *et al.* (2002) have modelled firm numbers in transgenic plant R&D – particularly soybeans – as exhibiting a cyclical pattern.

This chapter collects and describes general patterns in the data on transgenic plant R&D that bear on the empirically testable hypotheses of Oehmke *et al.* (2002). Given the relative newness of transgenic plant R&D, it is not possible definitively to characterize the evolution of firm numbers as exhibiting cyclical behaviour or not having cycles. Thus, a different testing strategy is required. Oehmke, Wolf and Raper (2002) present a set of corollary hypotheses, which can be examined with currently available data. In particular, we examine the number of firms, the level of innovative activity, industry concentration, and the level of M&A activity, for consistency with or falsification of the Oehmke *et al.* (2002) model.

We proceed by recapping the testable hypotheses to be examined. We then present currently available US data bearing on these hypotheses, for maize, cotton and soybean transgenic plant R&D (these crops represent essentially all the US commercial transgenic crop area and 95% of the global transgenic crop area (James, 2002)). Graphical exposition allows for examination of the testable hypotheses; correlation analysis provides statistical corroboration for the visual findings.

Application of the Model of Cyclical Behaviour in Transgenic Plant R&D

Testable hypotheses

Oehmke *et al.* (2002) present a neo-Schumpeterian model of investment in R&D races and apply it to the transgenic plant industry. The key assumptions of the model are that R&D is an uncertain venture, that greater R&D investment increases the probability of successfully innovating, and that the successful innovator earns a degree of monopoly profits. To drive the cyclical behaviour, Oehmke *et al.* assume that consumer recalcitrance toward biotechnology is an unanticipated, negative shock to expected monopoly profits.

The cyclical model behaviour is captured in a set of variable co-movements (Table 14.1). Of particular interest here are the following:

1. The number of firms moves in a cyclical pattern.
2. The level of inventive activity moves co-cyclically.
3. R&D industry concentration moves counter-cyclically.
4. M&A activity moves counter-cyclically.

Data

We focus on US maize, cotton and soybean transgenic R&D. As a measure of the level of inventive activity, we use the number of transgenic field trials. By law, the field trial represents an ‘environmental release’ of the transgenic crop, and as such is regulated by the USDA’s

Table 14.1. Empirically testable hypotheses generated by the model.

Economic phenomenon	Behavioural prediction
Number of firms	Moves in a cyclical pattern
Innovative activity	Moves pro-cyclically with number of firms
Time to discovery	Moves counter-cyclically to number of firms
R&D industry concentration	Moves counter-cyclically to number of firms
M&A activity	Moves counter-cyclically to number of firms
Duration of cyclic 'peaks' (periods of high firm numbers)	Short relative to periods of low firm numbers

Animal and Plant Health Inspection Service (APHIS). A publicly available data set provides a complete listing of all transgenic field trials conducted. The data contain information on the organism on which the trial is to be conducted, the phenotype, the institution conducting the trial, the date at which the application was approved or rejected, and the status of the application. Attention is restricted to applications which were 'issued' or 'acknowledged', i.e. where the field trial was allowed to go forward.

These data have both advantages and disadvantages. The major disadvantage is that they provide information on R&D activity at only one stage of the research-to-commercialization continuum, and so are not a comprehensive summary of R&D portfolios. The major advantages are that since all field trials are required to be registered, these data provide a complete census of R&D activity at the field trial stage; and that the information on the nature of the trial is sufficiently detailed so that good descriptions of the nature of the R&D activity and the industry structure affecting that activity are easily described. These data are used to construct time series representing the evolution of firm numbers over time, the level of inventive activity over time, and the concentration ratio. All 2002 data are preliminary.

The level of inventive activity for commodity c at time t is defined to be the number of field trials on that commodity at that time. The date of approval is used to assign a year to the field trial. The number firms in the industry at time t is defined to be the number of firms conducting at least one transgenic field on commodity c in year t.

Since transgenic field trials were conducted for a number of years before transgenic products were commercialized, a product-sales based concentration measure is uninformative. Following Brennan *et al.* (1999), we construct a CR4-like concentration ratio based on field trials. Specifically, the concentration for commodity c at time t is defined to be the number of field trials conducted on commodity c at time t by the four most active firms (in that commodity at that time), divided by the total number of field trials conducted on commodity c at time t. This measure lies in the interval, 0,100, in percentage terms, and has the same properties as the traditional, sales-based CR4 measure.

The M&A events we chose for this study are for years 1988 to 2001 and were screened from a database of financial actions, published by the Institute of Biotechnology Information (IBI, 1999), as well as exhaustively searching the Lexis–Nexis database for press releases and news reports about M&A activity. For the purposes of this study, an M&A event occurs anytime a company purchases 51% or more equity stake in another company or simply merges with the company. Licensing, research collaborations and joint ventures between two companies are not included. The M&A variable counts only those M&As in which both companies involved in the transaction also had field trials for the same crop in any period. This construction is selected to correspond to the model description of M&As occurring *within* the industry. The year in which the M&A was approved by regulatory and/or shareholders is assigned as the year in which M&A occurred.

Data Analysis

Maize

The evolution of transgenic maize R&D activity (number of field trials), firm numbers, concentration ratio, and M&A activity are presented in Fig. 14.1. As noted above, the time series is too short to examine directly for evidence of cycles,

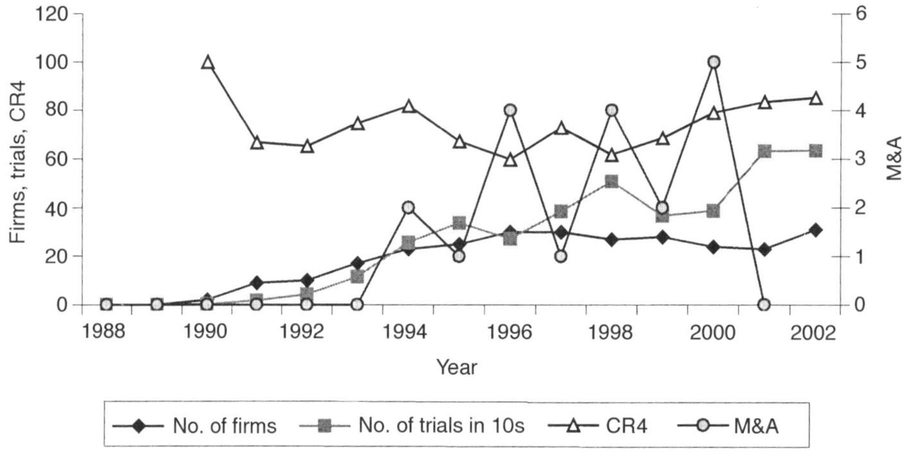

Fig. 14.1. Number of firms conducting transgenic maize trials, number of trials approved, CR4 concentration ratio, and number of mergers and acquisitions (M&A), 1988–2002. (Source: Constructed from APHIS, IBI data.)

so we search for co- or counter-movements of the type described by the model. To the eye, there appears to be a co-movement of the level of R&D activity with the number of firms, particularly in the earlier years. This is consistent with the model hypothesis. The M&A variable appears to bounce around with little direction. The concentration ratio does not exhibit much fluctuation, and appears to be relatively flat. The latter two variables appear to be inconsistent with the model hypotheses.

Correlation analysis (Table 14.2) shows that firm numbers and field trials are positively and statistically significantly (at the 1% level) correlated, with a parital correlation coefficient of 0.84. This is consistent with the model prediction. The CR4 ratio is negatively correlated with firm numbers, which is consistent with the model, but the coefficient is not statistically significant. The correlation between CR4 and field trials is practically and statistically zero.

The M&A variable is positively correlated with firm numbers (significant at the 5% level) and number of trials (significant at the 10% level) with coefficients of 0.65 and 0.52. The M&A findings are inconsistent with the theory.

Table 14.2. Pairwise correlations among number of firms conducting transgenic maize trials, number of trials approved, CR4 concentration ratio, and number of mergers and acquisitions (M&A).

Correlation (*P*-value)	No. firms	No. trials	CR4	No. M&A
No. firms	1			
No. trials	0.8441 (0.0001)	1		
CR4	–0.3411 (0.2541)	0.0214 (0.9447)	1	
No. M&A	0.6453 (0.0127)	0.5202 (0.0565)	–0.3200 (0.3105)	1

Cotton

The evolution of transgenic cotton R&D activity (number of field trials), firm numbers, concentration ratio, and M&A activity are presented in Fig. 14.2. The number of field trials and the number of firms appear to move together quite closely. The number of firms peaks in 1994; the number of field trials reaches an initial peak in 1995. Each series declines from the peak, and then begin to climb again in 1999, with the number of field trials reaching a new high in 2002. These co-movements are consistent with the model.

M&A activity in cotton is minimal. The little activity that does occur happens between 1997 and 2000 (with an additional event in 1991). This corresponds to a period of relatively low (sub-peak) firm numbers, which is consistent with the model. However, the data are too sparse to draw any conclusions.

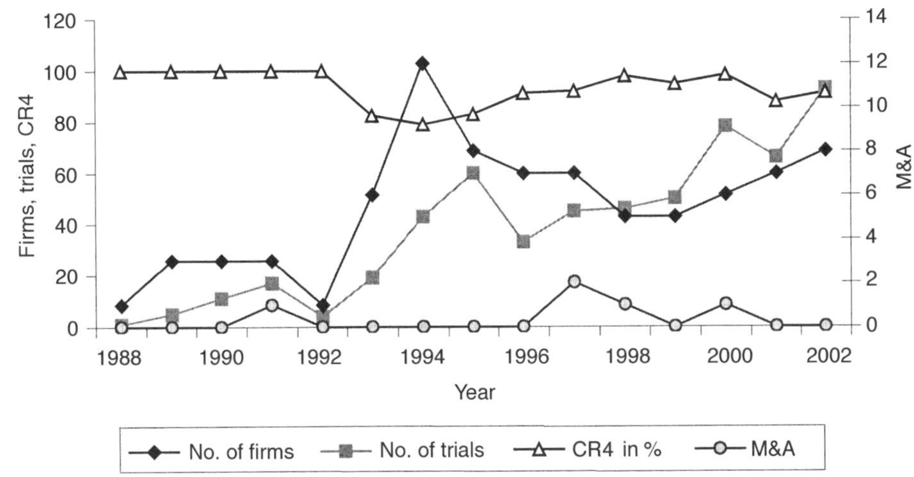

Fig. 14.2. Number of firms conducting transgenic cotton trials, number of trials approved, CR4 concentration ratio, and number of mergers and acquisitions (M&A), 1988–2002. (Source: Constructed from APHIS, IBI data.)

There is minimal variation in the CR4 ratio. It is interesting to note that the CR4 reaches its nadir simultaneously with the firm numbers reaching their peak, as the model would predict. However, again the data (and variability in the data) are too sparse to draw any conclusions.

Correlation analysis of the cotton data (Table 14.3) reveal that firm numbers and field trials are positively correlated with a coefficient of 0.67, significant at the 1% level. The CR4 ratio is negatively correlated with field trials and firm numbers, and the correlation coefficient of –0.86 between the CR4 and the number of firms is statistically significant at the 1% level. These findings are consistent with the model. The M&A variable is positively correlated with each of the other variables, but in no case is this correlation statistically significant.

Table 14.3. Pairwise correlations among number of firms conducting transgenic cotton trials, number of trials approved, CR4 concentration ratio, and number of mergers and acquisitions (M&A).

Correlation (*P*-value)	No. firms	No. trials	CR4	No. M&A
No. firms	1			
No. trials	0.6744	1		
	(0.0058)			
CR4	–0.8455	–0.3863	1	
	(0.0001)	(0.1550)		
No. M&A	0.0262	0.1653	0.2260	1
	(0.9263)	(0.5560)	(0.4180)	

Soybeans

The evolution of transgenic soybean R&D activity (number of field trials), firm numbers, concentration ratio, and M&A activity are presented in Fig. 14.3. The evolution of firm numbers and field trials follow very similar paths. The number of field trials reaches an initial peak in 1994 and a second peak in 1998. Each series dips into a trough between the peaks, and each is increasing over the 2000–2002 period.

M&A activity among soybean R&D firms is not large, but there is a noticeable peak in 1996, with additional activity primarily in the years 1994–1998. The peak activity corresponds to the dips in firm numbers and inventive activity, as predicted by the model. However, to the extent that firm numbers and inventive activity dip between 1998 and 2000, the model would suggest that M&A activity should be relatively higher in this period. The single M&A event in 2001 (with no events in 2000 and 2002 data unavailable) is insufficient to support the idea of a second M&A peak. However, this may be primarily owing to lack of currently available data.

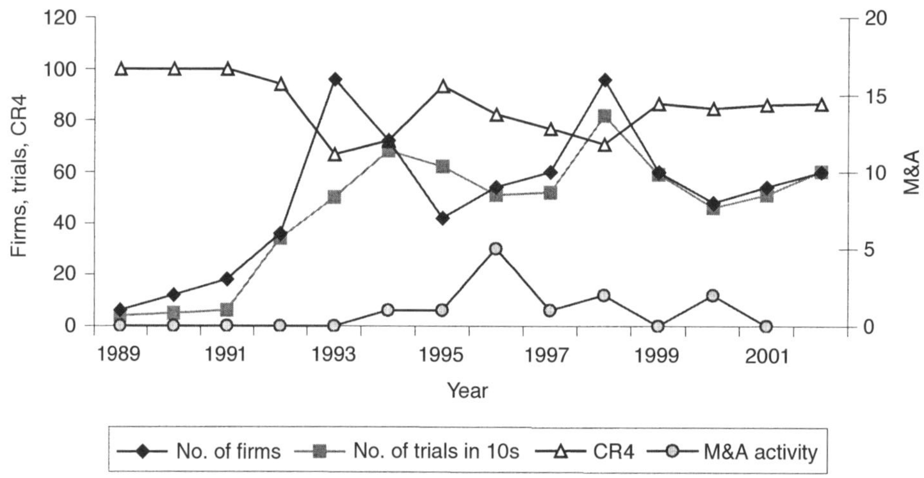

Fig. 14.3. Number of firms conducting transgenic soybean trials, number of trials approved, CR4 concentration ratio, and number of mergers and acquisitions (M&A), 1988–2002. (Source: Constructed from APHIS, IBI data.)

The CR4 ratio shows more fluctuation for soybeans than for maize or cotton. This ratio starts out high, and then declines to its nadir in 1993. It recovers to reach a secondary peak in 1995 – the same year that firm numbers reach the bottom of a trough. Then the CR4 ratio falls to a secondary nadir in 1998, the same year that firm numbers and inventive activity peak. The opposite movements of the CR4 ratio to the firm numbers and level of inventive activity support the model predictions.

Correlation analysis for the soybean data (Table 14.4) shows that firm numbers are positively correlated with the number of field trials, with a coefficient of 0.86, which is statistically significant at the 1% level. The correlations between CR4 and field trials and firm numbers are also statistically significant at the 1% level, with coefficients of –0.76 and –0.95, respectively. As with the cotton data, all correlations between field trials, firm numbers, and CR4 are consistent with the model. None of the M&A correlation coefficients are consistent with the model, although none are statistically significant.

Table 14.4. Pairwise correlations among number of firms conducting transgenic soybean trials, number of trials approved, CR4 concentration ratio, and number of mergers and acquisitions (M&A).

Correlation (*P*-value)	No. firms	No. trials	CR4	No. M&A
No. firms	1			
No. trials	0.8613 (0.0001)	1		
CR4	–0.9495 (0.000)	–0.7645 (0.0014)	1	
No. M&A	0.2621 (0.3870)	0.3876 (0.1907)	–0.3152 (0.2942)	1

Conclusions

The empirical analysis shows that the model performs best on soybeans and cotton, and less well on maize. For all three commodities, co-movement of number of firms and level of inventive activity is visually strong, with positive and statistically significant correlation coefficients. M&A activity performs poorly in all three models in terms of statistical significance, and usually has a sign opposite that of the model prediction.

Two explanations for the poor M&A performance come to mind. The first is that the model takes a rather naïve view of M&As – essentially an event is assumed to occur whenever firm numbers decline, on the assumption that even in case of exit the firm assets are somehow purchased by other firms. Doubtless, the complexities of actual M&A activity are not fully captured by such a model. The second explanation is that the poor empirical performance of the M&A variable could be due to failure to control for time trends or other common trends (this could also contribute to the extremely high negative correlations between the CR4 and number of firms and field trials for cotton and soybeans). The truth may well be that each of these explanations is a contributing factor.

The general consistency between the empirical findings and the model predictions for the number of firms, the level of inventive activity as measured by field trials, and the concentration ratio suggest that the cyclical model of biotechnology industry evolution is useful. More rigorous testing for further validation is clearly an item for future research, as is improving the explanation of M&A activity.

References

Aghion, P. and Howitt, R. (1992) A model of growth through creative destruction. *Econometrica* 2, 323–351.

[APHIS] Animal and Plant Health Inspections Service, US Department of Agriculture. Field Test Releases in the United States. Updated daily. Available online at http://www.nbiap.vt.edu/cfdocs/fieldtests1.cfm

Artuso, A. (1999) Life science industries: comparative analysis of structure, performance and regulatory policy in the agricultural biotechnology and pharmaceutical sectors. Paper presented at the conference *Transitions in Agbiotech: Economics of Strategy and Policy*, Washington, DC, June 1999.

Brennan, M., Pray, C. and Courtmanche, A. (1999) Impact of industry concentration on innovation in the U.S. plant biotech industry. Paper presented at the conference *Transitions in Agbiotech: Economics of Strategy and Policy*, Washington, DC, June 1999.

Cheng, L.K. and Dinopoulos, E. (1992) Schumpeterian growth and international business cycles. *American Economic Review* 82, 409–414.

Dasgupta, P. and Stiglitz, J.E. (1980) Uncertainty, industrial structure and the speed of R&D. *Bell Journal of Economics* 11, 1–28.

Dosi, G. (1988) Sources, procedures and microeconomic effects of innovation. *Journal of Economic Literature* 3, 1120–1171.

Folster, S. and Trofimov, G. (1997) Industry evolution and R&D externalities. *Journal of Economic Dynamics and Control* 21, 1727–1746.

Graff, G.D., Rausser, G.C. and Small, A.A. (1999) Agricultural biotechnology's complementary intellectual assets. Paper presented at the ICABR conference: *the Shape of the Coming Agricultural Biotechnology Transformation*, Rome, Italy, 17–18 June 1999.

Grossman, G.M. and Helpman, E. (1991) Quality ladders and product cycles. *Quarterly Journal of Economics* 106, 557–586.

[IBI] Institute for Biotechnology Information (1999) *Historical Actions Database*. Research Triangle Park, North Carolina.

Iosso, T.R. (1993) Industry evolution with a sequence of technologies and heterogeneous ability. *Journal of Economic Behavior and Organization* 21, 109–129.

James, C. (2002) Global review of commercialized transgenic crops: 2001. ISAAA Briefs No. 24: Preview. Available on the world wide web at http://www.isaaa.org

Johnson, S.R. and Melkonian, T.A. (1999) Policy and technology as factors in industry consolidation. Paper presented at the NE-165 conference: *Transitions in Agbiotech: Economics of Strategy and Policy*, Washington, DC, June 1999.

Kalaitzandonakes, N. and Hayenga, M. (1999) Structural change in the biotechnology and seed industrial complex: theory and evidence. Paper presented at the NE-165 conference: *Transitions in Agbiotech: Economics of Strategy and Policy*, Washington, DC, 24–25 June 1999.

Kamien, M.I. and Schwartz, N.L. (1976) On the degree of rivalry for maximum innovative activity. *Quarterly Journal of Economics* 90, 245–260.

Lee, T. and Wilde, L.L. (1980) Market structure and innovation: a reformulation. *Quarterly Journal of Economics* 94, 429–436.

Loury, G.C. (1979) Market structure and innovation. *Quarterly Journal of Economics* 93, 395–410.

Oehmke, J.F., Wolf, C.A. and Raper, K.C. (2002) *Cyclical Industry Evolution in Agricultural Biotechnology R&D*. Staff Paper, Department of Agricultural Economics, Michigan State University, East Lansing, Michigan.

Peña, I., Aldridge, J.T. and Boehlje, M. (1999) Collaborative agreements in the ag-biotechnology industry: the importance of transaction costs and investment strategy. Paper presented at the ICABR conference: *The Shape of the Coming Agricultural Biotechnology Transformation*, Rome, Italy, 17–18 June 1999.

Rauser, G.C., Scotchmer, S. and Simon L. (1999) Intellectual property and market structure in agriculture. Paper presented at the ICABR conference: *the Shape of the Coming Agricultural Biotechnology Transformation*, Rome, Italy, 17–18 June, 1999.

Reinganum, J.F. (1985) Innovation and industry evolution. *Quarterly Journal of Economics* 100, 81–99.

Scherer, F.M. (1967) Resource and development resource allocation under rivalry. *Quarterly Journal of Economics* 81, 359–394.

Segerstrom, P.A. (1991) Innovation, imitation and economic growth. *Journal of Political Economy*, 99, 807–827.

Segerstrom, P.A., Anant, T.C.A. and Dinopoulos, E. (1990) A Schumpeterian model of the product life cycle. *American Economic Review* 80, 1077–1091.

Vickers, J. (1986) The evolution of market structure when there is a sequence of innovations. *Journal of Industrial Economics* 35, 1–12.

15 The Innovation System in Agro-food Biotechnology – is it European?

Klaus Menrad and Thomas Reiss
Fraunhofer Institute for Systems and Innovation Research, Karlsruhe, Germany

Introduction and Theoretical Background

In recent years, increasing commercialization and growing of genetically modified (GM) plants has been registered, mainly in North and South America (James, 2001), while the de facto moratorium for GM products has prevented market approval of transgenic plants and derived products in the European Union (EU) since 1999. On the other hand, ongoing research activities, public funding of agro-food biotechnology, as well as some commercial activities (such as the founding of genomic-based agbiotech start-up companies or the use of biotechnology approaches in plant breeding) can be observed in almost all EU countries (Enzing *et al.*, 1999; PEW, 2001). Against this background the question arises whether national or sectoral systems of innovation are dominant in shaping the development of agro-food biotechnology within the EU.

There is a general agreement in modern innovation research that innovation is characterized by complicated feedback mechanisms and interactive relations, involving science, technology, learning, production, policy and demand (Edquist, 1997). Innovation processes occur over time and are influenced by many factors, i.e. commercial companies almost never innovate in isolation, but interact with 'organizations' of different types (e.g. suppliers, customers, research institutions) and their behaviour is shaped by 'institutions' as well (Edquist, 1997), which constitute constraints or incentives for innovation (e.g. laws, social rules, technical standards).

Due to their complex character, innovation activities represent an ideal area to use system theory approaches for the analysis of such processes at the level of a (national) economy. Since the 1980s, a series of systems approaches have emerged in this field and empirical studies can be observed. In this context 'national systems of innovation' are the most frequently used approach for understanding the complex relations of the innovation process (Lundvall, 1992; Nelson, 1993; Edquist, 1997; Lundvall *et al.*, 2002). Other analyses of innovation systems took a regional perspective (e.g. Saxenian, 1994) or were focused on specific sectors (Breschi and Malerba, 1997). The concept of 'technological systems' focuses more on the technology itself and its mediation and represents mainly dynamic knowledge and competence networks (Carlsson and Stankievicz, 1995).

There are no studies available in the economic literature that intend to analyse the innovation system of agro-food biotechnology in different EU member countries. In this context, the question arises whether sectoral features at a European level or national idiosyncrasies (which characterize national innovation systems) are the main driving forces of the sector. In order to analyse this question, the factors significant to

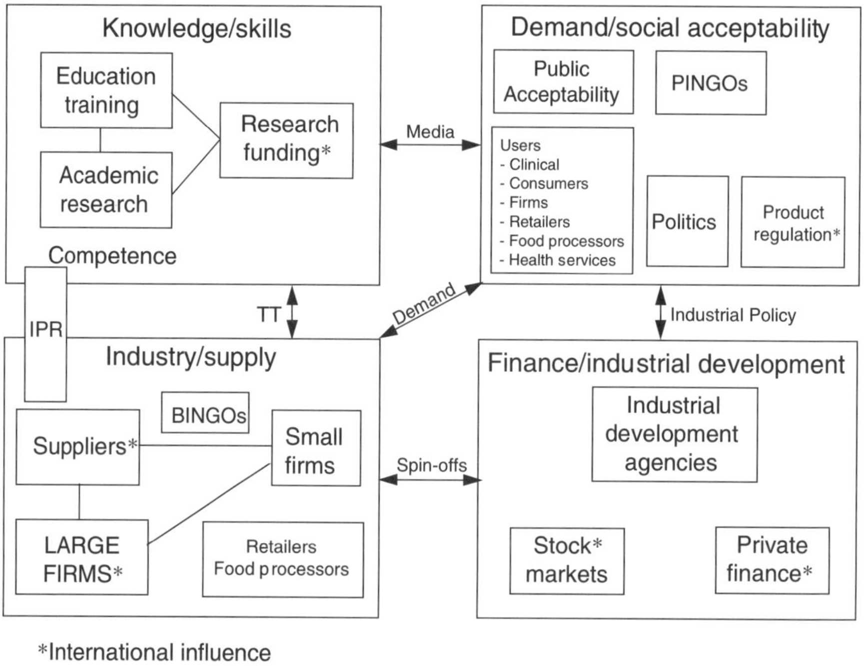

Fig. 15.1. Network of key factors influencing innovation. (Source: Senker *et al.*, 2001.)

innovation have to be considered. Figure 15.1 gives a simplified overview of these factors, the embedding networks as well as their interrelationships. The main components of the framework are networks of knowledge and skills, industry and supply, demand and social acceptability as well as financing and industrial development.

Methodology

The empirical data of this analysis is based on a European research project funded by the European Commission with the title 'European Biotechnology Innovation Systems' (EBIS) which was carried out between 1999 and summer 2001 by a team of eight European research groups and co-ordinated by the Science Policy Research Unit (SPRU) of the University of Sussex, UK. Fraunhofer ISI was the German partner in the project. Within this project a common framework was developed in order to analyse innovations in biotechnology in three sectors (biopharmaceuticals, agro-food, equipment and supplies) in the countries involved in the eight case studies: Austria, France, Germany, Greece, Ireland, The Netherlands, Spain and the UK.

The case studies on these countries provided information on the factors mentioned in Fig. 15.1 for biotechnology applications in the three analysed sectors (including agro-food). Additional sources of information (e.g. statistical data, other studies) were used for collecting background national or sectoral data. In each of the eight countries the research teams identified potential agro-food biotech companies and performed a company survey in the year 1999, including questions related to company structure, cooperations, and technology use, as well as product development and marketing. In the agro-food sector 162 companies filled in the questionnaire and participated in the survey. Due to the very limited number of completed questionnaires (up to six questionnaires per country), a detailed analysis cannot be made for Austria, Ireland and the UK. In the latter country the companies hesitated to participate in the survey because of a very conflicting situation concerning the use of agro-food biotechnology in 1999. In the other five countries between 12 (Greece) and 59 agro-food biotech companies (France) responded to the questionnaires. All data were entered into a common database which served as a basis for the statistical analysis of the sector. Additional information was obtained through interviewing experts and by screening the available literature on the sector.

Empirical Results

In the following, the results of the empirical analysis of the agro-food biotechnology innovation system in the eight analysed countries will be presented according to the four networks of key factors which form the framework of our analysis.

The knowledge/skills network

The current skills and knowledge base in each country is affected in part by when national policies first focused on the support of biotechnology. France, Greece, The Netherlands and the UK began to introduce policies to support a biotechnology research infrastructure around 1980. In contrast, Austria, Germany, Ireland and Spain started during the second half of the 1980s with biotechnology-specific policy measures. In the 1990s, the countries have developed differently in terms of public funding of agro-food biotechnology research. France, Germany and the UK are spending most public money for research in this sector (Enzing *et al.*, 1999; Giessler and Reiss, 1999). In France and Germany the annual research budget for agro-food biotechnology amounted to roughly €63 million between 1994 and 1998. The respective figure for the UK was €155 million. When related to the gross domestic expenditure on R&D, the German and French funding corresponds to between 0.1% and 0.2%, while the British contribution amounts to roughly 0.7%, indicating the rather high specific funding of agro-food biotechnology research in the UK.

On the other hand, Greece, Austria and Ireland spent the least on the science base during the 1990s with annual budgets below €3 million each (Enzing *et al.*, 1999; Reiss, 1999). Spain has an intermediate position with a fast growing science base and an average annual R&D budget for agro-food biotechnology between 1994 and 1998 of about €9 million (Senker *et al.*, 2001). The situation of The Netherlands is difficult to evaluate in terms of subsector-specific R&D expenditure because during the 1990s there was no specific programme directed towards certain sub-areas of biotechnology. Since the science base in The Netherlands can be considered as very well-developed, it seems reasonable to assume that a considerable amount of the total annual biotechnology budget of about €65 million during 1994 and 1998 flowed into agro-food biotechnology.

Looking at the actors generating and transferring scientific know-how for the agro-food biotechnology sector, different configurations can be identified in the eight countries. National universities are the most important actors in the knowledge/skills network in Austria and Greece. In Austria, 11 university institutes at six universities and one inter-university research centre are the main university-based actors in the system. In addition, two non-university institutes conduct agro-food biotech-related research. A similar situation can be stated for Greece, where six universities are the main actors in addition to few non-university institutions.

In France, Spain, the UK and Ireland, governmental research institutions are the main actors in the agro-food biotechnology knowledge/skills network. In France Institute National de la Recherche Agronomique (INRA) is the most important player, with about 500 full-time researchers in the agro-food biotechnology sector. In addition, 30 universities with various institutions and five Ecoles Nationales Supérieures Agronomiques are active with varying intensities in this sector. In Spain, among the governmental institutions, the Instituto Nacional de Investigacion y Technologia Agraria y Alementaria (INIA) with nine regional research centres and the Consejo Superior de Investigaciones Cientificas (CSIC) with 16 institutes are performing most of the agro-food biotechnology research. In the UK the eight institutes of the Biotechnology and Biological Sciences Research Council (BBSRC) are the key players in agro-food biotechnology research. In addition, various universities in Great Britain are performing related research. In Ireland a number of research centres are active in the sector, with the nine research centres of the The Irish Agriculture and Food Development Authority (TEAGASC) being the most important ones.

Finally, The Netherlands and Germany can be considered as a mixed system where universities and numerous non-university research institutes are performing agro-food biotechnology research. The non-university institutes in Germany belong to the Max Planck Society (five institutes), to the Helmholtz Society (one national research centre), and to the Leibniz Society (four institutes). In addition, five departmental research centres of the Federal Ministry of Consumer Protection, Food and Agriculture are key players in this area. In The Netherlands a very strong research base has been built up in agro-food biotechnology for many years. In addition to nine universities, the Wageningen Centre for Food Science, the Wageningen Research Centre, two institutes of the TNO and the Dutch Institute for Dairy Research of NIZO perform related research.

The industry/supply network

In total, 162 agro-food biotech companies were identified in the eight countries. A considerable increase in the annual number of agro-food biotech companies founded was observed in the second half of the 1980s. Similar developments occurred also in other biotechnology sectors such as biopharmaceuticals (Senker *et al.*, 2001). However, a downwards trend in foundation activities can be observed for the agro-food biotechnology sector in the 1990s. This is in contrast to the biopharmaceutical sector, where especially in the second half of the 1990s very intensive foundation activities were registered (Allansdottir *et al.*, 2001). By far most of the agro-food biotech companies (about 86%) were independent foundations. Only about 7% have been founded as spin-offs from research organizations, another 7% as spin-offs from industry. In other sectors, e.g. biopharmaceuticals or equipment and supplies, only about two-thirds of the companies were established independently (Senker *et al.*, 2001). The distribution of the agro-food biotech firms among the eight countries reveals a surprising picture: of the three larger countries only France is home to a considerable number of agro-food biotech firms, covering about 36% of the whole population. Germany and the UK on the other hand, have only a low number of such firms, whereby the very low share of the UK is most surprising. On the other hand, besides France some of the smaller countries (in terms of population), namely Spain, The Netherlands and Greece seem to focus on the agro-food biotechnology sector.

The analysis of the size distribution of the agro-food biotech firms in the eight European countries indicates that there are two main groups of firms: very small firms with fewer than 20 employees and medium-sized and larger firms with more than 100 employees. Both groups amount to roughly one-third of the total population (Fig. 15.2). Of those countries with a strong agro-food biotechnology sector, mainly the larger firms with more than 100 employees dominate the scene in The Netherlands and Greece. In contrast more than 60% of the agro-food biotech firms have fewer than 20 employees in France. Spain presents a different picture; the medium-sized and larger firms seem to be present in comparable numbers. Among the countries with the less developed sector, Ireland is interesting because it is home to the highest share of very small firms of all countries. This seems to indicate that the sector has started developing rather recently. In Germany the firms are distributed rather equally among the four size classes.

The finance/industrial development network

Most countries in the EU have established specific policies to support industrial innovation. Sometimes these policies focus on biotechnology in general, but Greece, Austria and The Netherlands (until the end of the 1990s) lack any specific focus on biotechnology. In the latter country initiatives were put into place to promote the creation of new start-up companies from 2000. Greek industrial policy is mainly connected to technology transfer initiatives. Since 1998 industrial policy in Austria has focused on the commercialization of results of public sector research, the promotion of knowledge transfer between public and private research, as well as the creation of start-up companies (Senker *et al.*, 2001).

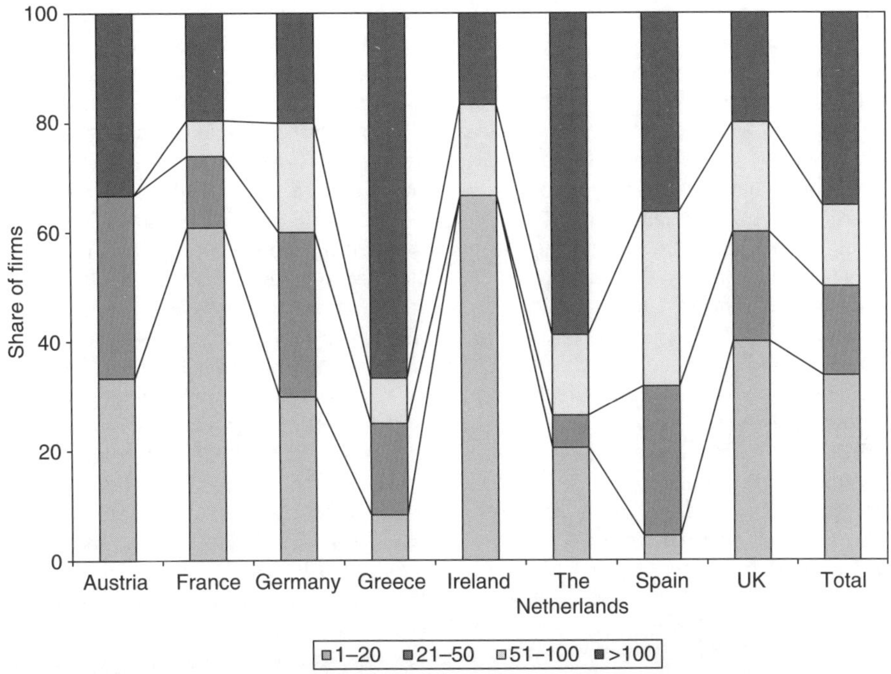

Fig. 15.2. Distribution of number of employees in agro-food biotech companies in different European countries ($n = 148$).

In the other countries industrial policy is more specifically dedicated to biotechnology. In Ireland the transfer of research results to existing biotech companies or new start-up firms is supported. German industrial policy tries to encourage firms to adopt new technologies and to raise industrial awareness concerning biotechnology. In addition, the founding of high-tech companies is supported by specific credit programmes and application-oriented research projects are financed by specific funds. French industrial policy provides tax credits for companies conducting in-house research (Senker *et al.*, 2001). In the UK, a wide range of different programmes exists to promote and fund industry–university projects as well as to encourage the creation of new start-up firms. In Spain, joint research projects by firms and public research organizations as well as development projects or activities aiming to adopt new technologies in companies are funded by specific programmes. In addition, there are credit programmes to support the creation of new companies (Senker *et al.*, 2001).

Although availability of venture capital (VC) for biotech firms has increased in the last years in the EU (excluding a decline in recent 2 years) much of the European investment in agro-food biotechnology research comes from public sources (e.g. European Commission, member states, regions), but not from private investments. In addition to a low availability of VC in some member states of the EU (e.g. Austria, Greece, Ireland), agro-food biotech companies have not generated enthusiasm within the investment community as agro-food biotechnology is regarded as high-risk business with very uncertain future market perspectives (Crowther *et al.*, 1999) and limited growth potential. Therefore, in particular agro-food biotech small and medium enterprises (SMEs) face significant difficulties in financing their business activities. This situation already existed before the recent collapse of the high-tech markets of European stock exchanges in 2001 and 2002.

Compared with Spain, Greece, Austria and Ireland, where all types of biotech companies face specific difficulties to find in particular 'early stage' or seed venture capital, the general availability of VC is much higher in the UK, Germany, France and The Netherlands (EVCA, 2002). The UK in particular has one of the most favourable legal and fiscal environments for private equity in the EU, providing the VC firms with a number of exit opportunities in specific segments of the stock exchange (Lessat *et al.*, 1999; Menrad *et al.*, 1999b). In Germany and The Netherlands, the VC market is well developed and several investment firms have specialized in biotechnology investments. However, many investments in agro-food biotech companies fail because start-up firms lack either the management capabilities demanded by VC firms (Menrad *et al.*, 1999a) or do not fulfil their high growth perspectives. In France a rapid growth of VC firms, which mostly invest in companies in the health sector, was observed in recent years. All these four countries have established specific segments in the stock markets in order to list high-tech companies and provide VC firms and other owners of start-up companies with specific exit opportunities (Menrad *et al.*, 1999b). However, only very few agro-food biotech companies have been listed on European stock markets in recent years. Given the collapse of these high-tech market segments in 2001 and 2002, it can be assumed that the private funding possibilities, in particular for agro-food biotech companies, will not improve in the coming years in the EU.

The demand/social acceptability network

Agro-food biotechnology products are mainly targeted to the markets of seeds, pesticides and other agro-chemicals, veterinary products as well as to the food market in general. Most of these markets represent large volume markets which are often separated in different segments. The commercial markets for agricultural seed and planting material in the EU range from €60 million in Ireland to €1.37 billion in France (ASSINSEL, 2002). The domestic food markets in the analysed countries have a considerably higher market volume and range from around €8 billion in Austria (Tradepartner, 2002) to €165 billion in Germany (BMELF, 2000). In general, the markets for seeds, pesticides or food products are not specifically regulated in the EU. However, in each country and at the European level, specific agencies control field trials with GM plants and are concerned with the impacts of these products on health, the environment and food security.

Public attitudes, regulatory aspects as well as the reactions of the food retail companies and food processing companies have a major impact on a potential demand for GM crops and food in the EU. Critical non-governmental organizations (NGOs) have played an important role in widening the debate about the use of genetically modified organisms (GMOs) in the agro-food sector and their views have been widely disseminated by the media in recent years. As a result, negative public attitudes to GM crops and foods are now widespread in the EU, although there are differences between the member states (Gaskell *et al.*, 2000). In Austria, Germany, Greece and the UK the public has a very negative attitude to GM crops and ingredients in food as well as very critical citizens and consumer associations which have taken an active part in the public debate in the last years. Food retailers which are highly concentrated in Austria, Germany and the UK have responded to these attitudes and partly removed GM ingredients from their products (Senker *et al.*, 2001). In Greece, food retailers and manufacturers have adopted the same position, while in Austria additional opposition came from the family-run farm sector.

In France, The Netherlands and Spain public opposition to GM crops and food was relatively moderate, compared with the other analysed countries. In France media interest in the GM debate has increased since 1997 and a growing distrust in GMOs can be observed on the part of the public. As shown in several studies, the Dutch public is more knowledgeable about biotechnology than the population in other European countries (e.g. European Commission, 1997, European Commission, 2001c). Interestingly, the Dutch attitudes towards biotechnology have not changed significantly during the 1990s. Although the Dutch government tries to ensure the freedom of choice of consumers with respect to GM products, the Dutch population also remains sceptical about applications of GM food. Compared with the other countries, there is relatively less public opposition to GM food and crops in Spain. The same applies to the Irish public, although GM food does not seem to attract high public support (European Commission, 2001c).

National, European and international regulations have relevance for the legal framework of agro-food biotechnology. The EU has the highest regulatory competency in this field, which partly determines the framework for national regulations as well. During the 1990s the use of genetic engineering in the agro-food sector in the EU was mainly regulated by directives 90/219 and 90/220 on the contained use and deliberate release of GMOs as well as regulation 258/97/EC (Novel Food Regulation). In February 2001 the European Commission adopted directive 2001/18/EC (which replaced directive 90/220/EEC), in order to regulate the deliberate release of GMOs into the environment. Market approval of GM products is restricted to 10 years, according to this directive. Furthermore, long-term monitoring as well as strict labelling activities are foreseen in this directive (European Commission, 2001b). In addition, the European Commission has proposed several other regulations, e.g. for food and feed consisting of GMOs as well as seed of GM plants (European Commission, 2001a). However, the de facto moratorium of new GMOs in the EU, which has been running since 1999 and prevents market approval of GM products, has not been repealed so far, although the European Commission is taking initiatives to change this situation (Meldolesi, 2002). In November 2002 the European Parliament also voted for the removal of the de facto moratorium on GMOs (EU Parliament, 2002).

The European regulatory framework related to agro-food biotechnology played an important, largely negative role for the development of GMOs in the EU in the last decade. During this time period increased regulatory oversight in agro-food biotechnology coincided with growing negative public opinion and diminished trust in public authorities and regulatory agencies (Senker *et al.*, 2001). In this context, in particular the continually changing regulatory environment has been criticized by industry, as well as the practical handling of existing regulations as being too slow, bureaucratic and causing high costs. During the interviews politics was criticized for not taking clear decisions regarding agro-food biotechnology and periodically intervening in regulatory processes. Due to their limited financial and personnel resources, in particular agro-food biotech SMEs face specific difficulties in the current regulatory environment of uncertainty in the EU.

The application for notification of field trials with GMOs under part B of directive 90/220/EEC gives information on the extent of field trials which have been carried out in the different member countries of the EU since 1991 (ICHP, 2002). Around 30% of all notifications on field trials with GMOs have been conducted in France, followed by Italy and the UK, in which 16% or around 12% of all notifications have been performed. In the other countries under consideration, a significant number of field trials have been carried out in Spain and The Netherlands, while in Germany only 115 notifications for field trials were reported. In Austria, Ireland and Greece a negligible number of field trials with GMOs was carried out during the 1990s.

Conclusions and Policy Implications

The development of agro-food biotechnology in the EU faces considerable constraints, mainly related to the high uncertainty in the regulatory framework as well as the practical handling of the existing regulations, very uncertain market perspectives due to low consumer acceptance of GM crops and ingredients in food, as well as lack of private financing opportunities for start-up agro-food biotech companies.

The countries that seem to have the best prerequisites to further develop their competencies in the agro-food biotechnology sector are France and The Netherlands (Table 15.1), based on their well-adapted science base, multinational companies as backbone of industrial activities, a relatively good financing environment as well as moderate public opposition to GMOs. Germany has a high number of domestic agro-chemical and seed multinationals as well as a large number of scientific institutions competitively active in agro-food biotechnology. Although the government tries to support commercial exploitation of biotechnology in general, future perspectives for companies in the agro-food biotechnology field are impeded by the low consumer acceptance of GM crops and food as well as corresponding reactions of food manufacturing and food retail companies, which try to avoid GM ingredients in their products or

Table 15.1. Overview of the innovation system in agro-food biotechnology in different member countries of the EU.

Country	Knowledge/ skills	Industry/ supply	Finance/industrial development	Demand/social acceptability
Austria	–	–	0	–
France	+	+	0	+
Greece	–	0	–	0
Germany	+	0	+	–
Ireland	0	–	+	0
The Netherlands	+	+	0	+
Spain	0	0	0	+
UK	+	–	+	–

+, positive; 0, neutral; –, negative.
Source: Estimations of Fraunhofer ISI, based on Senker *et al.*, 2001.

on their shelves. Spain's fast growing science base and moderate public opposition to GMOs (Table 15.1) are specific assets to further develop agro-food biotechnology in this country, which is, however, slowed down by the low private investments in R&D as well as a lack of VC to support the establishment of start-up companies in this field. The situation in the UK is characterized by a strong science base in agro-food biotechnology, as well as a high availability of VC funds and a national policy that tries to commercialize the existing scientific knowledge. However, strong public opposition to GMOs and GM food in particular since 1998 has resulted in a relatively low number of newly established start-up companies in this country (Table 15.1). Austria and Greece face specific difficulties in almost all areas which influence innovation activities in agro-food biotechnology. In Ireland commercial exploitation of agro-food biotechnology is mainly impeded by the weak industrial basis, as well as a limited science base (Table 15.1).

The findings of the study reveal significant differences in the innovation system in agro-food biotechnology between the eight EU countries studied. The great differences between these countries suggest that each country has its own pattern of innovation in agro-food biotechnology. The results of the study seem to indicate as well that countries with large home markets appear to have greater abilities than smaller ones to exploit the development of new markets for emerging technologies. The nature of demand and market seem to play a significant role in explaining different patterns of innovation activities by different countries (Senker *et al.*, 2001). In agro-food biotechnology the potential demand in the EU was negatively affected by a combination of public opposition to GMOs, media coverage, the response of food manufacturers and retailers and high regulatory uncertainty and led to relatively low innovation activities by firms in the last 5 years. All in all, it can be concluded that national factors have a high influence on commercial exploitation of agro-food biotechnology in the EU, which underlines the central role of national or regional innovation policies.

As the examples of Germany and the UK (where a favourable science base is not exploited commercially) indicate, there is an increasing need for taking a systemic perspective of the innovation process when designing and implementing research, development, technology and innovation policies. It is not sufficient to focus policy support only on individual sub-networks of the whole system. Rather, integrative policy approaches are needed which combine different policy functions to produce a more systemic policy concept.

References

Allansdottir, A., Bonaccorsi, A., Gambardella, A., Mariani, M., Orsenigo, L., Pammolli, F. and Riccaboni, M. (2001) Innovation and competitiveness in European biotechnology. *DG Enterprise Paper No. 7*. European Commission, Brussels.

ASSINSEL (2002) Seed statistics: world seed trade statistics. http://www.worldseed.org/statistics.html

Breschi, S. and Malerba, F. (1997) Sectoral innovation systems: technological regimes, Schumpeterian dynamics, and spatial boundaries. In: Edquist, C. (ed.) *Systems of Innovation. Technologies, Institutions and Organizations*. Pinter, London, pp. 130–156.

[BMELF] Bundesministerium für Ernährung, Landwirtschaft und Forsten (2000) *Statistisches Jahrbuch über Ernährung, Landwirtschaft und Forsten 2000*. Landwirtschaftsverlag GmbH, Münster.

Carlsson, B. and Stankiewicz, R. (1995) On the nature, function, and composition of technological systems. In: Carlsson, B. (ed.) *Technological Systems and Economic Performance: the Case of Factory Automation*. Kluwer Academic Publishers, Boston, Massachusetts, pp. 21–56.

Crowther, S., Hopkins, M.M., Martin, P., Millstone, E., Sharp, M. and van Zwanenberg, P. (1999) *Benchmarking innovation in modern biotechnology. Report to the European Commission, DG III*. Science and Technology Policy Research (SPRU), Brighton.

Edquist, C. (1997) Systems of innovation approaches – their emergence and characteristics. In: Edquist, C. (ed.) *Systems of Innovation. Technologies, Institutions and Organizations*. Pinter, London, pp. 1–35.

Enzing, C.M., Reiss, T., Schmidt, H. *et al.* (1999) *Inventory of public biotechnology R&D programmes in Europe. Volume 1: Analytical Report*. Office for Official Publications of the EU, Luxembourg.

EU Parliament (2002) Mehrheit für Aufhebung des GVO-Moratoriums. http://www.biosicherheit.de/aktuell/173.doku.html

European Commission (1997) *The Europeans and Modern Biotechnology. Eurobarometer 46.1*. Directorate General XII Science, Research and Development: Biotechnology. Office for Official Publications of the European Communities, Luxemburg.

European Commission (2001a) Proposal for a regulation of the European Parliament and of the Council on genetically modified food and feed. COD 2001/0173. http://europa.eu.int/comm/food/fs/biotech/biotech08_en.pdf

European Commission (2001b) Kommission verbessert Regeln für Kennzeichnung und Rückverfolgbarkeit von GVO in Europa, die dem Verbraucher die Wahl lassen und den Schutz der Umwelt sichern. *Pressemitteilung vom 25 Juli 2001*. IP/01/1095.

European Commission (2001c) *Europeans, Science and Technology*. European Commission, Brussels.

[EVCA] European Venture Capital Association (2002) *EVCA Yearbook 2002*. Brussels.

Gaskell, G., Allum, N., Bauer, M. *et al.* (2000) Biotechnology and the European public. *Nature Biotechnology* 18(9), 935–938.

Giessler, S. and Reiss, T. (1999) *Inventory and Analysis of Biotechnology Programmes and Related Activities in all Countries Participating in the Biotechnology Programme 1994–1998*. National report of Germany. Fraunhofer ISI, Karlsruhe.

[IHCP] Institute of Health and Consumer Protection, Joint Research Centre (2002) *Deliberate field trials notified under part B of Directive 90/220/EEC*. http://biotech.jrc.it/gmo.asp

James, C. (2001) *Global Review of Commercialised Transgenic Plants 2001*. ISAAA, Ithaca, New York.

Lessat, V., Hemer, J., Eckerle, T.H., Kulicke, M. *et al.* (1999) *Beteiligungskapital und technologieorientierte Unternehmensgründungen*. Gabler Verlag, Wiesbaden.

Lundvall, B.A. (1992) Introduction. In: Lundvall, B.A. (ed.) *National Systems of Innovation: Towards a Theory of Innovation and Interactive Learning*. Pinter, London, pp. 1–22.

Lundvall, B.A., Johnson, B., Andersen, E.S. and Dalum, B. (2002) National systems of production, innovation and competence building. *Research Policy* 31, 213–231.

Meldolesi, A. (2002) Political will lift the GMO moratorium emerging in Europe. *Nature Biotechnology* 20, 758–759.

Menrad, K., Kulicke, M., Lohner, M. and Reiss, T. (1999a) *Probleme junger, kleiner und mittelständischer Biotechnologieunternehmen*. Fraunhofer IRB Verlag, Stuttgart.

Menrad, K., Lohner, M., Estades, J. and Saviotti, P.P. (1999b) *New Possibilities for Accessing the Capital Markets for Small and Medium-sized Biotech Enterprises*. Office for Official Publications of the European Communities, Luxemburg.

Nelson, R.R. (ed.) (1993) *National Innovation Systems: A Comparative Analysis*. Oxford University Press, New York.

PEW (2001) *Harvest on the Horizon: Future Uses of Agricultural Biotechnology*. PEW Initiative on Food and Biotechnology. Washington, DC.

Reiss, T. (1999) *Inventory and Analysis of Biotechnology Programmes and Related Activities in all Countries Participating in the Biotechnology Programme 1994–1998*. National report of Austria. Fraunhofer ISI, Karlsruhe.

Saxenian, A. (1994) Regional advantage. Culture and competition in Silicon Valley and route

128. Harvard University Press, Cambridge, Massachusetts.
Senker, J., Zwanenberg, P. van, Reiss, T., Wörner, S. *et al.* (2001) *European Biotechnology Innovation System*. Final Report: EC TSER Contract No: SOEI-CT98–1117. SPRU, Brighton, UK.
Tradepartner (2002) Food and drink. http://www.tradepartner.gov.uk/food

16 How Firm Characteristics Influence Innovative Activity in Agricultural Biotechnology

Cassandra Klotz-Ingram,[1] David Schimmelpfennig,[1] Anwar Naseem,[2] John King[1] and Carl Pray[2]

[1]*Economic Research Service, United States Department of Agriculture, Washington, DC, USA;* [2]*Department of Agriculture, Food and Resource Economics, Rutgers, The State University of New Jersey, New Brunswick, New Jersey, USA*

Introduction

There has been a good deal of discussion concerning the causes and consequences of industry consolidation in agricultural biotechnology (agbiotech). One topic of particular concern has been the effect of mergers and acquisitions (M&As) on the future path of innovation (Kalaitzandonakes, 1999; Kalaitzandonakes and Hayenga, 1999; Brennan *et al*., 2000; Fulton and Giannakas, 2001; King, 2001; Schimmelpfennig *et al*., 2004). This chapter explores the relationship between agbiotech innovative activity and the financial characteristics of the companies conducting research related to the development of new plant varieties. We develop measures of inputs to and outputs from firm research and development (R&D) processes, and relate these measures to other firm performance characteristics for two classes of firms. We show that fundamental differences in innovative activity exist between smaller, specialized seed and agbiotech companies and the larger, more diversified chemical and pharmaceutical companies.

M&As and divestitures in the agbiotech industry are taking place almost daily, likely due to the search for an efficient scale for the production of new technology and research complementarities (Graff *et al*., 2002). Originally, it appeared that some companies were trying to create research synergies with pharmaceutical biotechnology, redefining themselves as large 'life-sciences' companies. Recently, many of these same companies have been taking steps to keep agricultural and pharmaceutical R&D separate, perhaps to isolate different shocks to shareholder equity due to uncertain agbiotech markets. Broader stock market fluctuations have also been taking place and these have influenced business strategies.

For economic analysis a tension exists between model richness and data availability. Oehmke *et al*. (2002) derive cyclical Schumpeterian relationships for agbiotech R&D, but as they readily admit, data do not exist to test their model. Our approach is closer to the less theoretical methods found in Kamien and Schwartz (1982) and reviewed by Cohen and Levin (1989) and Schmalensee (1988). Geroski (1998), in particular, provides guidance on the use of descriptive-type statistics in attempts to identify how differences in company characteristics influence firm performance. We take this approach one step further and relate these differences to the rate and form of innovation in agbiotech over the last 15 years.

The next section presents measures of innovative output. We then define associated measures of innovative inputs and present some analysis of firm characteristics. In the following section, we identify sources of firm heterogeneity and show how these influence the results. We conclude with a discussion of how firm assets and performance may have contributed to consolidation in the agbiotech industry and how consolidation influenced research.

Measures of Innovative Output

In the late 1990s there were approximately 1200 companies in the USA engaged in some aspect of biotechnology, from research and production, to providing services to the industry (Dibner, 1997). However, relatively few of these companies conducted any amount of research on agbiotech and even fewer on plant biotechnology, i.e. the specific application of biotechnology tools and techniques to plants. We characterize a company as being an innovative agbiotech company if it has at least one agbiotech patent and has conducted at least one field trial for regulatory approval of genetically engineered (GE) plants or organisms in the USA. While there are about 30 companies which meet these criteria, the list was reduced by the lack of available financial data for privately held companies to 15. Compustat (2000) has fairly complete financial data but only for publicly traded companies.

Patents have been widely used in economic analysis as a measure of innovation (Lanjouw *et al.*, 1998; Griliches, 1990). Growth in the patenting of biotechnology innovations has made patent counts an attractive gauge of innovation in the plant biotechnology industry (Klotz-Ingram and Day-Rubenstein, 2002). The US Patent and Trademark Office (PTO) classification system (US Patent and Trademark Office, 2002) was employed to identify the universe of plant biotechnology patents and company ownership. Our search strategy is similar to the one used by Foltz *et al.* (2001) who sought to identify patents that, broadly speaking, were cross-listed in primarily agricultural classes (800 and 47) with those in the biotechnology classes (435 and 536). The result is 2447 agbiotech patents awarded between 1985 and 2000.

Whereas patents indicate the successful completion of a research effort, field trials indicate on-going research activity. The Animal and Plant Health Inspection Service (APHIS) of the US Department of Agriculture (USDA) authorizes field trials and maintains records on approvals of field trials that is readily available (APHIS, 2002). Since the first field trial was registered in late 1987 there has been a steady growth in the number of field trials being conducted up to 2000. We also analysed counts of Plant Variety Protection Certificates (PVPC) from the Agricultural Marketing Service (AMS) of the USDA (AMS, 2002) as another measure of research activity. PVPCs are awarded for novel plant varieties and is another form of intellectual property protection.

The 15 firms we analyse fall naturally into two groups. Seven of the firms are small seed or plant biotechnology companies, and eight are large, diversified chemical or pharmaceutical firms. The companies and the years of available data are listed in Table 16.1. These companies hold 42% of all plant biotechnology patents, 46% of field trials, and 25% of PVPCs as shown in Table 16.2. The remainder of patents, field trials and PVPCs are primarily attributable to privately-held companies who do not publish (and often closely guard) their financial information, as well as non-profit organizations, universities, Federal labs and state agricultural experiment stations (SAES).

The companies with dominant positions in agricultural input markets are included in our sample. Pioneer Hi-Bred, Monsanto and DeKalb together accounted for 66% and 47% of 1997 North American maize and soybean seed market shares, respectively (Hayenga, 1998). Monsanto, Dow and DuPont acquired some of the smaller companies during the period of analysis. Monsanto acquired DeKalb and Calgene, while Dow acquired Mycogen, and DuPont bought Pioneer Hi-Bred in the late 1990s which is why the DeKalb, Calgene, Mycogen and Pioneer data end before 2000. Some important players in agbiotech are missing, such as Novartis (now Syngenta) and Aventis (now Bayer), because they have either only existed for a few years or are new to agbiotech. Companies that were established since 2000 do not have reliable innovation data because of approval lags in patents and PVPCs.

Table 16.1. Companies and years of available data.*

Company	Years
Agbiotech/seed	
Calgene	1985–1995
Crop Genetics International	1986–1993
DeKalb	1987–1998
Delta and Pine Land	1991–2000
DNA Plant Technology	1985–1995
Mycogen	1985–1996
Pioneer Hi-Bred	1985–1998
Large/diversified	
WR Grace	1988–1995
Amoco	1987–1997
Cyanamid	1985–1993
Dow	1985–2000
DuPont	1985–2000
ICI/Zeneca	1985–2000
Monsanto	1986–2000
Rhône-Poulenc	1985–1996

*Because data availability for each year is uneven, differing by variable and company, the table lists the years that data were available for either sales or R&D expenditures, whichever had more complete years of information. These financial variables typically were more complete than others. Data availability for the measures of innovative output depend on the years a company operated, but in general, adequate data were available as follows: patents (1985–1999), PVPCs (1985–2000), and field trials (1987–2000). Patents were truncated at 1999, reflecting the fact that it generally takes 3 or more years for a patent application to be approved or not. The first field trials were not approved until 1987. PVPCs, Plant Variety Protection Certificates.

Analysis of Firm Performance

A firm's success in technology development depends on factors influencing its ability to innovate, such as its stock of relevant knowledge, past technological achievements and R&D resources. However, other firm characteristics such as firm size, financial performance and market position also affect research success. The large number of M&As in this industry make the interpretation of firm characteristics for innovation quite challenging. In fact, only one of the smaller companies (Delta and Pine Land) was not acquired during this period. Several innovation-data assignment rules became necessary. First, we attributed the patents of a subsidiary to the parent company. For example, Agracetus patents go to WR Grace. Second, when a firm is acquired its new patents go to the new owner. Monsanto acquired Agracetus and is credited with all of the latter's measures of innovative activity after the acquisition. Third, data on the larger companies is only collected while they are involved in agbiotech research. The result is that the spin-off of an agbiotech division can result in a large company leaving the industry. Business segment data for firms is generally not available because regulations only require firms to file and publish data for their entire operation.

Several measures of innovative activity are taken from firm financial data. The level and growth of R&D expenditures,[1] and R&D intensity (calculated as the proportion of R&D expenditures to sales) can be interpreted as measures of inputs to technology development. Table 16.3 presents descriptive statistics for all the measures of innovation for both company groups. The table shows that the agbiotech and seed companies, on average, owned a greater number of agbiotech patents and PVPCs than the larger, diversified companies. Although the mean number of field trials was greater for the larger companies, the difference was not statistically significant. Within the group of large companies, Monsanto holds (by far) the greatest number of patents (170) and PVPCs (45) in our sample, and also conducted the greatest number of field trials (1035). This reflects the scope of Monsanto's agbiotech research efforts in this sample, a position that is widely accepted to be unique in the world. In fact, Monsanto is the only large multinational in our sample with the majority of its resources devoted to agriculture. Of the smaller seed and agbiotech companies, Pioneer Hi-Bred had the strongest innovation numbers with 460 patents, 322 PVPCs, and 582 field trials approved.

R&D expenditures by the larger companies are statistically greater (on average) in large part because separate data on their agricultural R&D

[1] R&D expenditures were deflated using an R&D price deflator constructed by Fuglie *et al.* (1996).

Table 16.2. Measures of innovative activity by company type.

Company type	Number of agbiotech patents (1985–1999)	Per cent of total agbiotech patents	Number of field trials (1987–2000)	Per cent of total field trials	Number of PVPCs (1990–1999)	Per cent of total PVPCs
Biotech/seed	738	30	1099	15	448	22
Large/diversified	297	12	2310	31	50	3
Total	2447	42	7442	46	2023	25

PVPCs, Plant Variety Protection Certificates.

Table 16.3. Descriptive statistics for innovative activity by company type.

	Agbiotech patents	Field trials	PVPCs	R&D exp. (million US$)	Annual change in R&D expend. (%)	R&D intensity (expend./sales)
Biotech/seed						
Mean	8.29*	12.35	7.00*	34.44*	11.07	0.45*
(SE)	(2.00)	(2.40)	(1.59)	(4.62)	(2.95)	(0.08)
Median	2.00	4.00	0.00	16.44	6.78	0.13
SD	18.86	22.67	12.72	40.57	24.67	0.70
Skewness	3.64	3.15	1.97	1.59	1.14	2.35
Kurtosis	16.75	13.79	5.89	4.52	5.83	8.17
Observations	89	89	64	77	70	70
Large/diversified						
Mean	2.78*	22.65	0.75*	897.02*	7.23	0.07*
(SE)	(0.72)	(6.82)	(0.23)	(67.09)	(4.52)	(0.005)
Median	1.00	1.00	0.00	800.33	0.13	0.06
SD	7.42	68.87	1.88	643.52	41.65	0.05
Skewness	6.40	4.34	2.61	1.47	4.02	0.91
Kurtosis	51.79	22.99	8.97	6.48	23.66	2.99
Observations	107	102	67	92	85	90

*The means are significantly different at the 5% level. The hypothesis test is H_o: $\beta_{small} = \beta_{large}$. Based on the F-statistic, we fail to reject H_o if the P-value > 0.05.
PVPCs, Plant Variety Protection Certificates.

is not available for these larger companies. Because the available statistics include non-agricultural activity, more informative innovation measures from firm financial data may be research growth and intensity. The average growth in R&D expenditures was statistically greater for the agbiotech and seed companies, which is not surprising given their smaller base level. However, the R&D intensity (i.e. R&D expenditures divided by sales) of the smaller companies was also significantly greater, 45% of sales compared with 7% for the larger companies. Of the companies analysed in the group of smaller firms, the agbiotech firms such as Calgene, DNA Plant Technology, Mycogen and Crop Genetics International had extremely high R&D intensity measures, ranging from 31% to 190%. These high levels of R&D intensity primarily reflect the low sales levels of the small, agbiotech firms. The agbiotech firms were mainly research companies with few products on the market.[2]

Two financial characteristics related to firm size are sales[3] and number of employees. As

[2] Throughout this chapter, we note that there are several financial differences between agbiotech and seed firms. It would be more informative to separate these firms based on their different lines of business, but insufficient observations prevented a separate analysis.

[3] Gross sales were deflated using the Gross Domestic Product implicit price deflator.

mentioned above this data is for all lines of business not just those related to agriculture. This particularly affects the interpretation of the large company results because they are more diversified than the small firms. The larger companies, of course, have higher average sales and number of employees than the smaller biotechnology companies, about US$17 billion in sales compared with US$362 million, and 61,000 employees compared with 1300 (Table 16.4). Sales growth and change in numbers of employees are also indicators of firm growth, and are more useful for comparisons of large and small companies. The growth rates for both sales and employees were significantly higher for the seed and agbiotech firms than for the larger, diversified firms. An examination of the individual annual means for the larger companies reveals that, except for Rhône-Poulenc, the diversified companies reduced their number of employees on average from the mid-1980s to the late 1990s. In fact, sales growth was negative for American Cyanamid, ICI and Monsanto during this period. This result was unexpected, especially for Monsanto, which acquired a number of firms during this period. However, at the same time that it was buying seed and biotech companies it was divesting itself of commodity chemicals, and nutritional products. All of the seed and agbiotech companies experienced growth in both sales and employees.

Additional insight on firm size may be obtained from examining changes in shareholder's equity and earnings per share (EPS) (Table 16.5). The equity growth rate in the table is the annual percentage change in shareholder equity interest in a firm, or firm net worth.[4] EPS is net income minus dividends divided by the average number of outstanding common shares. The agbiotech and seed companies experienced negative average changes in equity, about –15%, compared with positive (about 7%) growth for the larger, diversified companies. A similar result was found for EPS or profit earned per share from the annual percentage changes. The agbiotech and seed companies averaged about a 20% annual decline in EPS, whereas the mean for the larger companies was close to 52% growth. Of the smaller companies, the well established seed companies, DeKalb, Delta and Pine Land and Pioneer Hi-Bred, were the only firms that had equity growth and increases in EPS. This suggests that the overall financial well being of the agbiotech companies was uncertain. This, in part, may explain why the larger companies acquired many of the agbiotech companies, accessing the research assets of these companies at fairly reasonable prices compared with the prices being paid for seed company acquisitions (i.e. millions versus billions of dollars).

Another factor contributing to the more stable equity and EPS figures for the larger companies is that they have more options at their disposal to modulate and maintain their EPS than smaller companies. The larger companies often go to great pains (some in other industries have recently even violated the law) to show positive annual and quarterly results. This is partly because of the extreme effect that negative operating results have on share prices and also because executive compensation is often tied to performance, so companies use acquisitions, spin-offs and judgement calls in accounting for extraordinary items to put their best face forward. Of the larger, diversified companies, WR Grace was the only company with declining equity and EPS over this time period.

Gross profit margin (GPM) measures the markup on goods sold and is an indicator of a company's pricing strategy and financial health.[5] A high markup means that a company is better able to cover costs and other operating expenses through sales. The agbiotech and seed companies had close to a 29% markup, whereas the larger companies averaged about 42%. The lower GPM for the agbiotech and seed companies is an average (Table 16.4), which for several companies includes start-up years when more products were in the development pipeline and not yet commercialized. Additionally, from 1992 to 1995, Calgene's GPM declined rapidly, Crop Genetics International's GPM was very erratic and DNA Plant Technology had negative GPM (i.e. costs were greater than sales) in the last

[4] Equity is primarily measured as total assets minus total liabilities.

[5] GPM is calculated as net sales minus the cost of goods sold. This total is then divided by net sales and multiplied by 100.

Table 16.4. Descriptive statistics for firm characteristics by company type.

	Sales (million US$)	Annual change in sales (%)	Employees (1000s)	Annual change in employment (%)	Gross profit margin	Annual equity growth (%)	Annual change in EPS (%)	Debt-to-equity	Assets-to-equity	Debt-to-assets
Biotech/seed										
Mean	362.33*	27.52*	1.31*	15.80*	28.69*	–14.80*	–19.94*	0.32	1.65*	0.15*
(SE)	(62.42)	(8.38)	(0.23)	(7.71)	(6.06)	(8.71)	(19.56)	(0.05)	(0.07)	(0.02)
Median	115.40	6.93	0.45	1.48	41.04	–0.87	8.25	0.19	1.46	0.12
SD	525.96	67.06	1.82	58.72	52.85	73.35	161.26	0.36	0.64	0.12
Skewness	1.52	3.79	1.36	3.71	–3.87	–1.62	–3.24	1.64	1.44	0.96
Kurtosis	3.89	19.71	3.01	18.11	20.54	11.95	16.33	4.93	4.91	2.90
Observations	71	64	65	58	76	71	68	56	77	56
Large/diversified										
Mean	16,975.98*	2.53*	61.02*	0.89*	41.89*	7.01*	51.70*	1.33	3.80*	0.26*
(SE)	(1,200.03)	(2.83)	(4.28)	(2.96)	(1.36)	(3.27)	(27.54)	(0.51)	(0.94)	(0.01)
Median	16,279.21	1.56	46.13	–1.54	36.53	6.79	9.32	0.66	2.53	0.24
SD	12,060.18	27.71	40.11	26.65	13.50	32.53	255.37	5.11	9.44	0.10
Skewness	0.55	2.29	0.84	4.62	0.97	5.53	3.79	4.96	4.32	1.04
Kurtosis	2.29	15.96	2.47	28.80	2.70	55.27	18.85	35.39	32.15	3.99
Observations	101	96	88	81	98	99	86	101	101	100

*The means are significantly different at the 5% level. The hypothesis test is H_o: $\beta_{small} = \beta_{large}$. Based on the *F*-statistic, we fail to reject H_o if the *P*-value > 0.05.
EPS, earning per share.

2 years before being purchased by Savia. As we have noted, agbiotech company acquisitions were often made at basement sale prices, probably to get access to intellectual property and intangible assets such as scientific personnel. This is in contrast to many of the information technology acquisitions that were made at the height of the technology boom in the late 1990s that were made on the basis of projected future sales (that have since not materialized).

The more established seed companies, Pioneer Hi-Bred, DeKalb, and Delta and Pine Land, had higher and more stable GPMs than the agbiotech companies. Within the group of larger companies, American Cyanamid, Monsanto and Rhône-Poulenc drove up the average GPM for their group. One interesting point to note is that, for each year from 1996 to 2000, Monsanto had a GPM that hovered close to 60%. This is the time period that marked the first introduction of genetically modified crops. Prior to this time period, Monsanto's GPM was closer to 40% over several lines of business. Pharmaceutical companies often argue that long-term uncertain investments in R&D require a substantial markup. In contrast, we are finding that these agbiotech and seed companies were financing long-term investments in similar research at much lower profit margins. The Food and Drug Administration approval process can be quite lengthy for pharmaceuticals, possibly explaining some of this difference in margins.

Measures derived from company balance sheets can indicate firm financial stability and ability to recover from financial shocks and finance continuing operations. The leverage ratio, or the ratio of debt to shareholder equity, indicates the extent to which a company has financed asset acquisition through debt obligations. A leverage ratio greater than one implies that debt, rather than equity, provides a majority of asset financing. When this ratio is high, a company will have higher interest costs (all else equal) both because of the level and rate for its financing. A comparison of means shows that the larger companies financed a majority of their assets with debt rather than equity, whereas the smaller biotechnology companies did the reverse. The ratios are 1.33 and 0.32 (Table 16.4) respectively, a difference that is not statistically significant. The ratio of firm assets to shareholder equity is similarly larger for the big companies and this difference is significant. Debt-to-assets are also almost twice as large at the big companies.

Relating Firm Characteristics to Innovative Activity (Tables 16.5 and 16.6)

Correlations between firm characteristics and measures of research draw together the two major threads of this chapter. For the smaller biotechnology and seed companies the non-accounting measures of innovation, agbiotech patents (lagged 10 years),[6] PVPCs, and the number of field trials, were all positively correlated with sales, R&D expenditures, and number of employees. This suggests that for these companies, more research input may have led, quite appropriately, to more research output. Growth in sales and employees were negatively correlated with innovative output, perhaps indicating decreasing returns to scale in the production of technology within the group of smaller companies. Equity growth and the assets-to-equity ratio are also positively related to the production of new technology measured as patents. It is quite likely that as patent output increases it attracts investors, boosting the net worth of a firm. However, it is also possible that greater net worth allows more effective exploitation of research opportunities leading to increased research output. Assets-to-equity (leverage) is positively correlated with innovation suggesting the same kind of bi-directional causality. As a company wins more patents, its assets also rise, and greater assets may enable increased innovative activity.

It is also informative to consider how inputs into innovative activity, such as R&D

[6] No attempt was made to estimate the lag structure for this analysis. A simple lag was employed that is consistent with earlier work that found long lags between agricultural R&D and productivity (Pardey and Craig, 1989; Schimmelpfennig and Thirtle, 1999). This also is the longest lag that can be supported that retains any explanatory power given data availability.

Table 16.5. Selected correlation coefficients for seed/agbiotech companies: innovative activity vs. firm characteristics.

	Patents	Patents (10 year lag)	Field trials	PVPCs	ln R&D	R&D intensity	Growth in R&D
ln sales	–	0.811	0.486	0.708	0.869	–0.740	–
Growth in sales	–	–0.586	–	–	–	–	–
ln employees	0.503	0.897	0.494	0.762	0.865	–0.585	–
Growth in employees	–	–0.659	–	–	–	–	–
Gross profit margin	–	–	–	0.495	0.406	–0.420	–
Growth in equity	–	0.615	–	–	–	–	–
Assets-to-equity	–	0.501	–	–	–	–	–
ln R&D	0.550	0.799	0.564	0.767	1.000	–0.424	–

–The absolute value is less than 0.4.
*The R&D expenditure, sales and employee variables are represented as natural logs (ln) to reflect the non-linearity of these variables.
PVPCs, Plant Variety Protection Certificates.

Table 16.6. Selected correlation coefficients for large, diversified chemical/pharmaceutical companies: innovative activity vs. firm characteristics.

	Patents	Patents (10 year lag)	Field trials	PVPCs	ln R&D	R&D intensity	Growth in R&D
R&D intensity	0.438	–	0.433	0.460	–	1.000	–
ln sales	–	–	–	–	0.755	–0.531	–
Growth in sales	–	–	–	–	–	–	0.618
ln employees	–	–	–	–	0.636	–0.526	–
Growth in employees	–	–	–	–	–	–	0.540
Gross profit margin	–	–	–	–	–	0.788	–

–The absolute value is less than 0.4.
*The R&D expenditure, sales and employee variables are represented as natural logs (ln) to reflect the non-linearity of these variables.

expenditures and intensity, are affected by changes in business performance indicators. For the smaller firms, sales and the number of employees were positively related to R&D expenditures, but interestingly, not to R&D intensity. The first result was not unexpected because these different firm size measures usually vary together. The negative correlation with R&D intensity, however, suggests that the rate of increase of R&D expenditures is less than that of sales and employees. Perhaps this indicates that as firm size increases, within this group of smaller companies, R&D investments do not increase proportionally. This could suggest scale economies with increases in firm size or diminishing returns to scale for R&D, at least among the small companies. Furthermore, GPM was positively associated with R&D expenditures, but is negatively related to R&D intensity. Since larger margins are related to greater R&D investments, either one could reasonably lead to the other, but the negative relationship with R&D intensity suggests that once companies achieve higher sales they cut back on risky R&D projects.

Many of the correlation coefficients calculated for the group of larger, diversified firms demonstrate weak associations between innovative output and firm financial characteristics. This suggests problems related to the lack of agbiotech market segment information for the larger companies. Except for a positive association between the agbiotech innovation measures and R&D intensity, none of the other firm characteristics were correlated with the innovation output measures. For the large companies, the fact that R&D spending and innovation output go together could simply reflect the fact that all companies discontinue (with a lag) unproductive lines of research.

More revealing is that for the diversified firms, sales and employee numbers were positively related to R&D expenditures, and negatively related to R&D intensity as they were for the small companies. R&D scale economies may be implicated, but once again it could be firm size scale economies or decreases in R&D returns to scale. In contrast to the small companies, large company growth in sales and employees are positively correlated with R&D growth. Another contrasting result is that GPM is positively correlated with R&D intensity. There are any number of stories that could explain why profit and intensity (R&D divided by sales) go up together for the large companies but not for the small ones, but the important point is that sales and profits for the diversified firms increase with R&D investments. We cannot tell from this kind of analysis, whether profits or R&D come first.

Characterizing Firm Heterogeneity

Analysis of variance (ANOVA) can be used to provide at least two other kinds of information about the relationship between firm characteristics and innovative activity. When the large firms are grouped together, as are the small firms, ANOVA tells us that for most variables 80–90% of the variance is 'within' each group. The small remainder of variance is 'between' the two groups. We interpret this to mean that by analysing these groups separately we are considering most of the variability and have chosen our groups appropriately. This is not true for the size variables (R&D, sales and employees), which as indicated above gave some unexpected results. For size, perhaps all of the companies need to be pooled, or possibly other groupings are appropriate.

Once the companies are separated into large and small firms, the companies can be considered as individual units using ANOVA (this is a more standard approach). These results, presented in Table 16.7, show how much of the variance is 'within' each company and how much is 'between' companies (in that group). If there is a lot of within firm variation (more than between firm variation) it indicates that the firm as a unit of analysis is appropriate, firms should not be lumped together, and firm heterogeneity is important. We find that there is more within firm variation for patents and field trials, and more between firm variation in PVPCs for both groups. The firm growth variables, such as growth in sales, R&D expenditures, employees, EPS, and equity, had more within than between firm variation. This is also true for most of the leverage and liquidity ratios. These findings agree with Geroski's (1998) for other industrial classifications (see also Klette and Griliches, 2000) that within firm variability of growth is greater than between firm variation. Our results therefore also demonstrate his main points that a great deal of firm heterogeneity is to be found even within an industry, structural models need to be designed to exploit it, and pooling may not be appropriate. Only the size variables (and intensity) have most of their variance between firms within each group, indicating that the process of finding appropriate groups for the analysis of firm size begun in the previous paragraph will not end here.

Financial variables (GPM and assets-to-equity) were the only ones with different within/between splits for large and small company groups. For small companies most of the GPM variation is within (and large company is between), whereas for small companies assets-to-equity variation is between (and large company is within). These results reinforce our earlier conclusion that large companies maintain stable profits if at all possible, and have asset bases that vary considerably sometimes along with their M&As.

Conclusions

This chapter presents a preliminary analysis of 15 important biotech firms for which data on firm characteristics, innovative activity, and financial performance is available. Our goal is a better understanding of the factors that influence innovative behaviour and a better understanding of the mergers and acquisitions that took place in this industry.

The data on research inputs and outputs show important differences between the small

Table 16.7. Analysis of variance: by company in each group.

	For the agbiotech/seed companies		For the large/diversified companies	
	Percentage '*within*' variation	Percentage '*between*' variation	Percentage '*within*' variation	Percentage '*between*' variation
R&D variables:				
Agbiotech patents	70	30	76	24
Field trials	65	35	59	41
PVPCs	20	80	28	72
R&D intensity	32	68	37	63
Size variables:				
ln R&D	12	88	32	68
ln Sales	12	88	13	87
ln Employees	12	88	19	81
Growth variables:				
Growth in R&D	83	17	89	11
Growth in sales	76	24	98	2
Growth in employees	93	7	85	15
Change in EPS	88	12	97	3
Growth in equity	81	19	99	1
Other financial variables:				
Gross profit margin	77	23	13	87
Assets-to-equity	36	64	93	7
Debt-to-assets	67	33	71	29
Debt-to-equity	62	38	93	7

EPS, earnings per share.
PVPCs, Plant Variety Protection Certificates.

agbiotech and seed companies and the large, diversified chemical and pharmaceutical companies. The small firms are much more innovative by all measures of innovation and are far more research intensive. Their research budgets were also growing more rapidly than the budgets of the large firms. However, their financial situation was clearly not as strong as the large firms. All of the agbiotech and seed companies in our sample, except for Delta and Pine Land, were acquired by larger firms (some represented in our analysis here). One concern that these acquisitions raise, is that research inputs and research outputs will be reduced as these small innovative firms are acquired by their larger, less innovative rivals.

There are probably a number of reasons why the large companies bought up most of the biotech/seed companies. These smaller, specialized companies were attractive investments for the larger, diversified companies which had the financial assets and were trying to increase their market value, intellectual capital, seed market dominance and distribution networks, R&D complementarities, and scale and scope economies. The agbiotech companies were eager for the greater financial support they would expect to receive as part of a diversified, multinational company. The parent would be able to assume risks and costs of bringing products to market that might have been difficult to bear. Uncertainty in the markets for GE crops, regulatory costs for obtaining approval of GE crops and limited agricultural market knowledge, may have all entered the equation. The large companies may have seen lower average costs of research as they gained access to important research tools, facilities and human capital through consolidation (King, 2001). The data assembled for this chapter and the analysis of these companies will now be used to develop econometric models that will hopefully enable us to assess whether these mergers and acquisitions have had a positive impact on biotech innovation. We also intend to consider what factors most often lead to mergers and acquisitions.

Acknowledgements

The views expressed are those of the authors and are not necessarily those of the US Department of Agriculture. The authors are especially appreciative to Paul Heisey, Jorge Fernandez-Cornejo and Kelly Day-Rubenstein for their thoughtful comments and to Qiang Chen and David Lee for help with the data.

References

[AMS] Agricultural Marketing Service, US Department of Agriculture (accessed 2002) Plant Variety Protection Office. http://www.ams.usda.gov/science/PVPO/pvp.htm

[APHIS] Animal and Plant Health Inspection Service, US Department of Agriculture (accessed 2002) Field trial database. http://www.nbiap.vt.edu/cfdocs/fieldtests1.cfm

Brennan, M.F., Pray, C.E. and Courtmanche, A. (2000) Impact of industry concentration on innovation in the US plant biotech industry. In: Lesser, W.H. (ed.) *Transitions in Agbiotech: Economics of Strategy and Policy*. Food Marketing and Policy Center, University of Connecticut, Storrs, Connecticut.

Cohen, W.M. and Levin, R.C. (1989) Empirical studies of innovation and market structure. In: Schmalensee, R.L. and Willig, R.D. (eds) *Handbook of Industrial Organization*, Vol. II. Elsevier Science Publishers, Amsterdam.

Compustat PC Plus (2000) New York: Standard & Poor's. http://www.compustat.com/www/

Dibner, M. (1997) *Biotechnology Guide U.S.A.* Institute for Biotechnology Information, Research Triangle Park, North Carolina.

Foltz, J., Kim, K. and Barham, B. (2001) *A Dynamic Count Data Analysis of University of Ag-Biotech Patents*. Research Report 56. Food Marketing Policy Center, University of Connecticut, Storrs, Connecticut.

Fuglie, K., Ballenger, N., Day-Rubenstein, K., Klotz, C., Ollinger, M., Reilly, J., Vasavada, U. and Yee, J. (1996) Agricultural research and development: public and private investments under alternative markets and institutions. *Agricultural Economic Report* No. 735, May. http://www.ers.usda.gov/publications/aer735/

Fulton, M. and Giannakas, K. (2001) Agricultural biotechnology and industry structure. *AgBioForum* 4(2), Fall. http://www.agbioforum.org

Geroski, P.A. (1998) An applied econometrician's view of large company performance. *Review of Industrial Organization* 13, 271–293.

Graff, G.D., Rausser, G.C. and Small, A.A. (2002) Agricultural biotechnology's complementary intellectual assets. *Review of Economics and Statistics* 85(2), 349–363.

Griliches, Z. (1990) Patent statistics as economic indicators: a survey. *Journal of Economic Literature* 28(4), 1661–1707.

Hayenga, M. (1998) Structural change in the biotech seed and chemical industrial complex. *AgBio-Forum* 1(2), 43–55. http://www.agbioforum.org

Kalaitzandonakes, N. (1999) Biotechnology and Agrifood Industry Competitiveness. In: Amponsah, W., *et al.* (eds) *The Competitiveness of U.S. Agriculture*. Hayworth Press.

Kalaitzandonakes, N. and Hayenga, M. (1999) Structural change in the biotechnology and seed industrial complex: theory and evidence. Presented at the *Transitions in Agbiotech: Economics of Strategy and Policy Conference*, June 24–25, Washington, DC.

Kamien, M.I. and Schwartz, N.L. (1982) *Market Structure and Innovation*. Cambridge University Press, Cambridge, UK.

King, J. (2001) Concentration and technology in agricultural input industries. *Agricultural Information Bulletin, No. 763*, Economic Research Service, USDA. March. http://www.ers.usda.gov/publications/aib763/

Klette, J. and Griliches, Z. (2000) Empirical patterns of firm growth and R&D investment: a quality ladder model interpretation. *The Economic Journal* 110(April), 363–387.

Klotz-Ingram, C. and Day-Rubenstein, K. (2002) Agricultural research and development. *Agricultural Resources and Environmental Indicators*, Economic Research Service, USDA. http://www.ers.usda.gov/Emphases/Harmony/issues/arei2000/areich5_2.pdf

Lanjouw, J.O., Pakes, A. and Putnam, J. (1998) How to count patents and value intellectual property: uses of patent renewal and application data. *The Journal of Industrial Economics*. XLVI(4), 405–433.

Oehmke, J., Wolf, C. and Raper, K. (2002) Cyclical concentration and consolidation in agricultural biotech R&D. Presentation at the International Consortium on Agricultural Biotechnology Research (ICABR)'s 6th *International Conference on Agricultural Biotechnologies*, Ravello, Italy, presented 11 to 14 July, 2002.

Pardey, P. and Craig, B. (1989) Causal relationships between public sector agricultural research expenditures and output. *American Journal of Agricultural Economics* 71(1), 9–19.

Schimmelpfennig, D.E., Pray, C.E. and Brennan, M.F. (2004) The impact of seed industry concentration on innovation: a study of U.S. Biotech Market Leaders. *Agricultural Economics* (in press).

Schimmelpfennig, D.E. and Thirtle, C.G. (1999) The internationalization of agricultural technology: patents, R&D spillovers and their effects on productivity in the European Union and United States. *Contemporary Economic Policy* 17(4), 457–468.

Schmalensee, R. (1988) Industrial economics: an overview. *Economic Journal* 98, 643–881.

US Patent and Trademark Office (accessed May 2002) Searchable database, http://www.uspto.gov/patft/index.html

17 Dynamic Pricing of GM Crop Traits

Richard Perrin and Lilyan Fulginiti
Department of Agricultural Economics, University of Nebraska, Lincoln, Nebraska, USA

Abstract

The issue considered here is the retail pricing of patented crop traits such as Roundup Ready® herbicide resistance or *Bt* insect resistance. Our concern is not with the price of the seeds in which the traits are embodied, but rather with the implicit or explicit price for the traits themselves. Intellectual property rights are now available for traits, and while monopoly pricing of them has received some limited consideration in the economics literature,[1] no one has yet examined the possible implications of the durability of these traits as a factor in determining such monopolists' pricing behaviour.

Monopoly Pricing of Durable Goods

The theory of monopoly pricing of durables traces to Coase (1972). He noted that when the seller of a new durable good sets a price in the first period, a fraction of potential customers will buy, but the remaining fraction still remain as potential customers in the next period. At a lower price in that next period, a fraction of the remainder will buy, and similarly for the period after that. The seller clearly has a strong incentive to exploit this kind of price discrimination through time. However, buyers will probably anticipate this behaviour, and thus have an incentive to wait for next period's lower price. It is difficult for the seller to make a credible commitment that he will not reduce the price in the next period, given the obvious incentive to do so.

Thus Coase perceived a strategic game being played between the seller of the durable and his potential buyers. The seller's strategy for reducing future prices must be compatible with the buyers' incentives to wait for a lower price in the future. Buyers' incentive to wait can be weakened by a credible commitment that prices will not in fact fall in the future, but this credibility is difficult to establish. The outcome of the game, in terms of an equilibrium pricing strategy through time, is not obvious. Coase concluded that it is very likely that the equilibrium price will fall all the way to marginal cost (zero in the situations considered in this chapter) in every period. In this case the monopolist earns no rents, let alone the 'normal' monopoly rent obtainable by charging a single once-and-for-all monopoly price, or the even larger rent from intertemporal price discrimination. This conclusion has become known as the 'Coase conjecture'.

[1] See Perrin (1994), Moschini and Lapan (1997), Giannakas (2002).

 The Regulation of Agricultural Biotechnology (eds R.E. Evenson and V. Santaniello)

In this chapter we first discuss the durability of crop traits and how it is determined by technological considerations and by intellectual property rights. We then consider an equilibrium pricing strategy emerging from a specific formulation of a pricing game that is dependent on the nature of intellectual property rights.

Technology, Property Rights, and the Durability of Crop Traits

For purposes of this analysis, a durable good is an input that provides a flow of services for more than one production cycle. When seed is purchased, the producer acquires a bundle of traits, each of which can be thought of as providing a flow of services for the current crop year, and if the flow of services of a trait extends beyond that year, the trait may properly be considered a durable good. Varieties of crops such as soybeans and wheat are created by a recurrent selection process from which only phenotypically identical, self-replicating plants emerge. If seeds from such a crop are saved and replanted, the traits persist into subsequent years, and are thus durables. For crops such as maize, however, successful new cultivars are most often created by hybridization, produced by the crossing of two or more distinctly different genotypes. While the first generation of this cross is designed to be a highly uniform phenotypic population for the commercial crop, the traits expressed by subsequent generations can be disastrously heterogeneous. A trait expressed by a hybrid is therefore not a durable.

There are at least two other technological phenomena that may affect the durability of crop traits. The first is 'terminator' technology, such as the Technology Protection System (TPS) owned jointly by Delta and Pine Land Co. and US Department of Agriculture (USDA). Terminator seeds either produce a crop of sterile seeds, or a crop of seeds in which the trait in question is switched off, in either case insuring that the trait at issue is not a durable. The second technology is apomyxis, currently being developed by Pioneer and Centro Internacional de Mejoramiento de Maiz y Trigo (CIMMYT), also not yet commercially viable. Apomyctic seeds produce a crop of viable seeds that are genetically identical to the maternal plant. Seeds saved from an apomyctic hybrid crop will replicate the commercial hybrid, thus insuring that all of the traits in the crop are durables.

However, even if a crop trait is technologically durable, the seller may be able to exclude a customer from future use of the trait. This possibility is determined by the system of intellectual property rights and enforcement mechanisms to which a technologically durable crop trait is subject. The two systems of intellectual property rights that are relevant to crop traits are utility patents and plant breeders' rights (Plant Variety Protection Act or PVPA in the USA and Union for the Protection of Varieties or UPOV in much of the rest of the world). If a crop trait is protected by a utility patent, the seller will have the right to exclude the buyer from using the trait in subsequent years if he wishes to do so, whereas if it is protected by plant breeders' rights, the seller does not have that right (he only has the right to exclude the buyer from giving or selling the trait to other producers.) While utility patents are clearly the stronger form of property rights, they are not everywhere available, and they are more expensive to obtain. Within either system of property rights, however, the degree to which the seller is able to exclude future use of a durable trait depends on his enforcement effort and on the reliability and cost of the legal system through which enforcement takes place.

To summarize, if we have a trait that is technologically durable, the effect of patent protection is to allow the seller to exclude its use as a durable, while under breeders' rights the trait is a legal durable. We turn now to an analysis of how these alternatives might affect the sellers' choice of pricing strategy through time.

Property Rights and the Pricing of a Non-durable Crop Trait

We first consider the pricing of a non-durable trait, which is similar to the pricing problem facing any seller with a downward-sloping demand curve. Consider Fig. 17.1, for example, in which we present a demand curve that is derived from a schedule of heterogeneous users' valuations, υ, of the expected benefit of a

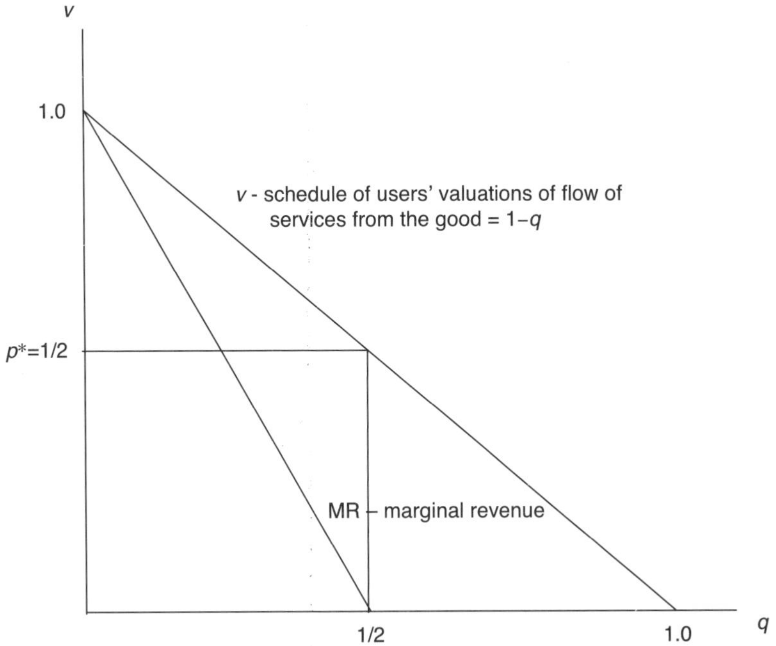

Fig. 17.1. User evaluations of annual benefit of a trait.

particular trait for one crop year on, say, 1 ha. We have scaled the function so that the valuation of the highest-valuation user is set at 1.0, and the total number of users (or hectares) deriving any benefit at all from the trait is also set at 1.0. The valuation curve $v = 1 - q$ can reasonably be considered to be the demand curve facing the owner of the trait. We assume that the marginal cost of incorporating a trait in seed for additional crop area is essentially zero.

In this stylized case with linear demand, the trait owner maximizes profit by setting the standard monopoly price every year, $p^* = ½$, resulting in adoption (purchase) of the trait by $q = ½$ of the potential users every year. The stream of monopoly rents realized is thus $r^* = ¼$, with present value $PV^* = k/4$, where k is the capitalization rate, presumed here to be the present value of a T-year annuity starting 1 year from the present, or $k = (1 - (1 + i)^{-T})/i$, where i is the discount rate.

Where property rights for the crop trait are not perfect and costless to enforce, the patent owner may not be able to exclude all pirating, or he may find it too costly to do so. The effective demand curve is not evident in this case. Deardorff (1992) and Perrin (1994) suggested that weakly-enforced property rights would result in payments only from some randomly-selected fraction θ of potential customers. This proportional pirating model implies that the quantity demanded is fraction θ of the quantity indicated by demand curve v, or line $v^{pp} = 1 - q/\theta$ in Fig. 17.2. The optimal monopolist price remains at p^*, but the optimal quantity to sell diminishes to $\theta/2$, the annual flow of rents falls to $r^{pp} = \theta/4 = \theta r^*$, and present value of rents falls to $PV^{pp} = \theta k/4 = \theta PV^*$.

Alternatively, Diwan and Rodrik (1991), followed by Perrin (1999), suggested that in the presence of weak property rights there is a limit royalty price, equal to some fraction ϕ of the valuation for each customer, above which piracy would occur. This demand curve has height equal to fraction ϕ of curve v, or line $v^{lp} = \phi - \phi q$ in Fig. 17.2. The optimal monopolist price falls to $\phi/2$, while the optimal quantity to sell remains at ½. The annual flow of rents is $PV^{pp} = \phi k/4 = \phi PV^*$, the same as for the proportional pirating case if the fractions ϕ and θ are equal. Giannakas (2002) suggests further that this limit pricing fraction ϕ may be determined by the customer's expected cost of being caught pirating the trait, which is in turn determined by enforcement costs

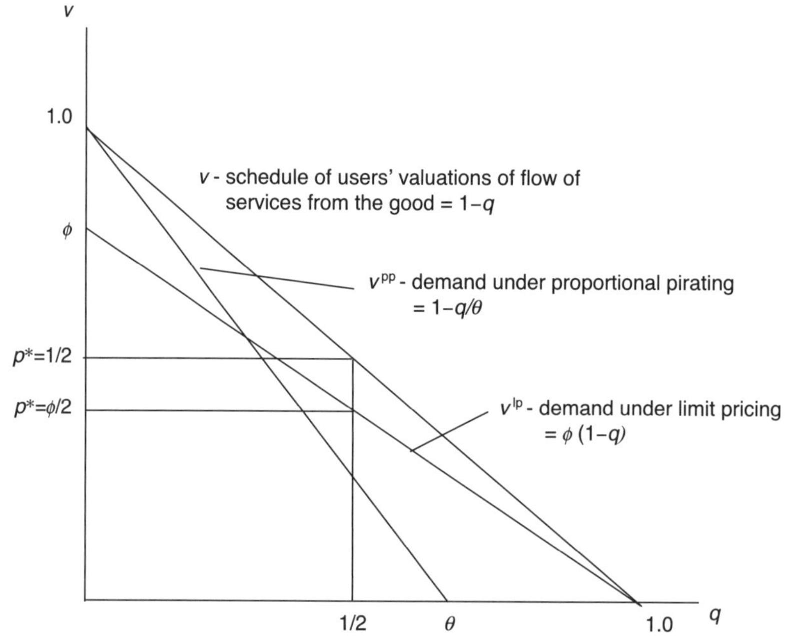

Fig. 17.2. Effective demand with imperfect property rights.

as well as by the nature of the patent system. This allows him to explore the static pricing of the trait within the framework of a regulatory game in which buyers, the monopolist, and the regulator are players.

The two theories above offer alternative explanations as to how a simple linear valuation schedule is transformed into the monopolist's derived demand curve when property rights are less than perfect. Neither is particularly persuasive, since it seems likely that potential customers' willingness to pirate is distributed in a way that is neither strictly random as implied by the first, nor strictly proportional to the expected benefit of pirating, as implied by the second. However, either approach is analytically convenient, and we will use the limit pricing approach in the analysis of durable pricing to follow.

Nash Equilibrium Pricing of Durable Traits

We now consider an explicit theoretical model of Coase's durable goods pricing theory to examine what intertemporal pricing strategies might emerge, and how they would be affected by property rights. Here we seek a Nash equilibrium solution to the game, which will insure the credibility of the resulting time path because by definition, none of the players in the game will have an incentive to behave otherwise. We convert the 1-year valuation curve of Fig. 17.1 to a valuation curve for the durable good using the capitalization rate $k = (1 - (1 + i)^{-T})/i$. The durable good valuation curve is $V = kv = k - kq$, shown in Fig. 17.3. Given this effective demand curve, the monopolist could charge some arbitrary price P_1 for the durable the first year, then in the second year charge the monopolist price for the remaining portion of the demand curve, $k/4$, etc., and in this manner extract most of the consumers' surplus.

However, buyers will anticipate this reduction in price, and a buyer with valuation kv and discount factor $\delta = 1/(1 + i)$ will have an incentive to wait until next year to purchase if $(V - P_t) < \delta(V - P^e_{t+1})$, where P^e_{t+1} is the price he expects to be charged for the durable next year. Given this buyer incentive to wait, just how much will the seller decide to charge in the first year and how fast will the price fall? A considerable number of papers have been published

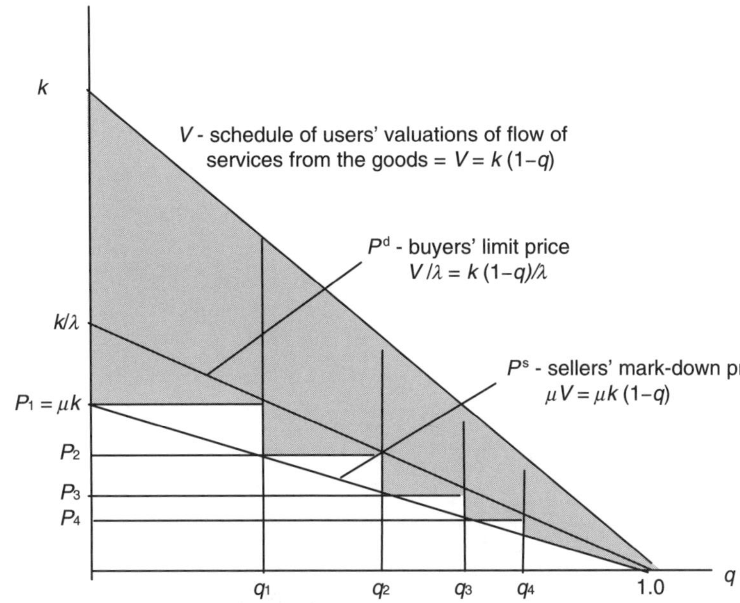

Fig. 17.3. Nash equilibrium pricing of a durable trait.

establishing conditions under which Coase's zero-profit conjecture would hold (see Tirole, 1988, Chapter 1). Here we adapt a relatively simple model that Tirole in turn adapted from Sobel and Takahashi (1983).

Consider first the case of plant breeders' rights with costless enforcement. If the monopolist could credibly establish that the trait would never be sold again, a one-time price $P^* = k/2 = kp^*$ could be charged, maximizing profits by selling only to the half of customers with the highest valuations. But it is difficult for the owner to assert credibly that the trait will never be sold again, and if so the initial price must be compatible with the buyer's incentive to wait for next year's lower price. We assume here that buyers' strategy is to identify an optimal limit price fraction λ such that they will purchase if $V = kv > \lambda P$. The effective demand curve is then $P^d = V/\lambda = k(1-q)/\lambda$ in Fig. 17.3, similar to the limit-pricing demand curve of Fig. 17.2. At the price marked P_1, buyers would purchase quantity q_1, realizing a surplus equal to the shaded area above the line $P_1 = \mu k$, leaving the monopolist the rent below it. We assume that the seller's strategy is to identify an optimal mark-down ratio, μ, such that if the buyers with valuations above $V = kv = k(1-q)$ have already purchased the trait and the others have not, then he will set the price at $P = \mu V$. This implies that the seller follows a pricing curve such as $P^s_t = \mu V_{t-1} = \mu k(1 - q_{t-1})$ in Fig. 17.3. This seller's behaviour implies that the seller will charge an initial price $P_1 = \mu k$. The buyers' behaviour implies that the initial quantity purchased will be $q_1 = 1 - \mu\lambda$, which in turn from the seller's behaviour implies that $P_2 = \mu\lambda P_1 = k\lambda\mu^2$ and $q_2 = 1 - \lambda P_2/k = 1 - (\lambda\mu)^2$, or in general, $P_t = k\lambda^{t-1}\mu^t$ and $q_t = 1 - \mu^t\lambda^t$ (here note that q_t represents the total quantity sold since the first period, $t = 1$).

The seller chooses the initial price to maximize the present value of future sales,

$$\begin{aligned} \text{PV} &= P_1q_1 + \delta P_2(q_2 - q_1) + \delta^2 P_3(q_3 - q_2) \\ &\quad + \ldots \\ &= P_1(1 - \lambda P_1/k) + \delta P_2(\lambda P_1/k - \lambda P_2/k) + \ldots \end{aligned}$$

Setting the derivative with respect to P_1 equal to zero yields $1 - 2\lambda P_1/k + \delta P_2\lambda/k = 0$, and since $P_1 = \mu k$ and $P_2 = \mu\lambda P_1$, then $1 - 2\lambda\mu + \delta(\lambda\mu)^2 = 0$. Solving this for μ we can obtain the seller's reaction curve as:

$$\mu = [1 - (1 - \delta)^{1/2}]/\delta\lambda \qquad (1)$$

For the marginal buyer at any point in time, $V = \lambda P$, and because he is indifferent to waiting, $V - P_t = \delta(V - P_{t+1})$. Given that $P_{t+1} = \mu\lambda P_t$, the marginal buyer's reaction curve is:

$$\lambda = (1 - \delta + \delta\mu)^{-1} \quad (2)$$

A Nash equilibrium under perfect information by both parties occurs when the reaction curves are mutually consistent, which occurs with

$$\mu = [(1 - \delta)^{1/2} - (1 - \delta)]/\delta, \text{ and} \quad (3)$$

$$\lambda = (1 - \delta)^{-1/2} \quad (4)$$

The time path of equilibrium prices under this solution is (μk, $\lambda\mu^2 k$, $\lambda^2\mu^3 k$, . . .). For $i = 0.10$, this time path of prices is (0.23k, 0.18k, 0.14k, 0.11k, 0.08k, . . .) and with a 5-year life cycle of the trait, $T = 5$, this becomes (0.88, 0.68, 0.52, 0.40, 0.31). For a discount rate of 0.20, the comparable numbers are (0.29k, 0.21k, 0.15k, 0.10 k, 0.07k, . . .) and (0.87, 0.62, 0.44, 0.31, 0.22). We show in Fig. 17.3 the first four prices in the sequence of equilibrium prices and quantities corresponding to the 20% discount rate. Buyers capture surplus equal to the shaded area, while the seller captures rent equal to the area beneath. The latter area, total revenue received, equals 0.51 for a 10% discount rate or 0.52 for a 20% rate. By comparison, the present value of returns from annual technology fees ($kp^* = k/2$) would be 1.89 and 1.50 for these two discount rates. This illustrates the 'problem' (from the monopolist's point of view) of the pricing of durables: he earns only about a third of the normal monopoly rent, let alone any additional gains from intertemporal price discrimination.

Now relax the assumption of perfect and costless property rights. Suppose, first, that only a randomly determined fraction θ of potential customers can be excluded from pirating the trait (i.e. from acquiring it from a supplier other than the patent owner or his licensee). Then the derived demand curve (analogous to v^{pp} in Fig. 17.2) is represented by a clockwise pivoting of the valuation schedule V through the point ($V = k, q = 0$) in Fig. 17.3, which would result in no change at all in the time path of equilibrium prices. The seller's revenues would fall, however, to the fraction θ of the level under perfect property rights. The seller's optimization problem would now include the amount to be spent on enforcement, if the fraction θ is affected by enforcement effort, but that problem is not directly relevant to questions addressed in this chapter.

Suppose, alternatively, that buyers set a limit price ϕV, above which they would choose to pirate the trait rather than purchase it from the seller. This would result in a counter-clockwise rotation of the valuation schedule V through the point ($V = 0$, $q = 1.0$). The Nash equilibrium price path through time would fall to the fraction ϕ of the level under perfect property rights.

Hence within the framework of a plant breeders' rights regime in which purchasers are permitted to re-plant the crop with the trait, this game-theoretic analysis results in an initial price only a quarter or so of the one-shot monopoly price, followed by prices that decline even further. Piracy would reduce the seller's returns proportionately below even these levels.

Will Buyers of a Durable Crop Trait Pay for a Durable?

To this point we have concluded that under a UPOV/PVPA breeder's rights regime, it is plausible that the seller of a technologically durable trait will charge a price that declines through time as suggested by Coase's conjecture. The height of this declining price path is clearly restricted by buyers' knowledge that the seller will in the future have an incentive to lower the price. However, in the case of a crop trait, today's customers are potential competitors of the monopolist – they will have the capability of selling the trait the next year. The entire crop of the first-year adopters could be used for seed the following year. Reproductive rates in small grains are on the order of 30 or more to one, so even a 3% adoption rate in year one would provide sufficient seed for the entire crop the following year. The price that the trait owner can charge the first marketing year therefore depends crucially on whether he can be credibly expected to exclude the future dissemination of the trait by those first-year buyers. Recall that as specified above, first-year buyers will only purchase if $P_1 < (1 - \delta)kv + \delta P_2^e = \delta v + \delta(P_2^e - \delta^T v)$. In the extreme case that next year's price, P_2, is expected to be zero, the buyer will pay no more for the trait than v, the value of its services for the coming year alone,[2] the owner would charge a price equal to the optimal rent, $P = r^*$, and sales would cease with this first year. Only if the owner could exclude all potential customers from this pirating activity could the time path of price and sales through time be as high as that derived in the previous section.

Conclusions

Crop traits are technologically durable if they are embodied in the seed of a true-breeding variety as opposed to a hybrid seed. If the trait is protected by a utility patent, the owner can be expected to charge the monopoly rental rate, or technology fee, each and every year for use of that trait. This rental rate should be in the vicinity of the median level of customers' valuations of the service of the trait for one year, with approximately half of the potential adopters choosing to adopt. However, if the trait is protected only by breeders' rights, the buyer retains the right to use the trait in the future, and the owner is selling a durable good to that buyer. Using an explicit game-theoretic model of trait pricing, we find that the Nash equilibrium path of prices through time falls in accord with Coase's conjecture about the pricing of durable goods. The price charged for the initial release of 5-year durable trait can be expected to be about 75% larger than the annual monopoly rental rate, but the present value of revenues would be only about 25% of those from annual monopoly prices. While this result holds for a trait protected by breeders' rights, a similar result could occur under utility patent protection, if enforcement costs under the legal system are sufficiently high.

References

Coase, R.H. (1972) Durability and monopoly. *Journal of Law and Economies* 15, 143–149.

Deardorff, A.V. (1992) Welfare effects of global patent protection. *Economica* 59, 35–51.

Diwan, I. and Rodrik, D. (1991) Patents, appropriate technology and north-south trade. *Journal of International Economies* 30, 27–47.

Giannakas, K. (2002) Infringement of intellectual property rights: causes and consequences. *American Journal of Agricultural Economics* 84, 482–494.

Moschini, G. and Lapan, H. (1997) Intellectual property rights and the welfare effects of agricultural R&D. *American Journal of Agricultural Economics* 79, 1229–1242.

Perrin, R.K. (1994) Intellectual property rights in agricultural development. In: Anderson, J.R. (ed.) *Agricultural Technology: Policy Issues for the International Community*. CAB International, Wallingford, UK.

Perrin, R.K. (1999) Intellectual property rights and developing country agriculture. *Agricultural Economics* 21, 221–229.

Sobel, J. and Takahashi, I. (1983) A multistage model of bargaining. *Review of Economic Studies* 50, 411–426.

Tirole, J. (1988) *The Theory of Industrial Organization*. MIT Press, Cambridge, Massachusetts.

[2] This would be technically true only for an asset with infinite life, which is effectively the case if a new asset can be acquired for free any time in the future. Given a T-year asset life as in the inequality here, the buyer expecting a zero price next year would pay even less than the value of current services, by the amount of the present value of services he would obtain in year $T + 1$ if he postponed purchase.

18 Identity Preservation, Segregation and Traceability: Marketplace Features and Uses

Stuart Smyth and Peter W.B. Phillips
Department of Agricultural Economics, University of Saskatchewan, Saskatoon, Saskatchewan, Canada

Introduction

Agricultural biotechnology has, in less than a decade, had a dramatic impact on the supply of and demand for agricultural food products. While biotechnology has offered few new food products to the marketplace, it has revolutionized the method of producing and delivering conventional food products. An increasing number of cereals and oilseeds are being differentiated to ensure their value or uniqueness is captured and maintained throughout the supply chain.

At present over 28 countries plus the European Union (EU) have either developed or publicly declared their intent to introduce mandatory labelling legislation for genetically modified (GM) products (Phillips and McNeill, 2000). If exporters in countries producing GM crops wish to retain those export markets requiring labelling, then systems of product differentiation will have to be established to ensure the continuity of exports to these concerned markets.

Formal governance structures for these differentiation systems are frequently lacking. Product differentiation systems can be imposed at the time of variety registration if the novel trait is deemed to harm food safety. More often than not differentiation systems are not required by government edict. Rather, they are undertaken to realize private objectives. However, private firms must take great care to ensure that the product differentiation systems they choose correlate with their objectives.

If the food industry does not adopt product differentiation systems, two alternatives are possible. First, the global food market could continue to divide into two distinct markets with only limited interaction between the two. Some markets, such as the EU, and some food processors, have decided to forgo GM technology for now, and are devoting increasing effort to securing adequate volumes of GM-free foodstuffs to satisfy their customers. Consumers in those markets for the most part do not have any opportunity to consume GM foods, as they simply are not available, even though a recent poll of British consumers found 40% were indifferent to consuming GM food (MORI, 2002). Other markets, such as in North America, have rapidly adopted the technology and for the most part do not offer a choice of GM-free food to their consumers. Trade between these two blocs has slowed dramatically in all product markets where GM varieties are grown. For example, American maize exports to the EU dropped to US$175 million in 1999 from US$574 million in 1995 and soybean exports dropped to US$1.1 billion from US$2.1 billion while Canada's oilseed rape (canola) exports to the EU dropped to Can$8 million from Can$204 million (Industry Canada, 2000).

Canada faces the risk of losing markets where GM traits have not been commercialized and likewise do not wish to have GM foods in any product lines. Kuntz (2001) has estimated that if GM wheat is introduced without a product differentiation system, Canada could lose as much as Can$185 million per year or 70% of the annual premium it earns in the global wheat trade. Meanwhile, food processors in places like the EU have diverted their purchases to markets where GM varieties are not produced, such as northern Brazil for soybeans and Australia for oilseed rape.

A second alternative would be for companies to shelve their new technology. In Canada, GM seed potatoes were withdrawn from the market in 2001 in response to food processor concerns while the developer of GM flax announced they had deregistered the variety because the UK, which imports approximately 70% of Canadian flax production, has not approved the variety for import.

Neither of these alternatives is desirable. If Canada, for example, adopts the technology and loses key premium markets, much or all of the benefits of the new technologies will be offset by market losses, with producers facing the greatest losses. Similarly, if Canada forgoes productivity or quality enhancing opportunities, its producers will face even stiffer competition in the residual commodity markets (see Phillips and Khachatourians, 2001, for further discussion of this).

Lehnert *et al.* (2000) succinctly makes the point about successful product introduction:

> Failures occurring during the establishment of a new product or a new process cause high correction costs. They may often lead to losses of market share and damage to the image of the supply chain. It is therefore reasonable to pay attention to potential failures in the early stages of establishment and process planning. (p. 409)

This linchpin concept defines a new approach required for the introduction of new GM crops – do no harm! Caution, diligence and concern for others must be the leading motto for all participants in differentiation systems. In the past, the focus has been on getting new products into the market, while adjustments to the supply chain were made as one went along. Clearly, this strategy is risky. Industry needs to identify and learn from the difficulties, successes and failures that occurred when introducing GM oilseed rape, maize, cotton and soybeans to ensure the successful introduction of new GM crops.

Definitions of Product Differentiation

The definition of product differentiation can have several nuances, depending on the justification for the differentiation. Frequently the terms identity preserved production and marketing (IPPM), segregation and traceability are used interchangeably in biotechnology and supply chain literature. This is creating misconceptions about the distinct role that each of these product differentiation systems has in the supply of food products. The purpose of this section is to identify definitions that exist in the literature to date and to suggest definitions where the literature is absent.

Identity preserved production and marketing

The first product differentiation system, identity preserved production and marketing, has evolved over time in the grain and oilseed industry. Purchasers of raw products became more demanding about the quality and purity of the product they were purchasing so the grain handling system gradually developed distinct channels to market the differing grades of grains and oilseeds. All grains and oilseeds are purchased through a grading system in today's marketplace and this grading system has premiums that rise as one moves from low to high grades. The relationship of premiums to differing grades for private market incentives is the defining feature of an IPPM system.

Identify preserved production and marketing systems are initiated by private firms in the grain and oilseed industry to extract premiums from a marketplace that has expressed a willingness to pay for an identifiable and marketable product trait or feature. An IPPM system is a 'closed loop' channel that facilitates the production and delivery of an assured quality by allowing identification of a commodity from the germplasm or breeding stock to the processed product on a retail shelf (Buckwell *et al.*, 1999;

Lin, 2002). These IPPM systems are predominantly voluntary, private, firm-based initiatives, ranging between systems that are loosely structured (i.e. malting barley) with high tolerance levels and those with rigid structures (i.e. non-GM European markets) with minimal tolerance levels. Firms operating in the minimal tolerance field achieve this by developing and adhering to strict protocols that specify production standards, provide for sampling and ensure appropriate documentation to audit the flow of product.

A survey of the literature on IPPM shows that while there is growing discussion about IPPM systems, there are very few working definitions. Lin (2002) suggests that an identity preservation system:

> is a more stringent (and expensive) handling process and requires that strict separation, typically involving containerized shipping, is maintained at all times. IP [identity preservation] lessens the need for additional testing as control of the commodity changes hands, and it lowers liability and risk of biotech and non-biotech commingling for growers and handlers.
>
> (p. 263)

This definition conflicts with the definition offered here as Lin sees IPPM systems as having a limited role in the movement of grains and oilseeds due to extremely low tolerance levels. Lin's definition of IPPM and segregation still deal with the same system, one that is initiated voluntarily by private firms in an attempt to capture premiums. It is shown below how IPPM systems differ from segregation systems.

The remainder of the literature on IPPM systems relates to theoretical and operational uses of IPPM systems. Bullock *et al.* (2000) and Bullock and Desquilbet (2001) discuss differentiation between GM and non-GM products and Herrman *et al.* (1999) examine the feasibility of wheat segregation. Bender *et al.* (1999), Bender and Hill (2000) and Good *et al.* (2000) have released a series of papers on handling speciality maize and soybean crops which are focused on costs. Additionally, Miranowski *et al.* (1999) offer some perspectives on the economics of IPPM, while Kalaitzandonakes *et al.* (2001) provide a solid theoretical model for examining the cost of identity preservation.

Numerous IPPM systems are operating in Canada and around the world. Some extend only between the breeders and the wholesale market or processor, while others extend right up to the retailer. Their structure depends on the attribute they are trying to preserve. Some novel oils, such as low linolenic oils that are more stable in fryers, only have value at the processing level while others, such as high oleic oils, have health attributes that can be marketed to consumers. Identity preserved production and marketing systems are important for providing information to consumers about the provenance of a product, as those attributes are not visible or detectable in the product itself.

There are a number of examples of IPPM systems operating in Canada. Organic products are one of the most noticeable IPPM products in today's marketplace. Cargill has an IPPM system in place to export oilseed rape to Japan. The oil of this oilseed rape variety gives off virtually no odour when used to fry food. General Mills is operating an IPPM system for a select variety of white wheat that possesses a special trait for 'flake curling' when processed into breakfast cereal. DowAgro Sciences is using an IPPM system to export the Nexera oilseed rape variety to Japan where it is sold into the speciality gift oil market.

Segregation

The second product differentiation system, segregation, has frequently been incorrectly applied to the grading of different classes of grains and oilseeds in order to receive a higher price for the commodity than if it were allowed to be commingled. Segregation systems have a formal structure and at times can act as regulatory standards. Segregation differs from IPPM in that the focus of the system is not on capturing premiums but rather on ensuring that potentially hazardous crops are prevented from entering supply chains where products are destined for human consumption.

Segregation can be defined as a regulatory tool that is required for variety approved and commercial release of grain and oilseed varieties that could enter the supply chain and create the potential for serious health hazards. Segregation systems often are developed as part of a variety registration process, where government regulators use contract registration to ensure that certain novel varieties will not enter the handling system of like varieties. The private firm seeking registration of the novel variety has to demonstrate that there is a segregation system developed to ensure the containment of the variety.

Lin (2002) defines segregation as the requirement 'that crops be kept separate to avoid commingling during planting, harvesting, loading and unloading, storage and transport' (p. 263). Segregation systems are used when potential food safety concerns exist over the commingling of the segregated product and all other like products. In short, IPPM is used to capture premiums and segregation is used to ensure food safety.

There are very few segregation systems presently operating in Canada. There is a small amount of *Brassica juncea* being segregated for the first time in 2002. The best known segregation system in Canada is for high erucic acid rapeseed. This variety of rapeseed has industrial value due to the high erucic acid content and is produced using a segregation system.

Traceability

The third product differentiation system, traceability, is commonly used in the food industry. Retail products found with unacceptable bacteria levels or intolerable levels of pesticide or chemical residues need to be quickly and completely removed from store shelves. Traceability systems allow for retailers and the supply chain to identify the source of contamination and thereby initiate procedures to remedy the situation.

The key focus of traceability is increasingly on food safety. For the purposes of this discussion, traceability will refer to systems that focus on ensuring food safety. Recently, the focus for developing traceability systems for new sectors of the marketplace has shifted from food safety towards extracting premiums from the marketplace. Extracting market premiums should never however, be the driver for developing a traceability system. In and of itself traceability systems do not define quality, they simply trace it. If market premiums are the driver, then the developers need to use an IPPM system as these systems are properly structured to capture premiums.

The International Organization for Standardization (ISO) has defined traceability as the 'ability to trace the history, application or location of an entity by means of recorded identifications' (p. 1) and the Codex Alimentarius Commission (Codex) has adopted this as their working definition for all Codex standards (Codex, 2001). The EU (2001) has defined traceability quite clearly in relation to GM products. Directive 2001/18/EC defines traceability as:

> . . . the ability to trace GMOs and products produced from GMOs at all stages of the placing on the market throughout the production and distribution chains facilitating quality control and also the possibility to withdraw products. Importantly, effective traceability provides a 'safety net' should any unforeseen adverse effects be established. (p. 2)

The economics literature from supply chain management defines traceability as the information system necessary to provide the history of a product or a process from origin to point of final sale (Jack *et al.*, 1998; Timon and O'Reilly, 1998; Wilson and Clarke, 1998). While Dickinson and Bailey (2001) suggest that their results from an experimental auction regarding features of meat traceability show there is willingness by consumers to pay premiums for traceability, in effect though they are buying the quality that is determined by the IPPM system. Before adopting traceability systems there has to be a clear indication of specifically what aspects of food safety can be improved by the adoption. Marginal improvements in food safety would be a dubious reason for proceeding – rather there must be a clear and evident improvement in the level of food safety which requires changes in production methods (hence IPPM).

Options for Product Differentiation

Each product differentiation system has features that are unique, while also possessing features that are common to one, if not both, of the other systems. Table 18.1 compares numerous features of product differentiation. These features are classified into those that apply to the

Table 18.1. Comparing identity preservation, segregation and traceability.

	IPPM	Segregation	Traceability
1. Overall management			
Objective	Revenue management	Liability management	Product safety
Status	Voluntary	Mandatory	Voluntary or mandatory
Lead stakeholder	Private company	Regulator	Commodity group, standards organization or regulator
Regulatory agency involvement	Consumer fraud	Regulatory oversight	Consumer fraud
Information	Asymmetric	Full	Asymmetric
Risk	Moral hazard	None	Moral hazard
Information flow	Two-way	Two-way	Two-way
Supply chain focus	Downstream	Downstream	Upstream
Penalties for failure in product market	Consumer fraud charges; lost brand value	Criminal prosecution; mandated product recalls	Consumer fraud charges; exclusion from product category
Testing/auditing	2nd party/brand owner	1st party/regulator	3rd party/standards organization
2. Production stage features			
Production arrangements	Formal production contracts	Regulation and contracts	Membership in quality standard
Production controls	In-season agronomic rules vary with product	Formal buffer zones; post production land use controls	Process standards adopted and record keeping
Premiums for producers	Short and long term	Short and long term	Short term
3. Processing stage features			
Enforcement	Private	Public	Collective
Quality criteria based on	Product standards	Regulations and or HACCP rules	Processes (e.g. ISO)
Tolerance levels	Variable	Set in law	Performance based
Testing/auditing	2nd party	1st party	3rd party
4. Retail stage features			
Provides access to	Branded product markets	Markets	Product categories
Information provided to	Consumer	Regulator	Regulator, retailer or processor
Final market price premiums	Yes	None	None
Labelling	Private brands	None	Quality standard

complete supply chain and those that apply to the three distinct stages of supply chains. The first stage involves features that are most commonly related to the production stage of the supply chain. Included in this stage are seed development firms, producers and grain handlers. The second stage of the supply chain is the processing stage. This stage includes all firms involved in the supply chain from the point when a raw ingredient is received to the point that a final product is shipped to the retailer. The third stage is the retail stage of the supply chain. This stage includes those firms that provide products to consumers, such as grocery stores and restaurants.

Overall supply chain management

The features in this stage are those that are important to the entire supply chain. Unlike the features in the following sections, these features span all sectors of the food industry and each participant in the supply chain must ensure that their commitment to these features is at least as strong as the other participants.

The objective of an IPPM system is revenue management. Premiums need to be available to attract participants and the efforts of participants will be directed towards receiving a share of the premium. Participation in these systems will be voluntary. The lead stakeholders in IPPM systems are private firms seeking to capture the increased value of special traits. The role of the regulatory body will be to ensure that industry standards are handled in such a way as to prevent consumer fraud from occurring. The information may be asymmetric as only the product seller can know with certainty what level, if any, of cheating has occurred in the delivery of the product. Moral hazards may be present due to the presence of premiums. Effective IPPM systems that span entire supply chains must have accurate two-way information flows. This means that information about purity and quality of the product flows downstream and that information coming from consumer demand flows upstream. While the information flow in IPPM systems is two-way, the focus of the system is downstream. Each participant in the system wants to ensure they extract a portion of the value of the special trait, whether from production, processing or retailing the product. This means that each participant will focus on the needs of the next participant in the supply chain. Market failure can result in fraud charges for mislabelling or improper labelling and also create awareness with consumers that certain brand names can not be trusted. Testing and auditing will be done by second parties acting on behalf of the brand owner or developer of the special trait.

The objective of a segregation system is to manage any and all risks that may arise through the production and processing of commodities. Participation is not optional – any producer or firm involved with segregated products will have to comply with standards that have been approved by the regulatory agency. The private firm will have the responsibility of developing the actual system, but the regulatory agency will be the final arbiter on approving the system for field use. Information will be fully disclosed because of the importance of protecting food safety, which will result in the minimization of risks in the system. Segregation systems must have two-way information flow due to the concern about food safety should commingling occur. The leading reason why the StarLink™ maize segregation failed is the lack of two-way flows of information. Aventis and their agents presented the principles of the stewardship programme they had designed to ensure segregation and did not incorporate any feature for information about the actual operation of the system to return back to Aventis. The focus of product delivery within a segregation supply chain will be downstream. Segregated commodities commonly have industrial value, so these products will be supplied to meet the criteria of the processor. The costs of market failure would most definitely see a complete recall of any products suspected of being affected. It may also result in criminal prosecution in the most severe instance. Testing and auditing will be vital features of segregation systems and will be conducted by agents of, or acting on behalf of, the regulator. This process will also reinforce the level of trust with foreign export markets.

The objective of traceability systems is ensuring that products available for consumption are as safe as possible. Participation in a traceability system can be voluntary, depending on where in the supply chain the participant is located. The closer the participant is to the start of the supply chain, the more likely it will be that participation is voluntary. The lead stakeholder may be a commodity group demanding greater clarity in or selection of food products, a standards council that is composed of industry representatives from all sectors of the supply chain or the regulator to ensure consumer protection. Information may be asymmetric due to the voluntary nature at the start of traceability supply chains. A moral hazard may also exist owing to the inability to test fully for some features of traceability. Traceability systems have information flowing two ways as these systems are designed to react quickly to food safety concerns. If a product is discovered to exceed any defined tolerance level at any point in the supply chain, traceability will be used to identify the source of the problem and to locate any and all retail products that may be affected. Information on food safety flows upstream while information on specific products flows downstream. This results in the focus of traceability systems being upstream. Market failures can also result in consumer fraud charges in addition to permanent exclusion from selling into that supply chain. Testing and auditing will be conducted according to the standards developed by third party organizations.

Production stage features

The production stage features are those at the front end of the supply chain and involve seed development firms, producers and grain and oilseed handlers. Historically, this has been the starting point for supply chains as seed development firms would commercialize a new crop variety. This push version of supply chains has had difficulty adapting to the demands of consumers for a pull supply chain.

Identity preserved production and marketing systems are voluntarily developed by private firms to ensure that all stakeholders in the supply chain for a specific product capture a share of the value from special trait varieties. Private firms may use technical use agreements to protect the intellectual property of the special trait or production contracts that have specific conditions that must be met in order to receive the premium. Grain companies typically organize and manage these contracts. These systems are typically developed for niche market products and are typified by small acreage and low volumes. There is presently some debate as to whether long-run premiums for producers are sustainable, as they may be bid away due to competition among producers.

Segregation is focused on ensuring that the integrity of the special trait is not allowed to commingle adventitiously with other products destined for the human supply chain. Production contracts would be used by private firms to ensure that all of the commodity being segregated is collected and that the producer retains no amount of seed. Buffer zones are required for segregation systems as a preventive measure for reducing cross-pollination. Producers may also have restrictions placed on what crop varieties would be allowed to be grown the following year on fields that produced segregated crops. Premiums would be available in both the short and long term to ensure that product supply is maintained.

Traceability is very fragmented at the producer stage. Production would be arranged largely through membership in an organization established to create and manage the industry. Production control would be through industry standards and stringent record keeping. The cost of initially becoming involved in a traceability system is usually offset by short-term premiums. Long-term benefits are often not available as the premiums evaporate when the desired number of producers become involved.

Processing stage features

Processing stage features are those that relate to firms involved in the manufacturing of food products. Most of these features contain aspects of quality assurance and industry-led standards.

Identity preserved production and marketing quality standards will be enforced by private commitment to industry standards, as the value of the product will be greater given higher purity and/or protein levels. The enforcement of standards is crucial as products that do not conform to the desired quality level will not be accepted. Tolerance levels will vary from product to product and also will depend on the preferences of the final consumer. Testing and tolerance levels will be important to ensure that the purity and the high quality levels of the product are maintained. Frequently, these tests will be conducted by second parties.

Enforcement of standards is very important in segregation systems. To ensure that products that could be a hazard to the human food supply chain are prevented from entering that supply chain, functional operating standards must be agreed to by all participants. The enforcement of these standards will need to be rigorous (it was the lack of standards enforcement by Aventis that resulted in the StarLink™ maize debacle). Quality will be defined in regulations or be created through the implementation of a hazard analysis critical control point (HACCP)-based system. Tolerance levels for commingling will be set by the regulator. Because of the importance of standards, the features of testing and tolerance levels will also be important. Testing will need to be conducted frequently to ensure that the commodity is being segregated properly and that none of the product is entering other supply chains accidentally. This will be done by agents of the regulator. It was the diligence of officials with the United States Food and Drug Administration that discovered unapproved material in test plots belonging to ProdiGene.

Traceability is very important at the processing stage, as this is where the supply chain begins to be rigorously applied. The lack of high standards and poor enforcement could result in costly recalls of products and losses for products in related categories. Therefore, the enforcement of standards will be done collectively. The focus will be on the production processes to ensure that the highest standards possible are maintained at all times. Nevertheless, tolerance levels exist for food safety reasons as no product can be entirely 100% free of potentially harmful effects. When tolerance levels are exceeded, the risk of harm to consumers develops and these products must then be recalled from the marketplace. The costs of recall can be substantial. Not only does the firm have the cost of gathering and disposing of the product in question, there may also be a loss of consumer trust in that brand name, which will require aggressive marketing campaigns to overcome. Testing and auditing of traceability systems are done by third parties.

Retail stage features

The final stage of the supply chain is the retail stage, involving those enterprises that sell food products to consumers. This is the stage of the pull supply chain that is now seen as driving many modern supply systems.

Identity preserved production and marketing systems are likely to play a large role in the introduction of new GM food products. New GM products may be introduced without complete international market acceptance and IPPM systems can be used to ensure continued access to these markets. An IPPM system will be able to provide information to the consumer about the uniqueness of the branded product that is being identity preserved. For an IPPM system to function properly, and ensure that all stakeholders remain committed to the process, final market price premiums must be available. If this premium is not available for the retailer, a disincentive is created for the retailer to continue to carry the product. Products of IPPM systems will need to be labelled to justify the final market premium. If the consumer has no ability to identify the product, the consumer will not pay a premium to purchase the product.

Segregation systems will also be used to ensure that market access is continually guaranteed. A co-ordinated education and marketing effort by the regulatory agency and the private firms involved can be effective in creating trust in foreign markets that production of potentially hazardous products will not jeopardize export food channels. (Most segregated production systems end at the processing stage and therefore, there is no need for final market premiums or labelling.)

Traceability is crucial for providing access to new categories of products. Many markets have demanded documentation regarding product composition before allowing market access. Traceability systems are designed to increase information regarding food safety to consumers and also provide information back up the supply chain to regulators and processors. Final market premiums are usually not available for traceability systems. Labelling is important for traceability, but in this case it is the special bar code and other features that allow two-way flow of information that relates to IPPM. While this is often perceived as creating consumer value, in practice it is a way of protecting value generated in an IPPM system.

The features common to product differentiation can be classified as to whether and how they pertain to identity preservation, segregation and traceability. As is evident, some features differ depending on the system in which they are applied. It will be important for those involved in product differentiation to examine what features are most commonly related to the product requiring differentiation and if and how those features overlap. This model of comparison will assist with determining which system best relates to the identified needs of the product being differentiated.

Conclusions

Biotechnology innovations in agriculture present a clear challenge to traditional marketing systems. Transactions for new, proprietary, novel-trait crop varieties require a more extensive set of institutions than for traditional commodity varieties. Companies assisted by governments and industry associations have developed product differentiation systems that handle both the risks and assist with capturing the returns from the introduction of new products with commercially valuable input and output traits. Spot markets are increasingly competing with proprietary, vertically integrated supply chains. The optimal structure and organization of these new supply chains has not evolved yet, but over time one would expect a more stable set of relationships to emerge.

Product differentiation offers biotechnology two immediate benefits. First, while it is expected that the use of product differentiation systems will be expensive to operate in the initial stages, these cost increases must be balanced against any expected or realized gains. Although research and seed development companies may lose some money due to the costs of product differentiation systems, they will gain significantly in terms of market adoption (Smyth and Phillips, 2002). Second, product differentiation offers biotechnology firms another means to protect their investment in new technology crop developments. The problem with the current package of intellectual property rights is that they do not fully control the use of a new technology once it is expressed in seed. Most GM crops can be propagated in subsequent years with seed from previous years. While regulations and private contracts attempt to manage that activity, many in the industry note that they are far from effective.

References

Bender, K. and Hill, L. (2000) *Producer Alternatives in Growing Specialty Corn and Soybeans (AE-4732)*. University of Illinois, Urbana–Champaign. Available at: www.stratsoy.uiuc.edu/specialty_corn_beans.pdf

Bender, K., Hill, L., Wenzel, B. and Hornbaker, R. (1999) *Alternative Market Channels for Specialty Corn and Soybeans (AE-4726)*. University of Illinois, Urbana–Champaign. Available at: www.ngfa.org/specialtybk.html

Bullock, D.S. and Desquilbet, M. (2001) Who pays the costs of non-GMO segregation and identity preservation? *Proceedings of the 5th International Conference of the International Consortium on Agricultural Biotechnology Research (ICABR)*, Ravello, Italy, 15–18 June.

Bullock, D.S., Desquilbet, M. and Nitsi, E. (2000) The economics of non-GMO segregation and identity preservation. Paper presented to the *American Agricultural Economics Association Annual Meeting*, Tampa, Florida, 30 July–2 August.

Buckwell, A., Brookes, G. and Bradley, D. (1999) *Economics of Identity Preservation for Genetically Modified Crops*. Report prepared for Food Biotechnology Communications Initiative. CAES 1745/GJB Wye, UK.

Codex Alimentarius Commission (2001) *Matters Arising from Codex Committees and Task Forces: Traceability*. Available at: ftp://ftp.fao.org/codex/ccexec49/al0121ee.pdf

Dickinson, D.L. and Bailey, D. (2001) Meat traceability: are U.S. consumers willing to pay for it? (UAES Journal Paper 7458). Personally requested.

[EU] European Union (2001) Regulation of the European Parliament and of the Council concerning traceability and labelling of genetically modified organisms and traceability of food and feed products produced from genetically modified organisms and amending Directive 2001/18/EC. Commission of the European Communities, Brussels. Available at: http://europa.eu.int/comm/food/fs/biotech/biotech09_en.pdf

Good, D., Bender, K. and Hill, L. (2000) *Marketing of Specialty Corn and Soybean Crops (AE-4733)*. University of Illinois, Urbana–Champaign. Available at: http://web.aces.uiuc.edu/value/marketingspecialty.pdf

Herrman, T., Boland, M. and Heishman, A. (1999) *Economic Feasibility of Wheat Segregation at Country Elevators*. Kansas State University, Manhattan. Available at: www.css.orst.edu/nawg/1999/herrman.html

Industry Canada (2000) *Strategis: Imports and Exports by Product*. Available at: http://strategis.ic.gc.ca/sc_mrkti/tdst/engdoc/tr_homep.html

Jack, D., Pardoe, T. and Ritchie, C. (1998) Scottish quality cereals and coastal grains: Combinable crop assurance in action. *Supply Chain Management* 3, 134–138.

Kalaitzandonakes, N., Maltsbarger, R. and Barnes, J. (2001) Global identity preservation costs in agricultural supply chains. *Canadian Journal of Agricultural Economics* 49, 605–615.

Kuntz, G. (2001) Transgenic wheat: potential price impacts for Canada's wheat export market. MSc Thesis, Department of Agriculture Economics, University of Saskatchewan, Canada.

Lehnert, S., Schmitz, T. and Peterson, B. (2000) Risk and weak point analysis in the range of chain-oriented data acquisition. In: Trienekens, J.H. and Zuubier, P.J.P. (eds) *Chain Management in Agribusiness and the Food Industry*. Wageningen, The Netherlands, pp. 409–416.

Lin, W. (2002) Estimating the costs of segregation for non-biotech maize and soybeans. In: Santaniello, V., Evenson, R.E. and Zilberman, D. (eds) *Market Development for Genetically Modified Foods*. CAB International, Wallingford, UK, pp. 261–270.

[MORI] Market & Opinion Research International (2002) British public supports EU Directive. MORI, London. Available at: www.mori.com/polls/2002/greenpeace.shtml

Miranowski, J.A., Moschini, G., Babcock, B., Duffy, M., Wisner, R., Beghin, J., Hayes, D., Lence, S., Baumel, C.P. and Harl, N. (1999) Economic perspectives on GMO market segregation. (Staff Paper 298). Iowa State University, Ames. Available at: http://agecon.lib.umn.edu/cgi-bin/detailview.pl?paperid=1768.

Phillips, P.W.B. and Khachatourians, G.G. (eds) (2001) *The Biotechnology Revolution in Global Agriculture: Invention, Innovation and Investment in the Canola Sector*. CAB International, Wallingford, UK.

Phillips, P.W.B. and McNeill, H. (2000) A survey of national labeling policies for GM foods. *AgBioForum* 3, 4. Available at: www.agbioforum.org

Smyth, S. and Phillips, P.W.B. (2002) Competitors co-operating: establishing a supply chain to manage genetically modified canola. *International Food and Agribusiness Management Review* 4(1), 51–66.

Timon, D. and O'Reilly, S. (1998) An evaluation of traceability systems along the Irish beef chain. In: Viau, C. (ed.) *Long-term Prospects for the Beef Industry*. Institute National de las Recherche Agronomique (INRA), Paris, pp. 219–225.

Wilson, N. and Clarke, W. (1998) Food safety and traceability in the agricultural supply chain: Using the Internet to deliver traceability. *Supply Chain Management* 3, 127–133.

19 Segmentation of GMO and non-GMO Soybean Markets under Identity Preservation Costs and Government Price Supports

Troy G. Schmitz,[1] Charles B. Moss[2] and Andrew Schmitz[2]
[1]Department of Agribusiness, Arizona State University, Tempe, Arizona, USA;
[2]Department of Food and Resource Economics, University of Florida, Gainesville, Florida, USA

Abstract

The introduction of genetically modified organisms (GMOs) has significant implications for the structure of marketing channels in the USA and worldwide. The introduction of GMOs results in differentiated market channels for GMO and non-GMO crops. However, the additional costs of identity preservation along with existing government subsidies, such as the US loan deficiency and target price programmes, affect the costs and benefits of introducing GMOs in the market channel. This study examines the conditions under which price premiums for GMO soybeans in US markets exist and obtains empirical estimates of the economic implications of introducing GMO soybeans in the US market.

Introduction

Biotechnology is defined as the use of biological organisms to address human problems (Caswell *et al.*, 1994). Genetically modified organisms (GMOs) comprise one of the most visible forms of technologies involving transgenic modification (moving genetic material between non-similar organisms) of agricultural crops. Initially, GMOs appeared to be a panacea for both consumers and producers. Consumers were promised foods with extended shelf life, reduced use of pesticides and herbicides and potential health benefits. Farmers anticipated decreased costs (mostly due to the reduction in the use of agricultural chemicals), increased yields and the possible emergence of new markets through the advent of medical foods. The new era promised by the biotechnology revolution has encountered a few pitfalls along the way. The largest barrier seems to originate from growing consumer concerns over GMO products in the European Union (EU) and elsewhere.

The question as to whether price premiums actually exist in US markets for non-genetically modified (GM) agricultural products was addressed by Lin *et al.* (2002). They found that immediately following the commingling of Starlink maize in the US marketing system, premiums for Starlink-free maize averaged between 7 and 12 cents/bu and in some rare instances reached as high as 15 to 20 cents. However, these premiums for Starlink-free maize soon dissipated over time. Premiums for non-GM soybeans exist under special contracts with certain processors who do not want the risk involved in having their processed output labelled as containing GMOs. However, these contracts are quite rare. Why have large premiums for non-GM crops in the US not materialized?

Moss *et al.* (2001) determined the conditions under markets for GM and non-GM maize in the USA. They found that premiums do not arise in maize markets because the demand for non-GM maize is so low relative to the overall supply of non-GM maize available. Moschini *et al.* (2000) estimated the welfare effects of adopting Roundup Ready® soybeans in global markets. However, their analysis emphasized the implications of market power derived from patents and spillover effects and did not explicitly include the additional identity preservation (IP) costs associated with segregating markets, nor did they take into account the effect of US price support programmes. Lapan and Moschini (2001) developed a theoretical model for the general effects of GMO labelling that included IP costs, but did not account for government programmes. Qaim and Traxler (2002) analysed the welfare effects of Roundup Ready® soybeans in Argentina, but since almost all soybeans in Argentina are now GM, they did not have to deal with the issue of IP costs. Moreover, they focused on the benefits of technology spillover from the USA.

This study examines the effect of the introduction of GMOs on agricultural marketing channels in the USA, focusing on the impact of the introduction of GM soybeans. We show that the introduction of these GM technologies has the potential to segment the market channel resulting in market premiums for the producers of non-GM products. However, the emergence of a market premium depends upon the level of IP costs and also depends on government price supports, such as the Loan Deficiency Payment Programme (LDP) or target prices. Our results indicate that sufficiently large IP costs, coupled with price supports, result in the loss of market segmentation opportunities.

The Existence of Price Premiums for non-GM Soybeans

The introduction of GM products into agricultural markets represents a product that is weakly inferior to similar non-GM products (Lapan and Moschini, 2001). For example, GM soybeans cannot be substituted for non-GM soybeans for export to countries that do not accept GMOs, nor can they be substituted for food manufacturers that require non-GM inputs. Given this lack of substitution in one direction, we develop a model of product differentiation based on the back-to-back model of trade flows. Product is allowed to flow from the non-GM market into the GM (or undifferentiated market), but product is not allowed to flow from the GM market into the non-GM market. As demonstrated in our discussion, this dichotomy (the ability of non-GM soybeans to be used in the GM market while GM soybeans cannot be used in the non-GM market) can give rise to a market premium only if the original non-GM supply is insufficient to meet the non-GM demand at the prevailing price in the GM market. However, if the non-GM supply exceeds the non-GM demand at the prevailing GM market price, no premium emerges. The equilibrium is further complicated by the imposition of transaction costs used to guarantee that output marketed as non-GM indeed contains no GM product and by government price supports which, when binding, essentially create a floor for the price received by US producers of both GM and non-GM soybeans.

Figure 19.1 presents the scenario in which there exists an excess supply of non-GM soybeans. Since non-GM soybeans can be substituted for GM soybeans in all uses, there will be no price premium at the producer level. However, consumers of non-GMs will have to pay a price premium over GMs due to the imposition of identity preservation costs in the grain handling system. In addition, US price supports, when binding, affect the final price realized by producers of both GM and non-GM soybeans.

In Fig. 19.1, the US supply of non-GM soybeans is denoted as S_N and the demand for US non-GM soybeans is denoted as D_N. D_N includes both domestic non-GM soybean demand and the excess demand for US non-GM soybeans in international markets (not shown). The excess supply curve for non-GM soybean is ES_N. Similarly, the US supply of GM soybeans is denoted as S_G and the demand for US non-GM soybeans is D_G. The excess demand curve for GM soybeans is ED_G. These curves all represent supply and demand curves before IP costs are considered. The price support (which is binding in Fig. 19.1) represents either the loan rate under the loan rate deficiency payment scheme, or the

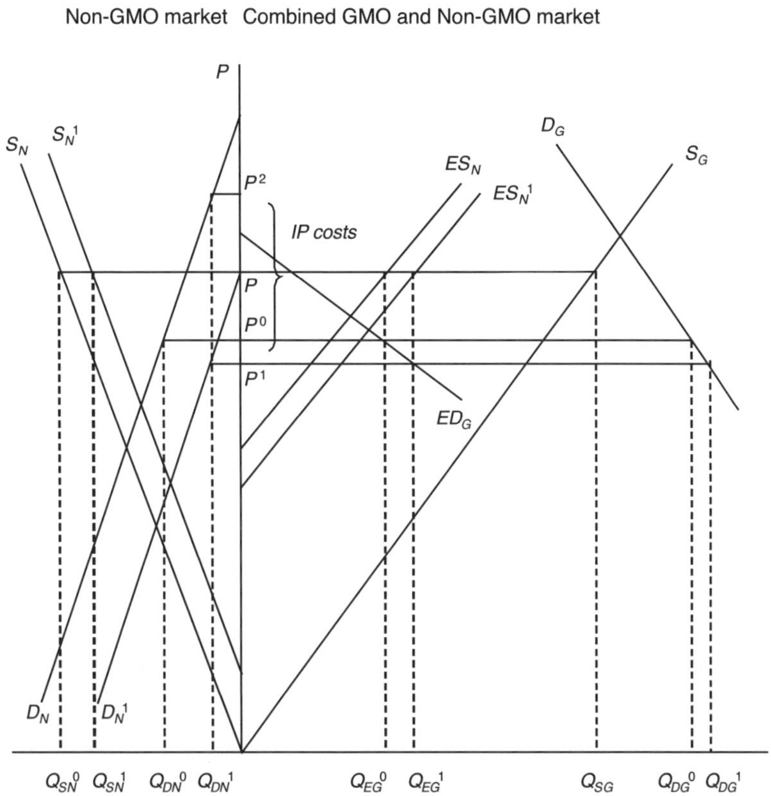

Fig. 19.1. Excess supply of non-GMO soybeans with price support.

target price under the farm bills before 1996 and after 2002. Since, until recently, the loan rate for soybeans has been binding, Fig. 19.1 reflects the actual market situation in the USA over the last couple of years. In the absence of IP costs and price supports, markets would clear at a price where the excess supply of non-GM soybeans (ES_N) intersects the excess demand for GM soybeans (ED_G) since non-GM soybeans can freely flow into the GM soybean market. However, a binding price support causes producers' realized price to remain constant at a price level equal to *P*. With the price support binding, non-GM producers will 'export' an amount of non-GM soybeans into the GM soybean market equal to the intersection of ES_N with the support price *P* (Q_{EG}^0), produce Q_{SN}^0 of non-GM soybeans, and Q_{SG} of GM soybeans. At the market clearing price (P^0) consumers will purchase Q_{DG}^0 of GM soybeans and Q_{DN}^0 of non-GM soybeans.

Now consider the imposition of IP costs at both the producer level and through the market channel until it reaches consumers. IP costs are comprised of two parts. The first costs are borne by producers because of on-farm cleaning, etc. The second part is borne by the rest of the market channel. These include testing costs and segregation of the product as it travels through the market channel. If producers want to sell their product as GM, they must incur additional costs, which essentially shift the supply curve for non-GM soybeans inward to (S_N^1). Similarly, processors, etc., incur additional costs that cause the demand curve to shift inward to (D_N^1). However, the demand and supply schedules for non-GM producers and consumers remain unchanged. The shifts in supply and demand cause the excess supply curve for non-GM soybeans to change from ES_N to ES_N^1. The direction of this shift depends on the allocation of costs between producers and processors. In Fig. 19.1, the excess supply curve shifts outward. A new equilibrium results in an amount Q_{EG}^1 'exported' into the GM market. The market price realized by producers drops from P^0 to P^1, which causes the quantity of non-GM soybeans demanded to decrease from Q^0_{DG} to Q^1_{DG} and the quantity of soybeans demanded in the GM market to increase from Q^0_{DG} to Q^1_{DG}. Now there are three different prices. The price that producers receive from the marketplace is P^1, the final price realized by producers is P^0, and the price paid by consumers of the end product rises to P^2 because the IP costs are added to the system. Although consumers pay a higher price due to IP costs, producers do not receive a premium for non-GM soybeans. Regardless of whether they sell GM or non-GM soybeans, their final realized price will still be P, which is the binding support price. The difference between the consumer and producer price is equal to the IP cost.

Now consider the situation depicted in Fig. 19.2, in which there is an excess demand for non-GM soybeans (as opposed to the excess supply situation in Fig. 19.1). The excess demand for non-GM soybeans equals ED_N and the excess supply of GM soybeans equals ES_G. First, consider a binding price support (P) where GM and non-GM soybeans are not differentiated so that no IP costs are incurred. If GM soybeans were allowed to flow freely into non-GM markets, the equilibrium market price realized by producers and paid by consumers would be P^0 and the government would pay the difference between P and P^0. The quantity of non-GM soybeans demanded would be Q^0_{DN}, the quantity of GM soybeans demanded would be Q^0_{DG}, and Q^0_{EN} would be 'exported' from the GM into the non-GM market.

Of course the situation described above can not occur because GM soybeans are restricted from entering the non-GM soybean market. In order for producers to sell non-GM products, producers must incur IP costs. In addition, if consumers want non-GM soybeans, they must incur IP costs as they are passed on through the marketing channel. Hence, the supply and demand for non-GM soybeans shift inward to S^1_N and D^1_N, respectively. Since the support price is still equal to P, producers of GM soybeans

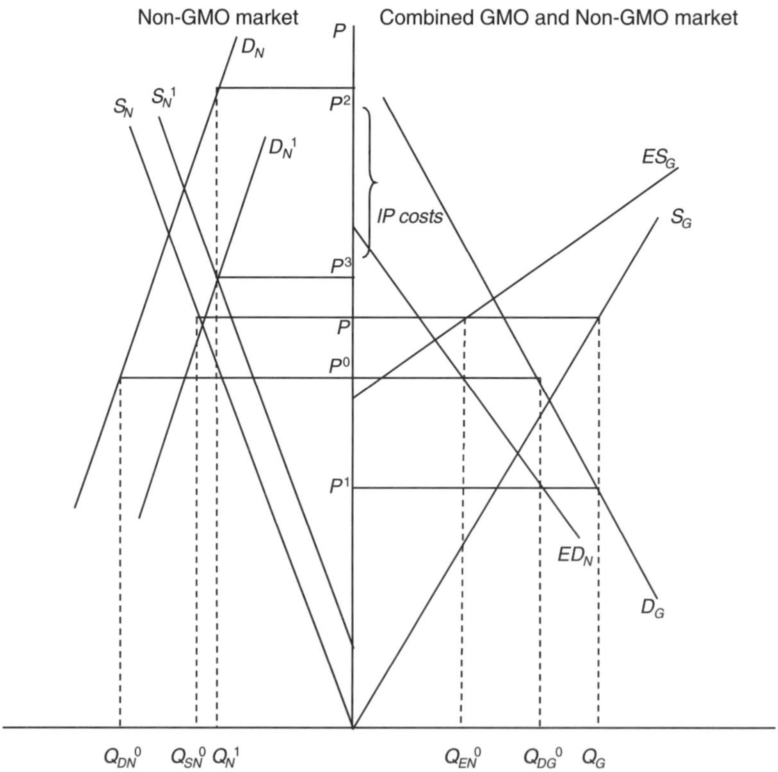

Fig. 19.2. Excess demand for non-GMO soybeans with price support.

will still produce Q_G. However, GM soybean consumers are now the only customers for GM products. In order to induce them to buy the full amount produced, the price must drop down to P^1, which is significantly lower than P^0. Producers of GM soybeans are still guaranteed the (much higher) price P by the government.

What happens to non-GM producers and consumers? It depends on the equilibrium price in the non-GM market that is determined by the intersection of S^1_N (the supply curve for non-GM soybeans once IP costs are incurred) with D^1_N (the demand curve for non-GM soybeans once IP costs are incurred). If the price resulting from the equilibrium is higher than the support price P (which it is in Fig. 19.2), then the price that producers receive for non-GM soybeans (P^3) will be higher than the support price, and producers will realize a premium ($P^3 - P$ in Fig. 19.2). However, different relative values for the elasticities of the supply and demand curve for non-GM soybeans could result in an equilibrium in which P^3 falls below the target price P. In such cases, non-GM producers will still not receive a premium.

To summarize, even if there is excess demand for non-GM soybeans, there will still not be a premium for GM soybeans if the price support is too high. If there is a premium (as in Fig. 19.2) then there will be four different prices. P^1 is the price that GM consumers pay, P is the final realized price that GM producers receive (once they receive payment from the government), P^3 is the price realized by non-GM producers, and P^2 is the price that consumers pay for non-GM soybeans. The difference between P^2 and P^3 in Fig. 19.2 is simply equal to the combined IP costs incurred by producers and by all participants in the marketing channel.

Model Parameterization

Roundup Ready®, or herbicide tolerant varieties are the most frequently used type of GM soybean seed. This innovation provides for a cheap and effective source of weed control. Given that GM soybeans reduce the cost of weed control, the technology may cause soybean production to become economically viable in states where soybeans were only marginally feasible under previous weed control technologies. As a result, these states may have average yields per acre below the national average.

Using the 1999, 2000 and 2001 crop years as a reference for production, Table 19.1

Table 19.1. Adoption of biotechnology in soybeans by state.

	Biotechnology crop				Total acres of crop	
	Per cent		1000 acres planted		1000 acres planted	
	2001	2002	2001	2002	2001	2002
Arkansas	60	68	1,740	2,006	2,900	2,950
Illinois	64	71	6,848	7,313	10,700	10,300
Indiana	78	83	4,368	4,731	5,600	5,700
Iowa	73	75	8,030	8,025	11,000	10,700
Kansas	80	83	2,280	2,324	2,850	2,800
Michigan	59	72	1,269	1,404	2,150	1,950
Minnesota	63	71	4,599	4,970	7,300	7,000
Mississippi	63	80	731	1,176	1,160	1,470
Missouri	69	72	3,416	3,384	4,950	4,700
Nebraska	76	85	3,762	4,165	4,950	4,900
North Dakota	49	61	1,054	1,495	2,150	2,450
Ohio	64	73	2,944	3,431	4,600	4,700
South Dakota	80	89	3,600	3,738	4,500	4,200
Wisconsin	63	78	1,008	1,131	1,600	1,450
Other states	64	70	4,925	5,406	7,695	7,723
United States	68	75	50,391	54,745	74,105	72,993

summarizes the adoption of GM soybeans in the USA. For example, the adoption of GM soybeans accounted for 75% of all soybean acres nationwide in 2001. Multiplying each state's adoption rate by the average soybean yield in each state results in a GM soybean yield of 2.085 million bushels for the average crop (between 1999 and 2001). This represents 72.2% of the total average soybean crop, or 2.767 million bushels. Thus, our baseline scenario assumes that 72.2% of soybeans produced are GM.

In addition to the planting data and average annual yield data, we use the USDA's 2002 agricultural baseline assuming that 2.945 million bushels of soybeans will be produced and consumed in the USA in 2002 and the projected market price at the farm level is US$4.35/bu (USDA, 2002). At this price, the average LDP would be US$0.57/bu under existing government programmes. In addition, we assume that the overall demand elasticity for soybeans is –0.40 with an elasticity of demand for non-GMO soybeans of either –0.35 or –0.20. The elasticity for GMO soybeans is then derived by holding the overall elasticity of demand constant.

A critical feature in modelling the impact of GMOs on agricultural marketing channels is the IP costs associated with separating non-GM output. As demonstrated in Fig. 19.1, if an excess supply of non-GM soybeans exists at the market equilibrium price, then the reduction in demand for non-GM output at the farm level (and hence the increase in excess supply of non-GM soybeans sold into the GM market) is due to these IP costs. As the cost of identity preservation increases, the economic loss due to the introduction of GM crops increases as well. Further, an excess demand for non-GM output increases the IP costs and reduces the price premium received by farmers (the net price premium $P^3 - P$ in Fig. 19.2).

A survey of the IP costs for maize and soybeans in the USA is presented in Table 19.2. Lin *et al.* (2000) find the IP costs of non-GM soybeans to be US$0.45/bu while Bullock *et al.* (2000) conclude the IP costs to be US$0.30/bu. In general, the IP costs for soybeans appear to be much higher for soybeans than for maize. We use the simple average of the three studies on soybean IP costs yields, which results in an estimated US$0.41/bu increase in costs caused by IP.

Empirical Effects of IP Costs and LDPs

The price received by US farmers and the market price paid by consumers of both GM and non-GM soybeans, given the IP cost estimates and the price floor created by the LDP programme are depicted in Table 19.3. Empirical results are computed for a range of original shares of non-GM demand of 2% to 35% of soybeans produced. The top half of Table 19.3 assume a demand elasticity of –0.35 for non-GM soybeans. If the share of non-GM demand is 2%, 10% or 25% (the first three columns in Table 19.3) then there is an excess supply of non-GM soybeans that get 'exported' into the GM market and the outcome is analogous to Fig. 19.1. For example, if the non-GM soybean demand is originally 25% of the total, the quantity of non-GM soybeans produced is 0.82 billion bushels but the demand for non-GM soybeans is only 0.72 billion bushels. Hence, there

Table 19.2. IP cost estimates for soybeans and maize. Source: author's compilation.

Study	Additional cost (cents/bushel)	Crop/characteristic
Bender *et al.* (1999)	17	Speciality maize
	48	Speciality soybean
Maltsbarger and Kalaitzandonkes (2000a)	16–27	High oil maize
	16.4	High oil maize
	16–15	High oil maize
Maltsbarger and Kalaitzandonkes (2000b)	16.4–36.6	High oil maize
Lin *et al.* (2000)	22	Non-GMO maize
	45	Non-GMO soybeans
Bullock *et al.* (2000)	30	Non-GMO soybeans
European Union	18.4	Non-GMO maize

is an excess supply of approximately 100 million bushels of non-GM soybeans that gets 'exported' into the GM market.

The final price realized by producers under a 25% market share for non-GM demand is US$4.92/bu in both the GM and non-GM market. This price is equal to the average price support level under the LDP programme. The demand price paid by GM consumers and received by US producers for GM soybeans is US$4.25/bu. The price received by US producers from buyers of non-GM soybeans is also US$4.25/bu. However, the price paid by non-GM consumers is US$4.67/bu, which is exactly equal to the non-GM price received by producers (US$4.25/bu) plus the IP cost of (US$0.41/bu). Hence, for market shares of 2%, 10% and 25%, there is no price premium received by US producers of non-GM soybeans.

If the share of non-GM demand is increased to 30% (column four in Table 19.3) there is an excess demand for non-GM soybeans and the outcome is like that depicted in Fig. 19.2, with the exception that the price received by US producers of non-GM soybeans is still at the loan deficiency price of US$4.92/bu. The quantity of non-GM soybeans produced is 0.82 billion bushels (because the price support is still binding) but the demand for non-GM soybeans becomes greater than 0.82 billion bushels. However, since GM soybeans cannot be 'exported' into the non-GM market, the markets separate so that the supply and demand for non-GM soybeans is 0.82 billion bushels and the supply and demand for GM soybeans is 2.12 billion bushels.

The demand price paid by GM consumers and received by US producers for GM soybeans drops to US$4.03/bu under a 30% market share for non-GM soybeans demanded. The price paid by US consumers for non-GM soybeans rises to US$5.24/bu. The market price received by US producers for non-GM soybeans rises to US$4.83/bu (the consumer price minus the IP cost). However, since the price received for non-GM soybeans is still lower than the support price of US$4.92/bu, the government makes up the difference. Hence, the final realized price by producers of both GM and non-GM soybeans remains at the LDP level of US$4.92/bu. Even though the markets separate in this case, there is still no price premium realized by non-GM producers.

Table 19.3. Price premiums for non-GM soybeans in the USA under IP costs and the Loan Deficiency Payments Programme. Source: author's computations.

	Original share of non-GM demand				
	0.02	0.10	0.25	0.30	0.35
Non-GM demand elasticity −0.35					
Non-GM soybean supplied[a]	0.820	0.820	0.820	0.820	0.878
Non-GM soybean demanded	0.057	0.286	0.717	0.820	0.878
Demand price for non-GM soybeans[b]	4.752	4.724	4.670	5.244	6.196
Supply price for non-GM soybeans	4.920	4.920	4.920	4.920	5.786
GM soybean supplied	2.125	2.125	2.125	2.125	2.125
GM soybean demanded	2.888	2.659	2.228	2.125	2.125
Demand price for GM soybeans	4.342	4.310	4.249	4.032	3.228
Supply price for GM soybeans	4.920	4.920	4.920	4.920	4.920
Non-GM demand elasticity −0.20					
Non-GM soybean supplied	0.820	0.820	0.820	0.835	0.916
Non-GM soybean demanded	0.058	0.289	0.724	0.835	0.916
Demand price for non-GM soybeans	4.756	4.740	4.709	5.551	6.771
Supply price for non-GM soybeans	4.920	4.920	4.920	5.141	6.361
GM soybean supplied	2.125	2.125	2.125	2.125	2.125
GM soybean demanded	2.887	2.656	2.221	2.125	2.125
Demand price for GM soybeans	4.346	4.330	4.299	4.074	3.406
Supply price for GM soybeans	4.920	4.920	4.920	4.920	4.920

[a]Quantities are in millions of bushels.

[b]Prices are in dollars per bushel.

Finally, consider the case in which the percentage of non-GM soybeans demanded is increased to 35% (column five in Table 19.3). In this case there is also an excess demand for non-GM soybeans and the outcome is exactly as depicted in Fig. 19.2. The markets separate so that the supply and demand for non-GM soybeans is 0.82 billion bushels and the supply and demand for GM soybeans is 2.12 billion bushels. The price that producers receive from sales of GM soybeans becomes US$3.23/bu, but the final realized price for GM soybeans is still US$4.92/bu (the loan rate). In addition, the final price realized by producers becomes US$5.78/bu for non-GM soybeans and consumers must pay US$6.19/bu for non-GM soybeans. Hence, there are now four different prices as in Fig. 19.2 (two prices for GM soybeans and two prices for non-GM soybeans). Producers receive a premium for non-GM soybeans equal to US$0.86/bu (US$5.78–$4.92).

The bottom half of Table 19.3 depicts the effect of a decreased elasticity for non-GM uses. The non-GM demand for soybeans in the USA and the export market is primarily driven by the demand for soybeans used directly for human consumption (as opposed to soybeans used as protein supplements in livestock feeds). Further, the demand for direct food use tends to be less elastic than the demand for livestock feeds, since other protein supplements such as cottonseed meal can be substituted for soybean meal in many cattle feeding operations. If the elasticity of demand is –0.20, an initial share of only 30% non-GM demand is sufficient to separate the markets (Table 19.3). Under these assumptions, the supply price of non-GM soybeans increases to US$5.14/bu causing the supply of non-GM soybeans to increase to 0.835 billion bushels. Producers realize a small premium of US$0.22/bu.

Discussion and Implications

The analysis demonstrates how the introduction of GMOs can lead to a segmentation of the soybean market in the USA. However, even if it exists, market segmentation alone may not result in a price premium for non-GM soybeans. In order for a price premium to emerge, the demand for non-GM soybeans must be greater than the supply of non-GM soybeans at the original price equilibrium. At the USDA baseline prices and quantities for 2002, separation of non-GM and GM markets requires that just less than 30% of the original demand represent non-GM uses. The results are further complicated by agricultural polices in the USA. In 2002, the loan rate for soybeans in the USA was US$4.92/bu. Thus, even when the markets separate, resulting in higher consumer prices for non-GM soybeans, the final price realized by producers may not be different, because of government price supports. Once producers begin to respond to market segmentation, the impact of market differentiation of GM soybeans increases with the GM soybean prices falling rapidly. This effect is driven by the differences in the demand elasticities in the differentiated market. For an overall demand elasticity of –0.40 and a non-GM demand elasticity of –0.35, the GM demand for soybeans is relatively more elastic (–0.42). Thus, prices must fall further in order to generate an equilibrium in the GM soybean market.

While the results show the potential for price premiums over a range of demand shares for non-GM soybeans, the actual share of non-GM soybean demand facing the USA is currently not known by the authors. Soybean markets appear to be different to maize markets (as studied by Moss *et al.*, 2001) with regards to the acceptability of GM output. While the EU has banned the importation of GM maize from the USA, it may not have a similar prohibition in the soybean market. This is, in part, due to the alternative forms of soybean imports. Most soybeans are imported by Europe as either soybean oil meal or soybean oil instead of raw soybeans. Thus, consumers may be less concerned about the risk implicit in processed products. If consumers are less concerned about the risk from processed soybean products, a high rate of GM soybean adoption may cause fewer problems than a similar rate of adoption for maize.

In order to perform a more detailed analysis on the international effects of the introduction of GM soybeans in the US market, competing countries such as Brazil and Argentina must be taken into account using a complete demand system for these markets. For example, while the US exported 8.41 million t of soybeans in 1998,

Brazil exported 10.65 million t and Argentina exported 12.9 million t (Moschini *et al.*, 2000). Brazil exports almost exclusively non-GM soybeans while Argentina exports almost exclusively GM soybeans. What are the reasons for the differences? Qaim and Traxler postulate that the main reason exports of soybeans from Argentina comprise over 90% GM is because Monsanto was not able to obtain the patent rights due to a national law that makes it difficult for patents to be issued for seed. As a result, producers in Argentina can readily purchase GM seed on the black market. Why does Brazil continue to remain relatively GM free? It may be that they can acquire a premium in countries that do not want to risk having GM soybeans enter their marketing system. A full analysis would be even further complicated if you were to allow for monopoly rents accruing to GM seed producers in international markets, along the lines of Moschini *et al.* (2000) or Lapan and Moschini (2001).

One more caveat, since soybean prices are above the loan rate as of December 2002, we expect the effect of market segmentation and IP costs to be more pronounced than under the USDA 2002 baseline projections, because producers are no longer shielded by the price floor set by the government. However, new US farm legislation passed in 2002 allows for the implementation of target prices that will be set at a price much higher than the loan rate. The analysis can be readily adapted to the US target price scheme by setting the price support at the target price level instead of the loan rate. Of course, acreage set-asides would then have to be incorporated that would shift the supply curve for soybeans inward by even more than the IP costs.

References

Bender, K., Hill, L., Wenzel, B. and Hornbaker, R. (1999) Alternative market channels for specialty corn and soybeans. AE-4726. Department of Agricultural and Consumer Economics, Agricultural Experiment Station. College of Agricultural, Consumer, and Environmental Sciences. University of Illinois at Urbana-Champaign (February).

Bullock, D.S., Desquilbet, M. and Nitsi, E.I. (2000) The economics of non-GMO segregation and identity preservation. Mimeographed paper, Department of Agricultural and Consumer Economics, University of Illinois, October.

Caswell, M.F., Fuglie, K.O. and Klotz, C.A. (1994) *Agricultural Biotechnology: an Economic Perspective*. Economic Research Service, United States Department of Agriculture, Agricultural Economic Report No. 687.

European Union, Directorate-General for Agriculture. *Economic Impacts of Genetically Modified Crops on the Agri-Food Sector* http://europa.eu.int/comm/agriculture/publi/gmo/fullrep/index.htm

Lapan, H.E. and Moschini, G. (2001) GMO labeling and trade: consumer protection or (just) protectionism. Presented at the Globalization, Technology, and Trade Conference of the International Agricultural Trade Research Consortium. Tucson, Arizona.

Lin, W.H., Chambers, W. and Harwood, J. (2000) Biotechnology: U.S. grain handlers look ahead. *Agricultural Outlook*. USDA/Economic Research Service, April.

Lin, W.H., Price, G.K. and Allen, E.W. (2002) StarLink: impacts on the U.S. corn market and world trade. *Agribusiness* 16(1), 33–55.

Maltsbarger, R. and Kalaitzandonakes, N. (2000a) Study reveals hidden costs in the IP supply chain. Economics and Management Center, University of Missouri-Columbia, September.

Maltsbarger, R. and Kalaitzandonkes, N. (2000b) Study reveals hidden costs in IP supply chain. *Feedstuffs* 72, 36.

Moschini, G., Lapan, H. and Sobolevsky, A. (2000) Roundup Ready soybeans and welfare effects in the soybean complex. *Agribusiness* 16(1), 33–55.

Moss, C.B., Schmitz, T.G. and Schmitz, A. (2001) Segregating GM and non-GM corn in a marketing channel. Paper presented at the International Agricultural Trade and Policy Consortium Meeting, Tucson, Arizona, December.

Qaim, M. and Traxler, G. (2002) Roundup Ready soybeans in Argentina: farm level, environmental, and welfare effects. Presented at the 6th International ICABR Conference. Ravello, Italy.

[USDA] United States Department of Agriculture, Interagency Agricultural Projections Committee (2002) *USDA Agricultural Baseline Projections to 2011*. Staff Report WAOB 2002-1.

20 EU Traceability and the US Soybean Sector

Gregory K. Price, Fred Kuchler and Barry Krissoff
Economic Research Service, United States Department of Agriculture, Washington, DC, USA

In 2001, the European Commission proposed regulations for labelling and tracing of biotech products, including bulk commodities, processed foods, and animal feed (Commission of the European Communities, 2001b). The proposed regulations also apply to items, such as soybean oil, in which genetically modified content is not detectable with current testing capabilities. Traceability systems are record-keeping systems where information about a particular product attribute is systematically recorded from creation to marketing. Traceability systems do not require segregation by product attribute. However, in practice, traceability systems are used primarily to keep foods with different attributes separate from one another (Golan *et al.*, 2002). The new regulations are not yet law, since they must be ratified by the European Parliament and all member countries. Given that nearly 70% of US soybean acreage was planted to herbicide-tolerant varieties in 2001 and about 30% of US soybean exports are destined for the European Union (EU), the proposed regulations could affect grain handling and marketing as well as food manufacturing in the USA (USDA-NASS, 2001; USDA-FAS, 2002).[1] However, since soybean exports to the EU are relatively small (less than 10%) compared with US production, the proposed regulations may not affect all players in the US market. While some may continue to supply the EU market, additional costs may cause others to channel their soybeans to domestic outlets or export to other countries.

The objective of this chapter is to investigate the potential costs of complying with the proposed EU traceability requirements as they pertain to biotech soybeans. These additional costs would only apply to those handlers and exporters that continue to supply the EU market after the implementation of the proposed regulations. Other suppliers may choose to find new markets for their soybeans. Thus, a redistribution in trade may occur. These costs could vary significantly, depending on a number of factors. Specifically, three cases are considered. One situation specifies the production of conventional varieties (traditional, non-biotech soybeans) as well as biotech GM varieties that are already approved in the EU.[2] Currently, Roundup Ready® is the primary biotech soybean GM variety produced in the USA, which is approved by the EU.[3] For those that export to the EU, this scenario would be easiest and least costly with respect to compliance since labelling and traceability schemes, but not segregation, would be required. Another case considers the hypothetical mass production of both EU-approved and

[1] US farmers intend to increase the adoption of herbicide-tolerant soybeans to 74% for the 2002/3 crop (USDA-NASS, 2002).

EU-unapproved soybean GM varieties by US farmers. This potential situation depicts an intermediate case, where compliance costs for suppliers of the EU market would be higher due to the additional requirement of segregation. The final scenario assumes that, due to consumer preferences, only non-biotech soybean and soybean products are exported to the EU and all biotech GM varieties are segregated out of shipments. This extreme case shows what could happen to costs if handlers, food and feed manufacturers, and exporters who continue to serve the EU market follow strict segregation, labelling, and tracing practices.

This chapter begins by summarizing the proposed EU regulations within the context of soybeans and soybean products (soybean meal and soybean oil). Subsequently, the costs of segregation are analysed for the purpose of estimating the potential cost of compliance in each scenario. Potential reactions to the proposed regulations both in the USA and the EU are considered. Finally, factors that determine who might bear or avoid the additional costs are assessed.

EU Labelling and Traceability Regulations

According to the proposed regulations, all biotech products (including seed, bulk commodities intended for processing, food items and animal feed) that contain genetically modified material must be labelled as such when being sold in the EU. The sellers of such goods would be required to disclose the unique codes of the events contained in the shipments. If the attachment of labels is not possible, biotech-related information must be included in accompanying documentation. Besides conventional varieties, only those events that have been officially approved by the EU would be accepted; EU-assessed and EU-unassessed events would not be permitted.[4] If a given product is intended for a specific use, such as animal feed, shippers may opt to include such notification on the label. Labelling must occur at the time when the items are first placed on the European market. Moreover, all biotech-related information must be transmitted to all future purchasers within the EU marketing chain and be retained for a period of 5 years. However, records of final consumers who purchase the products need not be maintained.

Under the current EU laws, Regulations (EC) No. 49/2000 and 50/2000, food products that contain more than 1% biotech content, ingredient by ingredient (including additives and flavourings), must be labelled (Commission of the European Communities, 2000a,b; Mitchell and Normile, 1997).[5] However, this rule only applies to food products in which modified DNA is present and detectable with current testing practices. If the new proposed regulations are passed, the EU will extend the labelling requirements to food and feed items that are derived from biotech commodities even though genetically modified material is not detectable with current methodologies (as in the case of soybean oil). Although the EU is proposing that animal feed be labelled as biotech if it is contains genetically modified material or contains ingredients

[2] EU-approved events are those that have been studied by the EU Scientific Committee for negative impacts on the environment and human health. These events have been officially accepted by the EU and are approved for sale and marketing there. By definition, conventional varieties are EU-approved.

[3] Roundup Ready® was not the only biotech event planted by US farmers in 2001. A small amount (some 3890 acres (1572 ha)) of high-oleic soybeans were grown under a tight, closed-loop production system that year. While high-oleic soybeans are pending approval in the EU, the closely monitored production system would eliminate compliance problems associated with the proposed regulations (American Soybean Association, personal communication 4 June 2002).

[4] EU-assessed events are technically unapproved by the EU but have undergone a risk assessment in terms of their potential to produce negative impacts on human health and the environment. EU-unassessed events have not been assessed or approved.

[5] In December 2002, ministers of the EU Environment Committee approved a threshold level of 0.9% for each biotech or biotech-derived ingredient contained in food and feed products (Dow Jones Newswires, 2002). This proposal must still be approved by the EU Parliament and Council before becoming law.

derived from biotech crops, meat that is produced from animals that had been fed biotech feed would be exempt.[6] The tracing of biotech-derived products presents great opportunities for fraud without third-party audit and verification. This is especially true when testing methods are not precise and when detection of biotech material is not possible. Contracting and identity preservation would become very important in preventing fraud (Golan *et al.*, 2002).

The proposed regulations for items that are produced from biotech products but do not contain genetically modified material are somewhat less stringent than for those that do contain such material. EU operators, including handlers, crushers, food and feed manufacturers, and importers, would have to indicate which ingredients (including additives and flavourings) are produced from biotech crops. However, the specific events would not have to be reported. If ingredients are not listed on a product's package, then the item would be labelled generally as being produced from biotech commodities. Recognizing the technical difficulty in segregating crops by GM variety, the proposed regulations allow for the presence of up to 1% EU-assessed GM varieties in food and feed items.[7] In those cases, operators must ensure that every effort is taken to exclude EU assessed GM varieties from the products. For other EU-unapproved varieties, zero tolerance remains. Biotech-related information must be passed along to all buyers in the EU up to the point of the final consumer. In addition, all parties must maintain the information for 5 years. An exception to this rule is provided for prepackaged products that are tracked by specific identification systems, such as lots or batch numbering systems. While purchaser data need not be maintained in these cases, the batch or lot information much be kept for 5 years.

The goal of proposed regulations is to facilitate product withdrawal from the market if there is an occurrence of some negative impact on human health or the environment. While the proposed regulations only require such systems be in place in the EU, US exporters are unlikely to bear the blame if a disastrous situation arises. For those that continue to sell to the European market, these rules would be extended to their suppliers (all the way back to the farm) regardless of whether or not some biotech GM varieties are approved by the EU. Record-keeping systems would aid handlers in those supply chains to keep track of particular GM varieties that are marketed, processed, and shipped within the USA. As a result, labelling and traceability systems would minimize risk exposure to negative occurrences by assigning accountability throughout the domestic side of the marketing chain.

Assessing the Costs of Traceability

Many factors affect the cost of implementing traceability programmes, as well as who ultimately bears the associated costs. Segregating, labelling, and tracing at each stage of the marketing chain would lead to a relatively long and complex set of activities for those US operators that choose to supply the EU market, each with their own costs. In order to analyse the potential costs of the proposed EU regulations, three possible scenarios are analysed to illustrate that the resulting costs may vary considerably. The first and least complicated case mirrors the current US soybean market. This scenario allows for the shipment of conventional varieties as well as EU-approved biotech events (such as Roundup Ready®). This case is likely to be the easiest and least costly since segregation would be unnecessary (Table 20.1).[8] However, this does not imply that compliance would be completely costless. Producers, handlers and manufacturers in the chain that supplies the EU

[6] The Environment Committee's decision to lower the threshold level would also apply to animal products, such as meat, eggs, and dairy, due to the limits on content of biotech or biotech-derived ingredients in feed.

[7] Ministers of the EU Environment Committee approved the lowering of the threshold level to 0.5% over the next 3 years (Dow Jones Newswires, 2002). This proposal must still be approved by the EU Parliament and Council before becoming law.

[8] There is a small segment in the EU that demands non-biotech soybeans when directly used in food products, such as tofu. Currently, about 100,000 t of non-biotech soybeans are exported from the USA to the EU (Toepfer International, Inc.). In this case, it would be necessary to segregate and trace food-bound non-biotech soybeans.

Table 20.1. Potential implications of complying with the proposed EU labelling and traceability regulations, if implemented.

Scenario	Actions needed for compliance	Likely relative cost	Potential US industry changes
1. US produces conventional (traditional, non-biotech) and EU-approved soybeans	Institute labelling and traceability schemes throughout the US oilseed marketing chain back to the farm level	Relatively low since all soybeans are EU-approved	Few since labelling and traceability would only require minimal management for record keeping
2. Hypothetical mass production of EU-approved and EU-unapproved soybean varieties in the USA	Same as No. 1. In addition, US operators would be required to segregate EU-unapproved GM varieties	Moderate, but the total cost depends on the number and volume of GM varieties that must be segregated as well as threshold levels for biotech content	May cause specialization among handlers that have basic facilities. Others may build facilities to segregate and accommodate multiple varieties. Niche marketers, especially those that handle several varieties, may not be greatly affected. Large companies may also find it relatively easy to assign particular varieties to certain elevators. Specialization or dedication of equipment at certain facilities may be sufficient to satisfy the proposed EU requirements and allow many elevators to continue handling commingled soybeans and avoid additional costs
3. USA only supplies non-biotech soybeans to the EU	Same as No. 2, but the strictness in segregation and tracing would be much higher. Containerized shipments may be required	Highest, assuming low tolerance for biotech content and strict segregation	Same as No. 2

would still institute labelling and traceability schemes to keep track of GM varieties and minimize risk exposure in case there are negative effects on human health and the environment. Supplying the EU would become more difficult and costly with a greater number of GM varieties.

The current situation with basically one EU-approved biotech GM variety may not continue in the future, and a scenario similar to biotech maize may evolve for soybeans where multiple GM varieties exist, with some not being approved by the EU. A second case considers the production of conventional soybeans as well as EU-approved and EU-unapproved GM varieties. The hypothetical mass production of EU-unapproved GM varieties would complicate compliance strategies for those that supply the EU market because of the need for segregation. In this potential situation, grain handlers would have to segregate to some degree, with the extent depending on the number and volume of EU-unapproved GM varieties. However, the task would probably not be as arduous and costly as supplying non-biotech soybeans.

A third scenario mandates that only non-biotech soybeans and soybean-derived products be accepted by the EU. Although this situation goes beyond the proposed regulations, it illustrates where the upper-limit of compliance costs may lie for those firms that supply the EU market. Those US operators would be forced to follow even stricter segregation practices to meet specifications and guarantee authenticity. Labelling and traceability schemes would become even more crucial than in the second scenario.

Segregating crops based on biotech traits is essentially an extension of the handling process for speciality grains and oilseeds (Lin *et al.*, 2000). Estimates of segregation costs incurred at different points along the marketing chain are based on several studies (Table 20.2). Ranges are given so as to allow for varying strictness in segregation. The low end is based on the segregation process for high-oil maize, while the high end refers to a more stringent process for synchrony-treated soybeans (STS).[9] If EU consumers fiercely oppose biotech crops, then an even stricter segregation system would likely be required and impose costs that are higher than those associated with STS.

Producers who grow non-biotech soybeans incur additional production costs stemming from higher harvesting, storage, transport, and marketing charges. They also incur greater market risks and have to test for quality. Additional on-farm production costs may range from US$0.01/bu to US$0.10/bu (Bender and Hill, 2000). Local elevators, subterminals, and export ports also realize additional segregation costs,

Table 20.2. Segregation, handling and marketing costs along the marketing chain.

	Additional per bushel costs of segregation, handling, and marketing US$	
Segment of the marketing chain	Low (high-oil maize)	High (synchrony-treated soybeans)
On farm (including elevator delivery)	0.01	0.10
Country elevator to export elevator	0.22	0.58
Inland transportation	0.00	0.45
Ocean freight	0.00	0.13
Total additional cost for soybeans	0.23	1.26
Soybean crushing plants	0.04	0.21

Sources: Good *et al.*, 2000; Lin *et al.*, 2000; Commission of the European Communities, 2001a; Bender and Hill, 2002.

[9] Segregation based on handling high-oil maize can usually meet a 5% threshold level for biotech content. However, the process may be grossly inadequate when attempting to segregate to meet a higher threshold level. Even though the STS process is stricter than that for high-oil maize, many STS farmers in the USA failed to meet the quality standards stated in their contracts. Thus, following the STS process does not guarantee that a stricter tolerance level for biotech content will be met.

with charges ranging from US$0.22/bu to US$0.58/bu, depending on the strictness of the segregation system (based on estimates reported by Lin *et al.*, 2000). The estimate of grain handlers' segregation costs accounts for extra storage, handling, and marketing costs; risk management; and testing. Extra charges incurred between country elevators and export terminals assume that bulk shipping is utilized. Stringent limits on biotech content may require containerized shipments, such as those used for food-grade soybeans. If this is necessary, then segregation costs from country to export elevators could increase significantly – up to US$1.83 as reported by Bender *et al.* (1999).

Transport costs could rise dramatically when small volumes are shipped or when low tolerance levels for biotech content mandate the use of individual rail cars or containers. According to one industry source, transport costs could double if the threshold level for biotech content is 1% or lower, typically requiring containerized shipments (Lin *et al.*, 2000). Good *et al.* (2000) estimated the additional cost associated with shipping food-grade soybeans domestically (usually containerized) to be as much as US$0.45/bu. Whether or not extra ocean freight expenses are incurred depends critically on the volume shipped. If the segregated volume exceeds 8000–9000 t, bulk handling can occur and there is no need for separate holds on vessels (Lin and Johnson, Chapter 21 this volume). For smaller shipments, separate holds would be required, which would add extra costs but would still allow for bulk handling. In the case of non-biotech maize shipments to Japan, Lin and Johnson, Chapter 21 this volume estimated that small shipments cost exporters an extra US$0.13/bu in ocean freight expenses, a value supported by Japanese grain trading firms.

Soybean processors that supply the EU market may not incur significantly higher costs if they produce soybean meal and soybean oil from a supply that contains only conventional varieties and EU-approved events. If traceability systems arise in the USA and crushers can source previously documented inputs, it should be a fairly straightforward task to label outputs and maintain records. Operational costs could be significantly higher for those firms in the presence of EU-unapproved GM varieties since obtaining EU-approved or conventional soybeans would require paying premiums to elevators for segregating, labelling and tracing those GM varieties or varieties. Sourcing strictly non-biotech soybeans would be even more costly for processors and may erode margins, which are already thin.

Owing to the large size of many operations, crushers with output being shipped to the EU are unlikely to specialize in processing particular kinds of soybeans. Doing so may cause underutilization of capacity. If the proposed regulations were to be implemented, there would be pressure to process different types of soybeans simply to operate at capacity. Such an effort would require separate storage facilities and equipment cleaning between batches. Processing costs could increase by between US$0.04/bu to US$0.21/bu if equipment cleaning, separate storage facilities, and identity preservation systems are required (Commission of the European Communities, 2001a). These additional costs would place downward pressure on the crushers' margins.

Adaptation and Cost Avoidance

Some firms are better positioned to adapt to, or avoid, cost than others. Grains and oilseeds in the USA are usually produced and marketed in bulk, although there are growing niche markets for specialized commodities. Speciality commodities may be supplied in bulk or containers, with the method being dependent on the volume and strictness of segregation. In these niche segments, buyers are willing to pay premiums for differentiated products, as in the cases of non-biotech food-grade soybeans in Japan and organic produce in the USA (Dimitri and Richman, 2000; Lin *et al.*, 2000). Companies that supply undifferentiated products are likely to find the proposed EU traceability regulations more difficult to comply with than firms that have experience in marketing differentiated products (e.g. high-oil maize). More traditional firms may choose to channel their soybeans to other markets (domestic or international) in order to avoid the added costs associated with the proposed rules.

The type of facility that a handler operates is critical in determining whether or not it would be relatively easy to comply with the proposed regulations. An elevator that has traditional, basic facilities is likely to have the most problems, particularly because it can only accommodate commingled, bulk commodities. Added costs of segregation may be avoided by specializing in certain GM varieties or non-biotech soybeans. Operators, either large or small, that deal with different varieties or speciality crops are likely to be much more adept in incorporating new requirements. This is because they have multiple pits and equipment dedicated to different varieties. Specialization and/or dedication of equipment by some operators would satisfy EU demands as well as permit other firms (particularly those with traditional bulk facilities) to continue to handle commingled soybeans without incurring additional costs. Large investments in infrastructure and equipment so as to comply with the proposed requirements may deter some traditional operators from doing business with the EU.

Large grain companies with multiple interior elevators could also accommodate the proposed regulations because they can designate certain facilities to accept particular soybean GM varieties. The choice of one or more elevators would be based on several factors, including the number of GM varieties to be segregated, local supply, and forecasted EU demand for certain traits. While dedicating facilities to particular GM varieties would significantly reduce segregation costs, elevators would still need to maintain paperwork for traceability verification. Testing may be employed to ensure the identity of GM varieties, although doing so is not required by the proposed regulations.

Using dedicated facilities does have some associated costs, including potentially higher transport costs due to non-optimal commodity movements and not being a part of a system with multiple delivery points. Thus, smaller elevators may be more ideal for segregation.

The proposed regulations may accelerate vertical integration in grain handling. Dominant firms operate country and export elevators as well as own their own railcars and barges. As a result, they can maintain control over commodities throughout large segments of the marketing chain. Thus quality standards can be maintained. A large scope with respect to in-house operations may also afford synergies in documentation. Because many of the major firms operate crushing facilities, such savings are likely to extend to the production of soybean meal and soybean oil. Since some activities are internalized, vertically-integrated companies may not need to pass large price discounts on to soybean producers if the regulations are implemented. In contrast, smaller, independent elevators that continue to serve the EU market (particularly those that do not specialize in certain soybean traits) may have to bear some additional costs brought on by compliance with the proposed EU regulations in order to remain competitive with the larger, vertically-integrated companies.

In terms of processed foods, manufacturers, particularly those with a strong presence and brand recognition in the EU market, may reformulate products to avoid labelling and tracing costs. Processors in Europe have already begun replacing biotech ingredients with ones that are not genetically modified to avoid labelling final products. However, such action is only taken when biotech ingredients are detectable with current testing capabilities. Items that are derived from biotech crops but do not contain genetically modified material (such as soybean oil) are still used in food manufacturing. More extensive reformulation would probably occur for biotech-derived ingredients if the proposed regulations are implemented. The new rules would make EU consumers aware of many more products that contain biotech ingredients and possibly encourage further opposition. In the case of soybean oil, manufacturers would be enticed to use other oils that do not yet have biotech counterparts in order to avoid the relatively high cost of non-biotech soybean oil. For those food makers that have relatively small sales in Europe may choose to withdraw from that market rather than incur higher manufacturing costs.

EU crushers are also likely to face additional costs as a result of the proposed traceability regulations. Since meat produced with biotech feed would not need to be labelled, crushers would have the option of continuing to import biotech soybeans and selling the soybean meal to feed manufacturers or livestock operators. However, biotech soybean oil would have to be labelled, although there would be no physical evidence to determine that it had come from biotech soybeans. Lower prices for the soybean oil may result due to EU consumers' aversion to products with biotech ingredients. As evidenced by current reformulation, food manufacturers and other users would probably avoid biotech soybean oil, forcing EU crushers to discard it or export the oil to other markets. Both situations would place pressure on crushing margins. Given that additional crushing capacity has arisen in a number of other countries, additional pressure on margins from biotech soybeans may place EU crushers in a difficult competitive position (Ash, 2000; USDA-ERS, 2001).

Market and Trade Impacts

The proposed regulations permit the import of EU-approved soybean GM varieties as well as the meal and oil obtained from crushing those varieties. Food and feed products containing either approved GM soybean material or ingredients derived from approved GM varieties would also be accepted. If only conventional varieties and EU-approved events are produced in the USA, segregation would not be needed but the products would be labelled and traced to ensure accountability. In this case, the cost of labelling and maintaining records would probably be small for those that continue to export to the EU market. As a result, only minimal adjustments in marketing and trade would be likely to occur. Those that do not want to incur the extra costs may channel soybeans and soybean products to domestic outlets or other countries.

In a case with both EU-approved and EU-unapproved GM varieties, compliance would be more difficult. Such a situation would require segregation to remove unapproved GM varieties from exports of bulk commodities and processed products to the EU. One factor that would influence compliance costs for suppliers to the EU is the volume of unapproved GM varieties relative to that of approved GM varieties and conventional varieties. With the existence of EU-unapproved events, those handlers and exporters must forecast EU demand for approved soybeans, meal and oil and isolate them for shipment. At the individual handler level, compliance costs are likely to be highest when the shares of approved GM/conventional varieties and unapproved GM varieties are evenly split. This is because the volume of 'approved' types would be large enough to involve significant segregation costs but not scale economies associated with bulk handling. Segregation costs for a particular elevator would be relatively small if the unapproved GM varieties comprise either a small or large share of the volume handled. In the case where EU-unapproved soybean volume is small, segregation costs may be inconsequential due to the small amount. Segregation costs may also be small when EU-unapproved GM varieties comprise a large share because individual elevators may be able to reap economies of scale and separate out EU-unapproved GM varieties in bulk. In addition, large-volume handling may help handlers that supply the EU market to economize on documentation. Costs would be likely to increase as the number of EU-unapproved GM varieties handled rises. Specialization in certain GM varieties or non-biotech varieties would eliminate segregation costs for those firms.

Besides labelling approved biotech products, handlers, processors and exporters that serve the EU would have to establish a tracing system to guarantee that shipments only contain conventional varieties or approved GM varieties. Segregation, labelling, and tracing activities would result in higher handling and marketing expenses (Table 20.2), raising the cost of supplying 'approved products' to the EU. These higher costs could be reflected in increased product prices for EU consumers, which may, in turn, lower EU demand for those items. As a result, trade between the USA and the EU may be adversely affected by the proposed regulations. However, this outcome may not necessarily lead to large losses for the USA since exports may be diverted to other markets and lead to a redistribution of global trade.

If the EU market begins to demand non-biotech products, segregation strategies for the US suppliers would become even more complex and costly. This specialization would require guaranteeing that the products are as claimed through stringent segregation and documentation (testing may also be utilized). The design and attributes of segregation and traceability schemes depend on the elasticity of demand in the EU and EU consumers' willingness to pay for segregated products. The total cost of compliance associated with this scenario hinges on the segregation strategies employed and the size of non-biotech demand in the EU. If most EU consumers are indifferent to biotech foods and the market for non-biotech soybeans is small, compliance costs are not likely to be exorbitant since there would not be a need for many handlers to segregate biotech from non-biotech soybeans. However, EU-unapproved GM varieties would still have to be removed and EU-approved soybeans would have to be labelled and documented. On the other hand, if most EU consumers do not want biotech foods, stringent segregation would probably lead to relatively high compliance costs for those suppliers of the EU market. The segregation, labelling, and tracing costs incurred in supplying the non-biotech segment in the EU would lead to higher prices for non-biotech products than those for 'approved' or conventional ones because of the increased stringency. The reduced demand resulting from higher prices could lead to greater disruptions in trade between the USA and the EU. But a redistribution of global trade may also arise, making up for lost exports to the EU.

Conclusion

The proposed EU traceability regulations may pose significant challenges for the US grain-handlers that serve the EU market. Potentially significant compliance costs may cause some handlers and exporters to channel their soybeans and soybean products to other outlets, such as domestic crushers or other foreign markets. In addition to higher costs, the proposed regulations present opportunities for fraud, which would have to be addressed through third-party verification. If only EU-approved GM varieties and conventional varieties are produced in the USA, compliance by firms that serve the EU would not necessitate segregation but would probably result in labelling and traceability schemes throughout their supply chain. Such systems would help to keep track of GM varieties as well as to ensure accountability, manage risk, and facilitate product withdrawal in the case of negative impacts on the environment or human health. Achieving compliance becomes more difficult if EU-unapproved GM varieties are produced in significant quantities or if EU consumers demand non-biotech products. Such situations would require segregation, imposing additional costs on elevators and processors. These costs could be very high if tolerance levels for biotech content are low and strict segregation processes (such as those with containerized shipments) are needed.

Handlers, large or small, that deal with multiple soybean varieties are likely to be in the best position to accommodate the proposed regulations, if implemented. This is because they have multiple pits and equipment that is dedicated to those varieties. In contrast, traditional elevators that handle bulk, commingled soybeans may need to either specialize in certain types or build new facilities in order to segregate. High costs may accelerate vertical integration so as to capture synergies with respect to commodity control and information transfer. US soybean processors are unlikely to specialize in processing certain kinds of soybeans because of higher input costs (caused by segregation at elevators), capacity underutilization, and extra storage and cleaning costs. Thus, the proposed regulations may place pressure on EU crushers' margins. Because biotech soybean oil would have to be labelled, it may be difficult to sell it to food manufacturers.

The market and trade impacts associated with the proposed regulations are likely to depend on the types of soybeans grown in the USA, their volumes, and the the stringency of segregation. Adjustments in trade flows may be smallest when US producers only grow EU-approved GM varieties and conventional varieties. In contrast, the trade impacts are likely to be the most severe if EU consumers demand non-biotech products. Adjustments in trade may fall somewhere in between if both EU-approved and EU-unapproved GM varieties are produced in the USA, with the effects depending on the number and volume of unapproved events as well as the strictness of segregation systems. However, losses in the EU market may be made up through a redistribution of trade.

References

Ash, M. (2000) Prices for bumper soybean crop hinge on China's imports. *Agricultural Outlook* AGO-274. http://www.ers.usda.gov/publications/agoutlook/sep2000/ao274d.pdf

Bender, K. and Hill, L. (2000) Producer alternatives in growing specialty crops. *Department of Agricultural and Consumer Economics Report No. AE-4732*. University of Illinois, Urbana, Illinois. http://www.ace.uiuc.edu/research/ae/papers/ae4732.pdf

Bender, K., Hill, L., Wenzel, B. and Hornbaker, R. (1999) Alternative marketing channels for specialty corn and soybeans. *Department of Agricultural and Consumer Economics Report No. AE-4726*. University of Illinois, Urbana, Illinois. http://www.ace.uiuc.edu/research/ae/papers/ae4726.pdf

Commission of the European Communities (2000a) Commission Regulation (EC) No. 49/2000 of 10 January 2000 amending Council Regulation (EC) No. 1139/98 concerning the compulsory indication on the labelling of certain foodstuffs produced from genetically modified organisms of particulars other than those provided for in Directive 79/112/EEC. http://europa.eu.int/comm/food/fs/gmo/legal_oj/reg49–2000_en.pdf

Commission of the European Communities (2000b) Commission Regulation (EC) No. 50/2000 of 10 January 2000 on the labelling of foodstuffs and food ingredients containing additives and flavourings that have been genetically modified or have been produced from genetically modified organisms. http://europa.eu.int/comm/food/fs/gmo/legal_oj/reg50–2000_en.pdf

Commission of the European Communities, Directorate-General for Agriculture (2001a) Economic Impacts of Genetically Modified Crops on the Agri-food Sector: A First Review. http://europa.eu.int/comm/agriculture/publi/gmo/full_en.pdf

Commission of the European Communities (2001b) Proposal for a Regulation of the European Parliament and of the Council Concerning traceability and labelling of genetically modified organisms and traceability of food and feed products produced from genetically modified organisms and amending Directive 2001/18/EC. http://europa.eu.int/comm/food/fs/gmo/biotech09_en.pdf

Dimitri, C. and Richman, N. (2000) Organic foods: niche marketers venture into the mainstream. *Agricultural Outlook* AGO-272: 11–14. http://www.ers.usda.gov/publications/agoutlook/jun2000/ao272f.pdf

Dow Jones Newswires (2002) EU Mins Set to Agree on GMO Labeling, Traceability Rules. 9 December.

Golan, E., Krissoff, B. and Kuchler, F. (2002) Traceability for food marketing & food safety: What's the next step? *Agricultural Outlook* AGO-288, 21–25. http://www.ers.usda.gov/publications/agoutlook/jan2002/ao288f.pdf

Good, D., Bender, K. and Hill, L. (2000) Marketing specialty corn and soybean crops. *Department of Agricultural and Consumer Economics Report No. AE-4733*. University of Illinois, Urbana, Illinois. http://www.ace.uiuc.edu/research/ae/papers/ae4733.pdf

Lin, W. and Johnson, D.D. (2002) Segregation of non-biotech corn and soybeans: Who bears the cost? Selected paper presented at the *International Consortium on Agricultural Biotechnology Research* conference in Ravello, Italy, 11–14 July.

Lin, W.W., Chambers, W. and Harwood, J. (2000) Biotechnology: U.S. grain handlers look ahead. *Agricultural Outlook* AGO-270, 29–34. http://www.ers.usda.gov/publications/agoutlook/apr2000/ao270h.pdf

Mitchell, L. and Normile, M. (1999) Consumer concerns elicit policy changes. In: *The European Union's Common Agricultural Policy: Pressures for Change*. US Department of Agriculture, Economic Research Service Report No. WRS-99-2. http://www.ers.usda.gov/publications/wrs992/wrs992.pdf

[USDA-ERS] US Department of Agriculture, Economic Research Service (2001) *Oil Crops Situation and Outlook Yearbook*. http://usda.mannlib.cornell.edu/reports/erssor/field/ocs-bby/ocs2001.pdf

[USDA-FAS] US Department of Agriculture, Foreign Agricultural Service (2002) *Foreign Agricultural Trade of the United States*. http://www.fas.usda.gov/ustrade/USTExFatus.asp?QI=

[USDA-NASS] US Department of Agriculture, National Agricultural Statistics Service (2001) *Acreage*. Cr Pr 2–5 (6–01). http://usda.mannlib.cornell.edu/reports/nassr/field/pcp-bba/acrg2001.pdf

[USDA-NASS] US Department of Agriculture, National Agricultural Statistics Service (2002) *Prospective Plantings*. Cr Pr 2–4. http://usda.mannlib.cornell.edu/reports/nassr/field/pcp-bbp/pspl0302.pdf

21 Segregation of Non-biotech Maize and Soybeans: Who Bears the Cost?

William Lin and D. Demcey Johnson
Economic Research Service, United States Department of Agriculture, Washington, DC, USA

The development of markets for non-biotech maize and soybeans is part of a more general trend toward product differentiation and market segmentation in grains and oilseeds. Markets for value-enhanced commodities, such as high-oil maize and food-grade soybeans, have been in existence for many years, but demand for non-biotech commodities is now adding additional pressure for product differentiation in both domestic and export marketing channels. Area devoted to 'traditional' value-enhanced maize increased from 3% of total US maize area in 1996 to about 8% in 2000 (US Grains Council, 2002). Maize marketed as 'non-biotech' accounted for an additional 2.6% of maize area grown.

Demand for non-biotech maize and soybeans has increased in recent years. First, changing preferences toward biotech products by consumers in some parts of the world, notably the European Union (EU) and Japan, have led to an increase in the use of non-biotech ingredients by food manufacturers in these countries. Second, consumers' perceptions about safety and health effects of biotech foods (which are not science-based), along with foreign governments' willingness to give consumers the right to choose between biotech and non-biotech foods, have led to the adoption of mandatory biotech food labelling and government regulations. Apart from the EU and Japan, biotech labelling requirements have been introduced in South Korea, Australia, New Zealand, China, and other countries. Food manufacturers in some of these countries have opted for procuring non-biotech ingredients to avoid having their products labelled as 'may contain genetically modified organisms (GMOs)'.

It is against this backdrop that the US grain and oilseed industries face the challenge of meeting the needs of foreign buyers through segregation or identity preservation (IP).[1] Suppliers' decisions about whether to segregate grains or oilseeds into biotech and non-biotech commodities depend crucially on the willingness of buyers to pay price premiums for non-biotech commodities. If price premiums cover the costs of segregation and provide sufficient incentives for producers, the US grain industry can accommodate foreign requirements.

This study focuses on the economics of segregating US non-biotech maize and soybeans for shipments to Japan, the primary non-biotech export market for US grains and oilseeds, as a

[1] IP or segregation, for purposes of this study, is broadly defined as a production–handling–distribution process by which crops are required to be kept separate to avoid commingling during planting, harvesting, loading and unloading, storage, and transportation so as to preserve the crops' identity in terms of end-use quality, genetic makeup, or unique production process, such as organic farming (Lin, 2001).

case study. The purposes of this chapter are twofold: (i) to ascertain or estimate price premiums that buyers in both the US domestic and Japanese export markets were willing to pay for non-biotech maize and soybeans for the 2000 and 2001 crops; and (ii) to examine who bears the cost of segregation.

Non-biotech Maize and Soybean Markets

At present, non-biotech maize and soybeans grown and marketed under IP systems remain niche markets. Non-biotech, IP maize and soybeans each account for about 2% of US production (Lin, 2002). Non-biotech maize and soybeans are basically an extension of value-enhanced crops, such as high-oil and white maize, which are typically grown by producers under contract with grain companies.

Japan is the primary export market for both US non-biotech maize and soybeans, according to grain trade sources (Table 21.1). In the marketing year of 2000/01, US exports of non-biotech maize totalled about 0.9–1.1 million t, 90% of which were shipped to Japan for food manufacturing (primarily starch). Similarly, US exports of non-biotech soybeans (totalling about 0.6–1.0 million t) were primarily to Japan for the production of processed foods, such as tofu, miso, and soy sauce. The 0.5–0.8 million t exports to Japan include non-biotech soybeans shipped as a bulk commodity (which are intended to meet a 5% threshold level of biotech content) or by containers, as well as organic soybeans (Table 21.1). Changing sentiment towards biotech products among some consumers and the decision of food manufacturers to avoid using biotech ingredients in the production of certain food items (26 products governed by the Japanese labelling regulations) are major factors behind the importation of non-biotech maize and soybeans into Japan.[2] In 2000/01, non-biotech exports accounted for 6–8% of US maize exports to Japan and 17–29% of US soybean exports to that country. Small volumes were exported to South Korea and the EU in the case of non-biotech soybeans and to South Korea in the case of non-biotech maize.[3]

Table 21.1. US exports of non-biotech maize and soybeans: 1999/2000 and 2000/01.

Commodity	Importing country	Volume exported (million t) 1999/2000	2000/01
Non-biotech maize	Japan	0.5–1.0	0.8–1.0
	South Korea	0.1	0.1
	EU	–	–
		0.6–1.1	0.9–1.1
Non-biotech soybeans	Japan	0.5–0.6	0.5–0.8
	South Korea	0.1	0.1
	EU	0–0.1	0–0.1
		0.6–0.8	0.6–1.0

Sources: US grain trade associations; Japanese grain trading houses; Toepfer International; and Japan Ministry of Agriculture, Forestry and Fisheries.

In addition to meeting the needs of export markets, US grain handlers also separate non-biotech from biotech crops through IP to meet the needs of some domestic food processors. Over the past few years, a few food manufacturers have decided to use only non-biotech crops in their operations. In July 1999, the Gerber and Heinz companies announced that their baby-food processing facilities would immediately stop using biotech ingredients. In January 2000, Bestfoods, Inc. decided to reformulate its food products destined for the EU, ending its use of biotech ingredients in order to avoid the labelling requirement (Howie, 2000). Then, in February 2000, Frito-Lay Inc. announced that it would replace biotech maize in its snack food manufacturing with non-biotech maize. The volume of non-biotech maize used annually by these food manufacturers is estimated to lie between 0.5 and 1.0 million t. Frito-Lay alone used about 21.4 million bu (0.5 million t) of maize in 1999 for processed foods production (Koenig, 2000).

[2] Japan began implementing its mandatory labelling regulations for bio-engineered foods in April 2001.

[3] According to Toepfer International (2002), market demand for non-biotech, IP food soybeans in the EU totalled about 100,000 t in 2000/01, mostly imported from the USA for making tofu.

Identity Preservation Systems and Costs

Because of limited demand for non-biotech maize and soybeans and the additional expense of maintaining separate handling and storage facilities, only a small to modest percentage of grain elevators segregate and market non-biotech products based on IP systems. The non-biotech niche market is defined by IP systems with contracts that specify the purity of non-biotech content in maize or soybean shipments so as to meet particular needs. Some US grain handlers are already segregating grain for certain export markets. For example, Cargill is segregating non-biotech maize under *Innovasure*, a so-called process-based IP system, for Japan, although without guaranteeing a specific tolerance level for biotech material. In many cases, the company provides its own process verification for the IP programme, similar to that used for high-oil (maize) corn (HOC). Patterning maize segregation after handling procedures for HOC can usually meet the 95% purity requirement of Japanese buyers (Lin *et al.*, 2000). Thus, the process has remained the core procedure for segregating non-biotech maize in recent years.

In contrast, the IP system for non-biotech soybeans has changed drastically in recent years. In the 2000 crop year, the Synchrony-Treated Soybeans (STS) IP system was a common segregation process with rigid purity requirements, ranging from 98% to more than 99%.[4] The STS IP system became less appealing to farmers in the 2001 crop year because of its very rigid requirements for maximum allowable biotech content and other quality attributes.[5] Many soybean producers did not receive the premiums promised in the contract because their shipments failed to meet the IP requirements. As a result, they moved to less rigid IP systems, typically resembling that for HOC.

Costs of Segregation

The cost of segregation depends on the requirements of the system used to preserve the identity of the product, including the purity level of non-biotech content. Hence, the cost of segregation is particular to a specific IP system. If an IP system is changed, such as the use of HOC system for 2001 non-biotech soybeans instead of the STS system, then the cost will change to reflect the new set of requirements.

An earlier Economic Research Service (ERS) study examines the costs of segregation for non-biotech crops, based on a survey of US grain elevators and speciality grain firms conducted by the University of Illinois (Bender *et al.*, 1999). The ERS study indicates that, on average, segregation could add about US$0.22/bu (12% of the average maize farm price) to the cost of handling non-biotech maize from country elevators to export elevators if segregation follows the HOC system (Lin, 2002). The ERS cost estimate is comparable to that (typically around US$0.20/bu) reported by other researchers or the grain handling industry (Miranowski *et al.*, 1999; Lence and Hayes, 2001; Moss *et al.*, 2001; Krejci, 2002). The ERS study also estimated that, on average, segregation could add about US$0.54/bu (12% of the average soybean farm price) to the cost of handling non-biotech soybeans from country elevators to export elevators, if segregation follows the STS system. In contrast, the cost of segregation could decline to US$0.18/bu (4% of the soybean farm price) if segregation follows the same handling process used for HOC.

In most cases, shipping non-biotech soybeans from US export ports to final destinations in Japan would not add extra ocean freight expenses. Non-biotech soybeans accounted for 17–29% of total US soybean exports to Japan, making it more likely to be shipped in larger volumes. A volume greater than 8000–9000 t

[4] STS were herbicide-tolerant, but non-biotech varieties developed by DuPont and marketed by ADM, Protein Technologies International, and other grain companies.

[5] According to Kim Nill of the American Soybeans Association, the sign up for the STS IP system was negligible for the 2001 crop. This statement was reaffirmed by analysts of the Protein Technologies International during a visit by the author in April 2002.

does not require separate holds or compartments on the vessel and thus does not lead to additional ocean freight expenses (Nishiyi). In addition, testing would be simpler because herbicide-tolerant soybeans are the only approved biotech trait. In contrast, shipments of less than 8000 t are more common for non-biotech maize, and testing for biotech content is more complicated for maize because of different biotech GM varieties and the existence of varieties that have not been approved by the importing countries (S. Sato, Nissho Iwai American Corporation, personal communication, March 2000; G. Martin, North American Grain Exporters Association, personal communication, May 2002). Shipments of non-biotech maize are thus more likely to have delays in loading on to the vessel, which adds extra opportunity costs (interest expense). Therefore, non-biotech maize shipments would typically add an extra US$5 per t (US$0.13/bu) of ocean freight expenses from US export ports to Japan, compared with no such additional expense for non-biotech soybeans (S. Sato, Nissho Iwai American Corporation, personal communication, March 2000).

The costs of IP also include segregation costs (including hidden costs) at the farm level, and incentives offered to producers to grow non-biotech crops. According to USDA's Value Enhanced Grain Survey conducted by NASS' *Illinois Market News*, non-biotech maize price premiums paid to producers by grain companies averaged about US$0.08–0.10/bu for the 2000 crop, and US$0.10 for the 2001 crop. In the case of non-biotech soybeans, the price premiums averaged about US$0.20 to US$0.25/bu for the 2001 crop, up from US$0.05 to US$0.10 for the 2000 crop. These premiums largely represent the incentives offered to producers to grow non-biotech crops.

The costs of IP vary depending upon the requirements of the system. In the case of non-biotech maize, the costs totalled around US$0.45/bu (24% of the average maize farm price) for the 2000 and 2001 crops, from US farm gate to final destination in Japan. The costs for non-biotech soybeans totalled around US$0.62/bu (14% of the average soybean farm price) for the 2000 crop and US$0.40 (9% of the average farm price) for the 2001 crop.

Price Premiums for Non-biotech Crops

If consumers in some export market segments (primarily the EU and Japan) show a preference for non-biotech food products, are they willing to pay premiums for non-biotech foods? This cannot be easily determined from retail data because food manufacturers opt to use non-biotech ingredients to avoid labelling. However, it is possible to determine whether Japanese buyers are willing to pay premiums for non-biotech maize and soybeans. In addition, it can be determined whether price premiums are large enough to cover the costs of segregation and to provide incentives to producers for entering into contracts with grain companies to grow non-biotech maize and soybeans.

Theoretical considerations

Before the investigation of price premiums that Japanese buyers are willing to pay for non-biotech maize and soybeans, it is appropriate to consider some conceptual issues. How does the market equilibrium for non-biotech maize or soybeans differ from that for biotech (and biotech-commingled) varieties? Under what market conditions would buyers pay price premiums for non-biotech maize or soybeans, or discount biotech products?

The conceptual framework divides the US maize or soybean market into two market segments: (i) the non-IP market, including biotech products and conventional varieties that are not segregated, and are thus commingled with the biotech products; and (ii) the IP market, which is reserved exclusively for non-biotech products grown under IP systems. The supply curve of the non-biotech product (but unsegregated) before the cost of IP being added is a 'kinked' supply curve *abf* (S_{nip}), as shown in Fig. 21.1, with the flat segment of line *ab* representing the fact that farmers have an option of selling their products at the conventional (but unsegregated) price (point *a*). After adding the cost of IP (distance of line *ik*) to the supply curve S_{nip}, the supply curve of the non-biotech, IP product shifts upward to *ecg* (S_{ip}). If the demand curve (D_{ip}) for the non-biotech product intersects the supply curve on its diagonal segment, farmers would choose

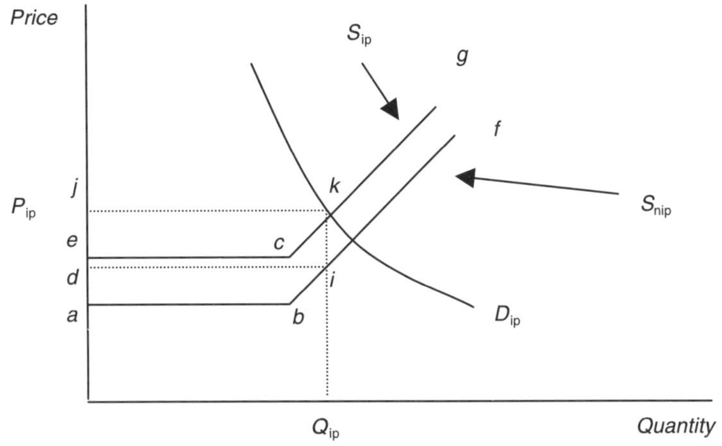

Fig. 21.1. Market equilibrium for non-biotech products in the IP market.

to supply the non-biotech, IP product rather than selling in the conventional, non-IP market. The market for the non-biotech product is cleared at the equilibrium price (P_{ip}), which is the vertical sum of the marginal cost of the non-biotech product production (point *i*) plus the cost of IP (distance of line *ik*, which is equal to the distance of line *ae*, or line *dj*). In other words, the non-biotech product IP market is in equilibrium when the market clearing price, which exceeds that associated with unsegregated production, is just sufficient to cover the cost of IP and marginal cost of production for the non-biotech product. The equilibrium quantity for the non-biotech product that corresponds to the equilibrium price is Q_{ip} (Fig. 21.1).

At the above equilibrium quantity and price, producers of non-biotech, IP commodities gain no excess returns beyond covering the marginal cost of production and the cost of IP. In other words, the price premium ($P_{ip} - a$) for the non-biotech product is sufficient just to cover the cost of IP and marginal cost of production. Given the assumption that producers can always sell at the unsegregated, non-IP price, the 'producer premium' for IP production, defined as the difference between the incentive price for IP production and the price received for unsegregated production, has a lower limit of zero. The incentive price for IP production refers to the effective price received by producers after subtracting the cost of IP (distance of line *ea*) from the price of IP commodities (point *e*) received by producers. Because the price premium ($P_{ip} - a$) can also exceed the IP cost, the 'producer premium' for IP production, in equilibrium, reflects differences in marginal cost between non-biotech, IP commodities and conventional (but unsegregated) ones, not pure profit for the producer.

The above 'producer premium' for the production of non-biotech, IP commodities, will disappear if the demand curve intersects the horizontal section of the supply curve *ecg* (inclusive of IP costs). Under this market condition, the market clearing price for non-biotech, IP commodities at the consumer level is at point *e*. The incentive price for producers is at point *a*, which is the same level as for production of conventional (but unsegregated) commodities. A portion of non-biotech maize or soybeans produced (after meeting its demand in the IP market) will be sold in the unsegregated market and commingled with biotech crops at this price level. Thus, the price premium for non-biotech, IP commodities, in equilibrium, is just sufficient to cover the cost of IP, leaving no 'producer premium' for IP production. Changes in the cost associated with IP (the vertical distance between supply curves) have different impacts, depending on whether the intersection with demand curve occurs on the 'flat' or 'diagonal' segment. If it occurs on the flat segment, an increase in IP costs will be borne by consumers. In contrast, if it occurs on the diagonal segment, there will be price changes for both consumers and producers.

The above graphic analysis has a limitation in that it assumes that IP cost can be represented as a fixed parameter. Related to this limitation is that the graph cannot show the incidence of IP costs for grain handlers or shippers. In addition, it assumes uniform cost structures for all producers of non-biotech, IP commodities. Individual producers may have higher or lower costs, depending on farm-specific characteristics and locational factors.

The above discussion is consistent with a theoretical development advanced by Lence and Hayes (2001), which suggests that whether price premiums paid for non-biotech crops exceed the costs of segregation depends on the relative size of the supply and demand for biotech and non-biotech crops. The market condition under which buyers are more prone to pay price premiums that are just sufficient to cover the costs of segregation is when the supply share of the non-biotech product is larger than the corresponding demand share (Lence and Hayes, 2001). At present, the supply share of non-biotech products grown in the USA exceeds their demand share (which is still a niche market, accounting for about 2% of domestic maize or soybean production). Hence, price premiums paid for non-biotech crops tend to cover just the costs of segregation. Non-biotech maize or soybeans, if exported, are primarily destined for Japan for food production, which is highly inelastic.[6] The inelastic demand for food maize and soybeans also makes it easier for US exporters to pass on the costs of segregation to Japanese buyers.

Foreign non-biotech premiums

Interviews with a few Japanese grain trading houses and the Japan Ministry of Agriculture, Forestry and Fisheries indicated that typical price premiums Japanese buyers were willing to pay for non-biotech maize ranged between US$0.40 and US$0.50/bu or US$15.7–19.7/t (c.i.f. price with average maize quality) for the 2000 crop. The price premiums (in Japanese yen per metric tonne) for the 2001 crop rose in part because the StarLink incident made Japanese buyers more cautious about biotech maize. However, a stronger dollar (the exchange rates averaged ¥121.6 per dollar in 2001, up from ¥107.8 per dollar in 2000) caused the premiums for non-biotech maize to remain largely unchanged when converted into dollars per tonne.

In the case of non-biotech soybeans, interviews with Japanese grain trading houses indicated that typical price premiums for non-biotech soybeans (c.i.f. price with average soybean quality) for the 2001 crop ranged between ¥1200 and ¥1500/t, or between US$0.27 and US$0.33/bu (US$9.90–12.10/t).[7] This premium range was a sizable decrease from the US$0.50–0.60/bu for the 2000 crop. Clearly, the use of more rigid STS IP system raised the cost of segregation, pushing up the price premiums for non-biotech soybeans for that crop year. In contrast, the use of an IP system patterned after HOC handling process is less stringent, leading to lower segregation costs for the 2001 crop.

The above price premiums for non-biotech soybeans estimated by Japanese grain trading houses are comparable to premiums obtained from the Tokyo Grain Exchange.[8] GM soybeans are specified in the contract as US No. 2 yellow soybeans of Indiana, Ohio and Michigan origin in the US. Monthly settlement prices of futures contracts expiring in December for GM and non-GM soybeans reflect the weighted average prices (in yen per tonne) of daily settlement prices for the contract expiring month. Non-GM

[6] According to Pekaric-Falak *et al.*, 2001, the price elasticity of demand for maize in food processing would probably be very inelastic, around –0.10 for the Western Hemisphere. Similarly, the demand price elasticity for food soybeans in Japan would probably be highly inelastic, around the same general magnitude.

[7] This range of premiums for non-biotech soybeans with average quality, which generally meet a 5% tolerance, is comparable with the US$0.35/bu price premium for the 2001 crop reported by the Clarkson Grain Company for shipment to Japan (Clarkson, 2001). In contrast, premiums for food-grade soybeans shipped by containers to the Japanese tofu market and organic soybeans for making tofu there were much higher – US$1.75/bu (US$64/t) for the former and US$11.5/bu (US$423/t) for the latter.

[8] Non-GM soybean futures were launched at Tokyo Grain Exchange (TGE) in May 2000. These futures prices and those for conventional, unsegregated soybeans permit the calculation of non-GM premiums. For details of contract specification for non-GM soybeans at TGE, the reader is referred to a study by Parcell, 2001 and TGE's web site: http://www.tge.or.jp/cgi-bin/monthly

soybeans refer to identity-preserved US No. 2 yellow soybeans of Iowa, Illinois and Wisconsin origin. The monthly settlement prices were converted to dollars per tonne or cents per bushel based on the mid-month exchange rate of the dollar. The December contract month is chosen for the analysis because it reflects new crop futures prices, which are more relevant for price premiums paid by grain companies to US non-GM producers based on seasonal sales patterns for soybeans. The weighted average price premiums for the 2000 and 2001 crops are computed by using monthly trading volume as the weighting factor, excluding December due to thin trading in that month. Price premiums for non-GM soybeans averaged US$0.31/bu for the 2001 crop, down from US$0.54/bu for the 2000 crop (Fig. 21.2).

The price premiums paid by Japanese buyers for non-biotech soybeans are comparable to costs of segregation from country elevators to export elevators for non-biotech soybeans reported in the earlier ERS study – US$0.54/bu for the 2000 crop and $0.18/bu for the 2001 crop and at the farm level. In the latter case, the costs of IP from farm gates to final destinations in Japan (US$0.38–0.43/bu) are slightly lower than those for non-biotech maize (US$0.45/bu) because testing for biotech content in herbicide-tolerant soybeans – the only biotech trait commercialized – is simpler.

Can the Incidence of Segregation Costs be Transferred?

Given the price premiums that Japanese buyers were willing to pay for non-biotech maize and soybeans, are the premiums large enough to cover the costs of segregation and to provide producers incentives to grow non-biotech crops? In other words, who bears the cost? Is it Japanese buyers, or US producers, grain handlers, or some combination?

In this analysis, the incidence of segregation costs is addressed by comparing price premiums paid by Japanese buyers with the costs of segregation. However, the results are drawn from *ex post* data on price premiums, which may differ from an *ex ante* analysis. In an *ex ante* context, farmers are assured of a selling price that is no less than that for unsegregated production. In other words, producers and handlers of non-biotech maize or soybeans will only be induced to grow and market these IP crops if price premiums are sufficient to cover the costs of segregation.

The price elasticity of demand for the commodity plays a key role in the distribution of the costs of segregation. If the commodity is less price elastic, either because of strong consumer preference for non-biotech products, a lack of substitutes, or some other aspect of market demand, then suppliers are in a better position to

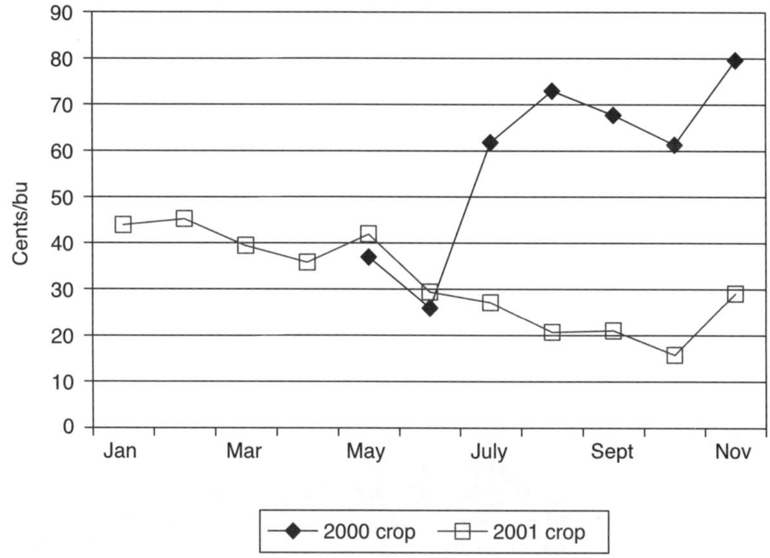

Fig. 21.2. Price premiums Japanese buyers were willing to pay for non-GM soybeans.

pass on the costs of segregation to consumers. The demands for both food-grade maize and soybeans in Japan are highly inelastic, which augments the Japanese consumers' willingness to pay for non-biotech maize and soybeans. In addition, Japanese consumers regard soybeans from Brazil and Argentina, which have a reddish tint, to be inferior in making tofu. This distaste for South American soybeans for tofu manufacturing makes it easier for US non-biotech soybean suppliers to pass on the costs of segregation in the form of higher prices to consumers.

It is clear that in the case of non-biotech maize, Japanese buyers have been willing to pay price premiums ranging from US$0.40/bu to US$0.50/bu in recent years. This level of non-biotech premiums is large enough to cover virtually all IP costs, including additional ocean freight expenses (due to the small shipment volumes that were commonly less than 8000–9000 t), segregation costs from country elevators to export elevators, and price premiums paid to producers (Table 21.2). Price premiums paid by Japanese buyers were able to cover 93% to 111% of the cost of segregation and producers' non-biotech premium for the 2000 crop. Similarly, about 89% to 111% of the cost of segregation and producers' non-biotech premiums were covered by premiums paid by Japanese buyers for the 2001 crop. In other words, the incidence of the segregation costs ultimately falls on Japanese buyers. US producers, grain handlers and exporters were able to pass on all the costs to Japanese buyers.

Price premiums paid by Japanese buyers for non-biotech soybeans also covered the bulk of the cost of segregation and incentives offered to producers for growing non-biotech soybeans. For the 2000 crop, price premiums that Japanese buyers were willing to pay for non-biotech soybeans ranged from US$0.50/bu to US$0.60/bu, which covered 85% to 94% of the cost of IP. Lower price premiums for 2001 non-biotech soybeans, due to the implementation of a less stringent IP system, resulted in a lower coverage range. However, these lower price premiums still covered about three-quarters of the cost of IP (Table 21.2). This lower coverage range might be attributed to three potential factors. First, in the event that NASS' *Illinois Market News* overstated producers' non-biotech premiums, the range of IP costs covered by

Table 21.2. Price premiums paid by Japanese buyers for non-biotech maize and soybeans and the costs of identity preservation (IP) 2000 and 2001 crops.

		Non-biotech maize		Non-biotech soybeans	
Premium or IP cost	Unit	2000 crop	2001 crop	2000 crop	2001 crop
(a) Premium paid by Japanese buyers	US$/bu	0.40–0.50	0.40–0.50	0.50–0.60 (0.54)[a]	0.27–0.33 (0.31)[a]
IP costs:					
Ocean freight exp.	US$/bu	0.13	0.13	–	–
Segregating cost from country elevators to export elevators	US$/bu	0.22	0.22	0.54	0.18
Price premium paid to producers[b]	US$/bu	0.08–0.10	0.10	0.05–0.10	0.20–0.25
(b) Subtotal	US$/bu	0.43–0.45	0.45	0.59–0.64	0.38–0.43
(a)/(b) % of IP costs covered by Japanese buyers' premium[c]	%	93–111	89–111	85–94	71–77

[a]Price premiums for non-GM soybeans based on Tokyo Grain Exchange futures prices for contracts expiring in December.
[b]Source: USDA-NASS, Value enhanced grain survey. *Illinois Market News* 8 Feb., 2000 and 6 Feb., 2001.
[c]The percentage of IP costs covered by Japanese buyers' premium for non-biotech maize or soybeans is calculated by matching lower end (or upper end) of the price premium range with lower end (or upper end) of the IP costs. Different ranges of the percentages could follow if one mixes the comparison between lower end of one with upper end of the other.

Japanese buyers' premiums would be understated. However, this scenario is regarded as unlikely because the Value Enhanced Grain Survey covered about 75–80% of the non-biotech soybean, IP market in Illinois (English, personal communication, 2002). Second, while the –0.10 demand price elasticity for food soybeans contributes to the transfer of segregation costs to Japanese buyers, it is less than perfectly inelastic. Hence, the transfer of segregation costs to Japanese buyers would probably be less than 100%. Third, although price premiums paid by buyers are reported for a specific crop year, sales are spread over several months after harvest. In the event that market prices for non-biotech products decline owing to unforeseen market forces, price premiums paid by Japanese buyers could be lower than US producers and grain handlers were expecting at the time of harvest. Given limitations of the available data (which include a number of industry cost estimates and price averages but not data for a complete set of actual transactions within a supply chain), these results should be viewed with some caution.

Conclusions

The IP system used to segregate non-biotech maize and soybeans from biotech varieties or biotech-commingled crops is an extension of the handling process that has been employed for value-enhanced commodities. The system evolves over time as the US grain and oilseed industries adapt to the needs of foreign buyers for these niche products.

The costs of IP vary depending upon the requirements of the system. As the grain and oilseed industries adapt to the needs of foreign buyers, more cost-effective IP systems will be adopted by US producers and handlers. In addition, the costs of IP might vary over time, as in the case of non-biotech soybeans, owing to the adoption of less costly IP systems.

At present, non-biotech maize and soybeans grown and marketed under IP systems remain a niche market – the share of non-biotech maize or soybean supply exceeds their corresponding demand share. This market condition induces foreign buyers, such as those in Japan, to pay higher prices for non-biotech products sufficient to cover the costs of segregation. In addition, the inelastic demand for food maize and soybeans in Japan, which is the sole use of US non-biotech maize and soybeans destined for that country, together with a lack of viable substitutes, make it easier for US grain handlers and exporters to pass on the costs of IP to Japanese buyers.

Evidence available to date suggests that price premiums that Japanese buyers were willing to pay for non-biotech maize and soybeans imported from the USA, by and large, can cover the costs of IP systems. In other words, the incidence of the segregation costs largely falls on Japanese buyers. This is especially true for non-biotech maize in both the 2000 and 2001 crops and for non-biotech soybeans in the 2000 crop. Even in the case of the 2001 non-biotech soybean crop, where the premiums did not cover the costs of IP as fully as in other cases, the premiums still covered about three-quarters of the IP costs. The remaining quarter of the costs was borne by US non-biotech, IP producers, grain handlers, exporters, or some combination of the three. However, these results have to be interpreted in an *ex post* context. In an *ex ante* sense, farmers are assured of a selling price that is no less than that for unsegregated production.

Current IP systems are not primarily designed to meet a 1% threshold level of biotech content as required for EU biotech food labelling regulations. Instead, they are designed to meet a more lenient threshold level, such as the 5% level for exports going to Japan. While these IP systems appear to meet the needs of Japanese buyers, their applicability to the EU is uncertain. There is already trade with the EU in non-biotech products; this is now a small market niche, but one that could grow as traceability and labelling regulations are put into effect. The costs of meeting a more stringent EU standard could decrease over time as a result of economies of scale and commercial adaptation of IP systems by producers and handlers, but these changes are hard to quantify. Likewise, the EU consumers' willingness to pay for non-biotech products under a more stringent standard is not known with certainty.

Acknowledgements

Insightful comments or suggestions from Gregory K. Price, Barry Krissoff, Joy Harwood, Mary Bohman, Carol Goodloe, Elizabeth Jones and Jay Mitchell are gratefully acknowledged. The authors are agricultural economists of the Economic Research Service, US Department of Agriculture. The views expressed herein are those of the authors, which do not necessarily reflect official USDA policy.

References

Bender, K., Hill, L., Wenzel, B. and Hornbaker, R. (1999) *Alternative Market Channels for Specialty Corn and Soybeans*. Department of Agricultural Economics, University of Illinois at Urbana–Champaign, AE-4726.

Clarkson, L. (2001) Identity preservation, in the growing and export of US soybeans. Presented at an INTSOY training course. 18 June, 2001.

Howie, M. (2000) Food industry may prefer non-GM grain instead of labels. *Feedstuffs* 72(1), 1.

Koenig, D. (2000) No Biotech Corn for Frito-Lay: Decision Angers Farmers, Please Environmentalists. The Associated Press, 1 February, 2000. Available at: http://abcnews.go.co...ess/DailyNews/fritolay000201.html

Krejci, D. (2002) Feasibility and cost of marketing identity-preserved crops: challenge to changing the infrastructure. Speech presented at USDA's Agricultural Outlook Forum 2002 in Arlington, Virginia, 21–22 February.

Lence, S. and Hayes, D. (2001) *Response to an Asymmetric Demand for Attributes: an Application to the Market for Genetically Modified Crops*. Working Paper, Iowa State University, Ames, Iowa.

Lin, W. (2001) Identity preservation for nonbiotech corn and soybeans in the United States. Paper presented at the *Argentine Seed Association Identity Preservation (IP) Seminar* in Buenos Aires, Argentina, 11 July.

Lin, W. (2002) Estimating the costs of segregation for non-biotech maize and soybeans. In: Santaniello, V., Evenson, R.E. and Zilberman, D. (eds) *Market Development For Genetically Modified Foods*. CAB International, Wallingford, UK, pp. 261–270.

Lin, W., Chambers, W. and Harwood, J. (200) Biotechnology: U.S. grain handlers look ahead. *Agricultural Outlook* AGO-270, ERS-USDA, Washington, DC, pp. 29–34.

Miranowski, J.A., Moschini, G., Babcock, B., Duffy, M., Wisner, R., Beghin, J., Hayes, D., Lence, S., Baumel, C.P. and Harl, N.E. (1999) *Economic Perspectives on GMO Market Segregation*. Working Paper, Iowa State University, Ames, Iowa.

Moss, C., Schmitz, T. and Schmitz, A. (2001) The economics of GMO market segmentation. Paper presented at the *International Agricultural Trade Research Consortium Meeting* in Tucson, Arizona, 14–16 December, 2001.

Parcell, J.L. (2001) An Initial Look at the Tokyo Grain Exchange Non-GMO Soybean Contract. *Journal of Agribusiness* 19(1), 85–92.

Pekaric-Falak, I., Meilke, K. and Huff, K. (2001) *The Trade Effects of Bt Corn*. CATRN Paper 2001–02, Canadian Agrifood Trade Research Network. Available at: web site http://www.eru.ulaval.ca/catrn/

Toepfer International (2002) Concerning agricultural policy: Marketing of genetically modified agricultural products. *Market Review*, April 2002, pp. 7–12.

[USDA-NASS] United States Department of Agriculture (2001) Value enhanced grain survey. NASS, *Illinois Market News*, 8 February, 2000 and 6 February, 2001.

US Grain Council (2002) *2000–2001 Value-Enhanced Grains Quality Report*. Available at website http://www.vegrains.org/documents/2001veg_report

22 Future Impact of New Technologies: Three Scenarios, their Competence Gaps and Research Implications

Hanne Harmsen, Anne-Mette Sonne and Birger Boutrup Jensen
The MAPP Centre, Aarhus School of Business, Aarhus, Denmark

What will the impact of science in the food industry be 10 years from now? Large or overwhelming most people will probably agree. But if we want to be more specific, we might start by defining, what type of technology we have in mind and secondly, what kind of impact we are talking about. Since there are many technologies whose impact can be relevantly studied from a variety of different angles, this is indeed a challenging task if we want to end up with an overall picture of the impact of technology.

In this chapter, we consequently do not start by looking at any particular technology and discussing what impact it will have in the future. Our approach is to construct a number of likely scenarios of the future and then look at the role and impact of technology in each of the scenarios.

We do this by using an industry-level scenario technique, in which we rely heavily on expert and industry participants, who represent both 'technology push' and 'market pull', which we feel are both very important basic driving forces.

The aim is to get an idea of the very different roles science or technology can play in the near future for a specific industry, in this case the Danish food industry, and present a methodological approach that might accomplish this aim.

How Can We Take a Closer Look at the Future?

If we want a closer look at the future, there are various methodological approaches that can be taken. We can use forecasting, which usually is tightly linked to predicting the development in previously identified quantifiable factors, in which a trend can be detected (Wright and Ayton, 1987). Or we can develop one or more scenarios of the future, which, in brief, involves identifying the factors that are expected to affect the issue concerned, separating the largely certain from the largely uncertain ones and the important (large impact) from the less important (small impact), and drawing the various scenarios based on looking closely at variations in the uncertain factors and combining these with other less uncertain, but at the same time important factors in a logical pattern (Schnaars, 1987; Von Reibnitz, 1988; Schwartz and Ogilvy, 1998; De Jouvenel, 2000). Scenario construction therefore is not the same as forecasting. Its advantage lies in areas where we cannot expect continuity, but where a number of driving forces can still be identified. Since scenarios are not predictions, they cannot be used to the same extent as forecasting as a basis for planning. Rather, their contribution is to shed light on the extremes of possible future developments and thereby challenge conventional wisdom and stimulate visionary thinking both on industry and organizational level.

Scenario Construction

Shoemaker (1995), Godet and Roboulat (1996), Van der Heijden (1996) and Fahey and Randall (1998) recommend the following criteria for the creation of scenarios. They should be:

- plausible, i.e. the scenarios should be possible and credible;
- have internal consistency, i.e. events in the scenario cannot be mutually exclusive (e.g. it is difficult to imagine a scenario describing a situation with a high rate of inflation and a low interest rate);
- challenging, i.e. scenarios should challenge people's mindset and stretch their perception of the future;
- relevant, i.e. connect with the mental maps and concerns of the users and be relevant to the issue concerned; and
- archetypal, i.e. describe generically different futures rather than variations on a theme and highlight competing perspectives.

Our scenarios were developed through the following research process. We combined a future backwards approach, where Danish and international academics in food research developed vignettes of the future with validation and expansion of this material through workshops with representatives from four sectors of the Danish food industry (dairy, meat, fish, fruit and vegetables), and a Schwartz-inspired (Schwartz, 1991) scenario construction method (the methodological approach and the theoretical frame of reference is explained in more detail in Harmsen *et al.*, 2001). As participants were asked to think about the future related to the market currently served, the result is scenarios that focus geographically mainly on Western Europe and larger industrialized markets, as these are the most important export markets of Danish producers. The scenarios are therefore not global and the discussions of impacts of technology are limited to the same area.

This process resulted in three descriptive and qualitative scenarios, representing three equally justified future end states. The scenarios are archetypal or borderline scenarios. The relevance and plausibility of the three scenarios was validated through industry workshops, while internal consistency was ensured via influence diagrams and several rounds of analysis. The time frame chosen is approximately 10 years, which is regarded as realistic, balancing the ability for creative thinking while still maintaining a high degree of uncertainty.

Three Alternatives for the Food Industry Year 2010

Naturalness – the first scenario

This scenario is based on consumers' mistrust of conventional food production and rejection of genetic modification and other biotechnological advances. This, together with a deteriorating environment and a series of food scandals, has resulted in organic farming being the dominant production method. Consumers show great interest in sustainability from farm to table and consider organic products to be more wholesome.

Production and processing have undergone radical restructuring, which has helped restore confidence in the food sector. The food sector has responded swiftly to consumer demands for naturalness, fewer additives and increased transparency. Product development and the quality improvement of organic products, together with larger volumes, have lowered the price of organic products compared with conventional products, which in turn have contributed to the success of these products.

Organic production has developed in two directions. One trend, which represents an attempt to meet the retail trade's requirements for homogeneous and stable supplies at relatively low prices, is based on specialization and large-scale sales to the industry. The other trend is based on small local units with a versatile and safe production. The few remaining conventional farmers try to market their products as 'light green' with a minimum use of pesticides.

Danish exports are targeted at foreign consumers who demand 'credence characteristics' as, e.g. ecology, and who are prepared to pay for them. Denmark – as opposed to other countries – has declared itself a genetically modified (GM)-free zone, which has reinforced its green image.

Though Danish consumers have rejected functional foods per se, producers have focused heavily on the optimization of natural health-promoting substances, e.g. by making fruit juices from fruits and berries that contain a high level of vitamins. Most people think that good health depends on a wholesome and varied diet. The focus is on food quality and high-quality raw materials. Home cooking has a high priority, especially at weekends, though convenience foods are still in demand when it comes to whole organic meals or choice components.

The number of small speciality shops is again increasing. These outlets offer advice on how to prepare and cook the products, and enjoy the confidence of consumers. To a large degree the retail trade stocks locally produced food products. Within certain product categories retail chains no longer require large-scale, uniform, nationwide supplies, which has enabled small producers to supply supermarkets. Standard products are sold via the Internet, and Net stores sell organic products, which are delivered directly from the farm. In general, there is great interest in more direct contact between producers and consumers.

Recruiting staff is a problem in the processing industry, which is therefore beginning to pay more attention to the work environment. The agenda of value-based management includes good working conditions, which the ordinary consumer also has in mind when shopping for food. Having been won over to environment-friendly food products, consumers are now turning their attention to social consciousness and improved working conditions.

Technology-driven health – the second scenario

There are three driving forces of this scenario: consumer acceptance of new technology (such as functional foods, genetically modified foods and intelligent materials), a hitherto unseen consumer focus on personal health, and more liberal legislation – especially as regards labelling and research.

Consumers have confidence in food production – they feel well informed and believe they have a realistic picture of modern food production. Consumers have gained confidence in the quality of food products through systems that control traceability and documentation, such as intelligent packaging, which continually check product quality. The large brand manufacturers carry out their own laboratory checks and contribute to the development of traceability systems. In primary production, changing those elements that attracted the most consumer criticism, has restored confidence. For example, factory calves and battery hens no longer exist.

Food products and the health sciences are more closely linked than ever before. Since documented health effects have been allowed on product labels, many companies have started producing functional foods. Products that are assumed to have a beneficial effect on such widespread diseases as cardiovascular disease, obesity and diabetes are especially popular, e.g. low-fat products and fibre- and vitamin-enriched products.

Consumers are very knowledgeable about which food products are healthy for them, and therefore demand that producers provide relevant information. This has led to increased individualization of products, e.g. adaptations of the same basic product to fit the health and nutrition requirements of various target groups.

Research has shown that genetic modification not only offers manufacturers advantages, but that it may also be a vital element in combating animal and human disease, and can even effectively curb certain types of pollution. This has persuaded consumers to accept genetic modification, and gene technology is now also permitted in Danish primary production and processing. Vast amounts of public research funds are spent to ensure that genetic modification does not cause imbalances in the ecosystem and to monitor long-term effects.

Today's lifestyles have left little time for cooking, which has resulted in a heavy demand for convenience products and take-aways. Part of this rise in demand can be attributed to improved quality, which has been achieved through the use of better raw materials and improved packaging and manufacturing processes that ensure product freshness with a reduced use of additives.

Organic farms are few and far between. The demand for organic products is constrained by the fact that organic farmers reject the use of genetic modification and the use of unnatural substances, which has resulted in relatively more expensive products.

The retail chains give priority mainly to well-documented brands, downgrading their own products as manufacturers' brand strategy has become dependent on research in control and health. E-commerce has become normal, and consumers buy standard goods and foods, as well as more individualized products, on the Net. Net outlets store information about consumers' individual preferences, and the home is run by means of information technology, e.g. the fridge automatically re-orders groceries from the preferred supplier via the Internet.

The small cohorts entering the labour market have intensified the struggle for manpower. The food industry automated monotonous and strenuous manual work as soon as it became technically possible, but is still experiencing problems with employees leaving for other industries.

Tight spending – the third scenario

This scenario is based on lower disposable real income and extensive internationalization. Price has become the main criterion of choice of both Danish and foreign consumers. For example, many consumers only demand convenience if it is cheaper than cooking from scratch.

Danish food products are under increasing price competition from Eastern Europe. Eastern European products are cheaper because they do not have to meet the same environmental requirements, they are not subject to the same level of duties, and Eastern European wages are lower. However, Eastern European products do meet basic EU requirements, and as more Eastern European countries have entered the EU, consumer scepticism has faded. More and more consumers are thus demanding these low-priced products, which are also of fair quality.

High-quality products and soft quality parameters have a limited market, which is largely characterized by the intensified competitive situation. While the main competitive focus is on price, producers also have to meet rigorous demands on quality and food safety in order to enter these markets, which has therefore made this market considerably less attractive.

Primary production has become international. Parts of the Danish food industry buy cheap raw materials abroad, while others own Eastern European companies that process regional food products. Danish primary producers set up outside the country to avoid duties and environmental requirements, leading to a slump in Danish primary production. Some of the former primary producers now make a living from selling know-how to Eastern Europe, among others. Organic products are no longer in demand, prices being far too high for price-conscious consumers.

Powerful international chains dominate the retail trade. Most domestic chains are owned by, or work in close collaboration with, multinational retail chains. All retail chains are members of global buying organizations, which generate an extremely high and stable demand for products. The power this confers is used to put pressure on producers as regards price, as well as to gain the value added from own labels. Consumer trust is shifting towards the retail chains and their labels, which, among other things, guarantee food safety. As a countermove, some brand producers are trying to launch joint e-businesses in an attempt to reach consumers directly.

Health is still a top issue. There is a large international market for cut-price functional foods, which are regarded as an inexpensive way of staying healthy. The Danish food industry has not enjoyed much success with functional foods, partly due to restrictive labelling regulations and partly due to minimal R&D investments in this area.

Denmark is at a crossroads with regard to genetic modification in food production, a situation that has not been made easier by the fact that consumer attitudes towards genetic modification have become less clear-cut. Some consumers have begun to accept GM products, whereas others stick to GM-free supermarkets. This has therefore resulted in intensive research in high-yielding GM crops and products.

The recession and resulting unemployment have alleviated the recruitment problem – at least in the short term. However, companies with old-fashioned attitudes to the work environment are still experiencing problems in recruiting young staff.

Future Impact of New Technologies – What do the Scenarios Show Us?

The scenarios are broad pictures of a potential future, in which biotechnology, specific technologies like genetically modified organisms (GMO) or science in general has not had any predefined role or impact. From the three pictures of the future, various characteristics for the role of technology and its implications can however be deduced, which will be done in the following.

Depending on how broadly we define the concept of technology, its impact across the scenarios differs. Looking at technology from a detailed perspective a large number of technologies have been discussed. Examples include production technology ('careful processing'), new packaging technologies (edible coating, permeable barriers) and new conservation technologies. However, in the following discussion we will focus on the more general role of technology, covering new technologies such as GMO, biotechnology in general and functional foods (despite the very unclear technology component in this area, industry participants tended to include functional food under the heading of 'new technology').

Scenario 1 is to a large extent characterized by technology scepticism. There are two major reasons for this scepticism. One is related to consumers' attitudes. The background for the attitude towards technology is a number of negative experiences from various food scandals. It is not specified in which areas these scandals have been, but it is well known, that in the case of food scandals consumers can form rather general attitudes, not only related to the specific incident or technology. A lesson is, therefore, that things happening with technologies in one area can have a large impact on your own specific application, since consumers seem to think in rather broad terms regarding 'technology'. We also see in this scenario that consumers conclude what is safer or better than the previous approach to food production. In this case organic production. At the workshops held with industry participants many argued that this was simply not true, since there are a number of risks and unsolved problems related to this production form and a number of benefits from new technologies even within a 'sustainability' perspective. The lesson an 'illogical' scenario element like this can teach us is that consumers are not experts and will react to various incidents in ways that might not be correct, but nevertheless will form a major pressure and influence the implementation of technologies.

The second reason for consumer scepticism in this scenario is concerned with the altruistic thinking that to some extent can be identified in the scenario. Consumers are concerned with sustainability, not so much because of direct effects on their personal health, but because of a more general concern with the environment and the perceived negative impact of 'new' technologies.

The impact of technology is naturally also linked to legislation, and the core of this scenario is that legislation has moved in the direction of consumer attitudes (as opposed to industry's attitudes). One other aspect of technology impact stands out in the scenario. In this case, the absence of technology, i.e. being a GMO-free zone, actually enables a specific – in this case regional – product differentiation. This clarifies that the impact of technology can also be indirect and an inability to implement new technological possibilities, whether political or economic, can actually turn out to be an advantage.

Turning to Scenario 2, we see a very different picture when it comes to the role and impact of technology. There is a wide acceptance of new technologies, again in the form of a general attitude rather than an attitude towards specific technologies. This general positive attitude is triggered by success stories. A transparent, well-documented industry, which further has had the benefit of real advantages in relation to fighting animal and human diseases. The positive attitude is further strengthened by the fact that consumers personally like the implications of the new technologies. We see a much more individualistic positive attitude 'this can do good things for me and my personal health' more than an altruistic orientation towards the environment and other people. Blurred borders between industries, e.g. the pharmaceutical and the food industry, are also a consequence of the application of technologies in this scenario and something that can strengthen the positive consumer attitudes. Consumers have traditionally not been very sceptical towards technologies for developing and producing pharmaceuticals. A trend towards functional food with documented health effects can contribute to these borders being blurred as consumers transfer their less technology sceptical attitude from pharmaceuticals to food. It is also an underlying hypothesis in the scenario that consumers will transfer their positive attitudes from existing technologies to new and thereby become eager or at least early adopters of 'really' new technologies.

A condition for the largely technology driven advantages in this scenario are large investments in knowledge related to the possibilities and effects of the technologies. This is not a direct impact, but more of a precondition, which requires a lot of publicly funded research, that can give companies a basis for technology application and enable monitoring of effects on individuals and nature. It also requires privately funded research, which will enable companies to turn the combination of publicly available knowledge and company-internal knowledge into competitive advantages. In the workshops, many participants argued that this type of development favours large companies. An indirect impact of pursuing competitive advantages through application of advanced and new technologies in the food industry will thus perhaps be an increased concentration of industry structures. Further, a question regarding technology and industry structure that is unveiled in this scenario is the question of who will get the economic benefit of the new technologies. Several people argued, that sectors related to, but currently not directly part of the food industry will be better equipped for the new developments, because of their more knowledge-intensive production today. Examples are again pharmaceutical companies and ingredient companies. An impact of technology can thus be changes in industry structures.

In the last scenario, Scenario 3, the role and impact of technology are more fuzzy than in the others, where new technologies are positioned in each end of a spectrum from negative to positive. Here, technology does not have a key role in forming consumer attitudes. Consumers are generally not for or against new technology, but look at what technology can do to make them achieve their goals of primarily low-cost food. If a certain technology can make prices lower, it is accepted, otherwise not.

Here we see a more rational or utilitarian and less emotional attitude towards technology. This consumer attitude may be very relevant for consumers on very tight budgets, e.g. from developing countries, regardless of the likelihood of the scenario as a whole.

Turning from consumers' perception of competition and the performance effects of new technologies, this scenario shows that the timing dimension – related to both strategic decisions in companies and legislation – is key. Economic advantages are related to cumulative knowledge generation both in research and in companies with the result that the first movers get the advantage.

Summing up, the scenarios can be seen as speculative suggestions concerning the future. This implies that the deduced roles and impacts of technology are just as speculative. But just as scenario construction does not have the aim of prediction, but rather of stimulating thought of the future and visionary decision making, so do the impacts presented and discussed here. Across the scenarios it is clear that the impact of technology depends on the interplay between legislation, consumer attitudes, industry structure, resources for public and private research as well as every single success or scandal related to new technologies.

Further Implications and Limitations

Scenarios can be used in several ways. Here we have used them to discuss the potential impact of technology in the near future. Scenarios can also be the starting point for identification of competencies that are important for competing successfully in the various futures. Scenarios can, more traditionally, be used as a basis for making long-term decisions and identifying major threats and opportunities (Schriefer,

1995). Merely by discussing the scenarios, the personal beliefs of a management group or other interest groups will become much more articulate and transparent, and both agreements and disagreements will be important for developing some sort of shared perception of where the company or organization will or should be in the future. It is also important to note that the various actors in an industry influence the future. Thus, the scenarios can also be used as a starting point for discussions about the ideal direction the company should take in the future, and what it can actively do to influence this. In the case of technology impact, a long-term definition of a technology strategy and related R&D priorities and objectives is one way of actively approaching the future. Another option is actively to influence the political system. Whether opting for adaptation or proactively influencing the future, an important step is to identify the main events that have to take place for the specific scenario to become a reality, and to track these signals in order to monitor the direction in which the industry is moving.

This study is subject to several limitations, of course. The scenario process is highly qualitative, and results are dependent on the participants that have been included in the process. While we have tried to increase the plausibility of the scenarios by holding several rounds of group discussions, it can be argued that, in this way, we have eliminated the wildest ideas that could really challenge current thinking.

This study has been based on the Danish food industry. Therefore, it is impossible to say anything about the extent to which the scenarios are relevant for (food) sectors in other countries. The scenarios are international to the extent that we have asked industry participants to consider their relevant markets in the process. The dairy, meat and fish industries are heavily export oriented, while fruit and vegetables are less so. The scenarios are therefore a mixture of more and less international perspectives.

Scenarios can be developed at different levels of analysis. Often, they are developed for a specific company, which means that, among other things, competitive positioning in the industry and the existence of different consumer segments would be taken into account. At the other end of the spectrum, global scenarios can be developed, taking into account all kinds of macroeconomic factors. Our scenarios are industry-level scenarios, and do not include either the variation within specific industries or considerations about factors such as global population growth, global environmental problems or global political developments.

We have tried to develop scenarios for the year 2010. However, we found it very difficult to stick precisely to this time perspective. Participants seem to have a different view of the time frame of 10 years – some see it as just around the corner, while for others it is far in the future. This makes it difficult to evaluate whether the various inputs actually relate to the same point in time. This is a problematic aspect of the scenario method, which requires further methodological developments.

References

De Jouvenel, H. (2000) A brief methodological guide to scenario building. *Technological Forecasting and Social Change* 65, 37–48.

Fahey, L. and Randall, R.M. (1998) *Learning from the Future*. John Wiley & Sons, New York.

Godet, M. and Roboulat, F. (1996) Creating the future: the use and misuse of scenarios. *Long Range Planning* 29, 164–171.

Harmsen, H., Jensen, B.B. and Sonne, A.M. (2001) Scenarios for the food industry in 2010. *Agrarwirtschaft – Zeitschrift für Betriebswirtschaft, Marktforschung und Agrarpolitik* 50(6), 314–322.

Schnaars, S.P (1987) How to develop and use scenarios. *Long Range Planning* 20(1), 105–114.

Schoemaker, P.J.H. (1995) Scenario planning: a tool for strategic thinking. *Sloan Management Review* 36(2), 25–41.

Schriefer, A.E. (1995) Getting the most out of scenarios: some questions and answers. *Planning Review* 23(6), 6–37.

Schwartz, P. (1991) *The Art of the Long View: Planning for the Future in an Uncertain World*. Doubleday, New York.

Schwartz, P. and Ogilvy, J.A. (1998) Plotting your scenarios. In: Fahey, L. and Randall, R.M. (eds) *Learning from the Future, Competitive Foresight Scenarios*. John Wiley & Sons, New York, pp. 55–80.

Van der Heijden, K. (1996) *Scenarios: the Art of Strategic Conversation*. John Wiley & Sons, Chichester, UK.

Von Reibnitz, U. (1988) *Scenario Techniques.* McGraw-Hill, New York.

Wright, G. and Ayton, P. (1987) *Judgmental Forecasting.* John Wiley & Sons, Chichester, UK.

23 *Ex Ante* Welfare Effects of Agricultural Biotechnology in the European Union: the Case of Transgenic Herbicide Tolerant Sugarbeet[1]

Matty Demont and Eric Tollens
Department of Agricultural and Environmental Economics, Katholieke Universiteit Leuven, Leuven, Belgium

Introduction

In the USA, the first published *ex post* welfare studies reveal the distribution of the benefits from agricultural biotechnology. However, no parallel *ex ante* study has yet been published for the European Union (EU). The EUWAB-project (European Union Welfare effects of Agricultural Biotechnology, http://www.agr.kuleuven.ac.be/aee/clo/euwab.htm) tries to fill this gap in the literature by estimating the impact of biotechnology innovations in the EU and its distribution among member countries, producers, consumers and input suppliers.

Until now, the available *ex post* studies are applied on typical US export crops such as cotton (Falck-Zepeda *et al.*, 2000b) and soybeans (Falck-Zepeda *et al.*, 2000a; Moschini *et al.*, 2000). The 1998 *de facto* moratorium on transgenic crops in the EU has to be analysed using an *ex ante* approach. Studying the *potential* welfare effects associated with agricultural biotechnology in the EU reveals the *benefits forgone* or *costs* of the *de facto* moratorium. To illustrate these potential benefits, a representative case study is selected. Since most of the recent agricultural biotechnology innovations are embedded in the seed, a crop is chosen that is representative and important for the majority of EU member countries in terms of production and export. The case of sugarbeet is in line with our criteria.

Sugar

Sugar is one of the most heavily protected agricultural commodities (International Policy Council, 1996), implying for our study that these market interventions distort the flow of benefits from R&D in agriculture, such as biotechnology research. However, the sugar industry is facing a slow but steady progress towards greater liberalization of global trade. Over the last 40 years, real world sugar prices have fallen, on average, by between 1.5% and 2.0% per year (Duff, 1999). Even in the case of the highly protected European beet industry, growers are paid a fixed 'green rate' price, i.e.

[1] Research financed by the Flanders Interuniversitary Institute for Biotechnology (VIB) in the frame of the Technology Assessment programme.

 The Regulation of Agricultural Biotechnology (eds R.E. Evenson and V. Santaniello)

not corrected for inflation. This means that they have to compete continuously against an annual real price decline of 1.9% via technological progress. This is actually the only way benefits of technological progress end up being passed on to consumers in the EU sugar sector (Thirtle, 1999).

These arguments provide a powerful economic rationale for enhancing competitiveness by exploiting any cost savings that can be achieved, e.g. through the use of biotechnology. However, Table 23.1 reveals that, although most European countries have sufficient research experience in genetically modified (GM) beets, no authorization or commercialization of these crops is expected before 2002–2005 (Krick, 2000). Yet, most sugar industries have not even adopted a strategy for this technology. Nowhere in the world has transgenic sugarbeet been adopted on a commercial scale yet. Clearly, it is wise for the EU to take a careful science-based look at all the economic, agricultural and environmental issues involved. This justifies the elaboration of an *ex ante* study about the potential welfare effects of agricultural biotechnology in the EU sugar sector.

Transgenic Sugarbeets

Effective weed control is essential for economic sugarbeet production in all growing areas of the world. This was recognized as soon as the crop was first grown (Achard, 1799). The postemergence herbicides glyphosate and glufosinate-ammonium provide a broader spectrum of weed control in sugarbeet than current systems, while at the same time reducing the number of active ingredients. Glyphosate was first introduced as a herbicide in 1971. The gene that confers tolerance to glyphosate was discovered in a naturally occurring soil bacterium. This bacterium produces an enzyme, which prevents glyphosate from attacking another enzyme called EPSPS that controls the production of essential amino acids in the plant, and without which the plant would die. The gene was isolated using microbiological techniques, and introduced into the beet genome using the gene transfer technology. Glufosinate-ammonium was discovered in 1981. The gene that confers tolerance to glufosinate was also discovered from a naturally occurring soil bacterium and introduced into the beet's genome, accompanied by an antibiotic 'marker' gene (Dewar *et al.*, 2000). Two commercial transgenic sugar beet varieties resulted: Monsanto's Roundup Ready®, tolerant to glyphosate and Aventis' Liberty Link™, tolerant to glufosinate-ammonium. These combinations of transgenic seed combined with a post-emergence herbicide, offer farmers broad-spectrum weed control, flexibility in the timing of applications, and reduce the need for complex compositions of spray solutions. For most growers, this translates into cheaper weed control (Jassem, 2000). Moreover, these innovations are entirely coherent with the ongoing trend towards post-emergence weed control and reduced tillage techniques and the sharpening of the legal constraints for the application of herbicides (Schäufele, 2000). Both herbicides have a low toxicity and metabolize fast and without residues in the soil. Hence, the introduction of herbicide tolerant (HT) sugar beet varieties could be an approach to sustainable sugar beet cultivation (Märländer and Bückmann, 1999).

The Model

To analyse the welfare effects of the adoption of HT sugar beet in the sugar industry, we need to choose an appropriate spatial model. Therefore, a preliminary look at the geographical distribution of production and trade of sugar is in order. Table 23.2 reports the average global production and utilization of sugar during 1996–2000. A differentiation into sugarcane and sugar beet appears logical, accounting for respectively 71% and 29% of global sugar production. The sugar beet region can be further divided into the EU and the rest of the world (ROW), both responsible for half of global beet sugar (Table 23.3). In this ROW beet region, non-EU Europe is dominant (58%), followed by the US (21%). Hence, we believe that we can adequately capture the essence of production and trade in the global sugar market with a three-region model: EU, ROW beet, and ROW cane.

Table 23.1. Situation and research regarding transgenic sugarbeet by country.

Country	Are any GM beet varieties currently being tested?	For what characteristic(s)?	When can the first GM beet variety authorization be expected?	When can one expect a GM beet variety to be actually grown?	What strategy is to be adopted with the sugar industry regarding processing of GM beet?
Belgium	Yes (1989)	RR & LL	Not before 2005	Not before 2005; use will depend on acceptance by consumers and the interprofession	No strategy adopted yet No GM beet in sugar factories nor in the animal feed market
Denmark	Yes (1996)	RR	Not before 2002–2003	Very uncertain	As long as authorization of GM beet is not expected, the question of processing is not relevant; there will be no GM beet in sugar factories until industry is sure that consumers accept sugar from GM beet
Germany	Yes (1998)	RR & LL	Impossible to assess	Impossible to assess	No GM beet in sugar factories before 2001; only GM varieties of cultural value (better than conventional ones) will be authorized
Greece	Not yet; Novartis and Agrevo have filed applications for trials of GM beet varieties	RR & LL	Trials may start in a few years	Not foreseeable at present	The sugar industry has reservations, but will consider the issue if other countries grow GM beet
Spain	Yes (1997)	RR & LL	Unknown	Unknown	Continue work with close collaboration between the administration, beet growers, sugar industry, seed producers and research institute (AIMCRA)
France	Yes (1995)	RR & LL, resistance to nematodes and various diseases	Not foreseeable	Not before 2003–2005	Market and consumer acceptance of sugar from GM beet is a precondition; no double sector (GM and non GM); GM beet on negative lists (exempt from labelling).

continued

Table 23.1. *Continued.*

Country	Are any GM beet varieties currently being tested?	For what characteristic(s)?	When can the first GM beet variety authorization be expected?	When can one expect a GM beet variety to be actually grown?	What strategy is to be adopted with the sugar industry regarding processing of GM beet?
Ireland	Yes; research trials	RR	Not before 2005	Not before 2005	No strategy adopted yet
Italy	No, not even for research trials				No strategy adopted yet
The Netherlands	Yes (1997)	RR & LL	Not before 2003	Not before 2003	Any acceptance of GM beet for processing will be decided on a case by case basis; a strict surveillance of the production chain is indispensable
Austria	No; growing of GMOs, even just for pure research, is forbidden		Unlikely even in the medium term	Difficult to foresee	No strategy
Portugal	No		–	–	–
Finland	Yes (1998)	RR & LL	2002	Not before 2003	No strategy adopted yet
Spain	Yes (1998)	RR & LL	Not before 2002–2003	Not before 2002–2003	No GM sugar beet to be grown unless there is full consumer confidence
UK	Yes	RR	Not before 2003	Unknown	Continue work with close collaboration between the administration, beet growers, sugar industry, seed producers and research institutes; market and consumer acceptance of sugar from GM beet is a precondition
Switzerland	No		Unknown	Unknown	No strategy; sugar industry does not want GM beet
Hungary	No	–	–	–	–
Poland	No	–	–	–	–
Czech Republic	Yes	RR & LL	2002	Not before 2003	No strategy adopted yet
Slovak Republic	Yes	RR & LL	2002	2002	No strategy adopted yet

Source: Krick (2000).
GMOs, Genetically modified organisms; RR, Roundup Ready®; LL, Liberty Link™; GM, genetically modified.

Conventionally, research benefits were estimated assuming that the research is publicly funded and innovated inputs competitively sold in the input market. In contrast, most of the recent agricultural biotechnology innovations have been developed by private firms protected by intellectual property rights (IPRs), such as patents, which confer monopoly rights to the discoverer. Monopolistic prices are higher than competitive ones. Therefore, Moschini and Lapan (1997) complete the conventional framework by including welfare measurement in the input market. However, in a more recent paper Moschini *et al.* (2000) adapt their methodology to a model that is closer to the actual working of the herbicide tolerance innovation and apply it to the case of Roundup Ready® soybeans. Our model is inspired by the latter. The spatial dimension is defined by 16 regions *i*: the ROW cane ($i = 0$) and beet growers ($i = 1$), and 14 production blocks in the EU ($i = 2, \ldots, 15$). Belgium and Luxembourg are united in one block. The temporal dimension includes seven agricultural seasons *j*: one 'benchmark year' 1996/97 without adoption ($j = 0$), five sequential years of adoption ($j = 1, \ldots, 5$): 1996/97, . . ., 2000/01, and one 'evaluation year' 2001/02 ($j = 6$) to which the welfare effects, are actualized and aggregated. Average profit per hectare is a function of sugar price and adoption rate $\rho \in [0,1]$ (Moschini *et al.*, 2000) (Equation 1, see bottom of page):

Supply of land to the sugar industry by country *i* in year *j* is written in constant-elasticity form as a function of average land rents, which depend on output price and the adoption rate:

$$L_{i,j}[\bar{\pi}_{i,j}(p,\rho)] = \lambda_{i,j}[\pi_{i,j}(p,\rho)]^{\theta_{i,j}} = L_{i,j}(p,\rho) \quad (2)$$

with $\lambda = 16 \times 6$ matrix of scale parameters. Multiplying the land supply function by the (optimal) yield function results in a region- and year-specific supply function incorporating four technology-specific parameters, acting as shifters and enabling parameterizion of the herbicide tolerance innovation in detail:

$$Q_{i,j}(p,\rho) = L_{i,j}(p,\rho)(1+\rho\beta_i)G_{i,j}p^{\eta_i} \quad (3)$$

Aggregation of supply functions will allow us to model the effect on world sugar prices of the interaction between two aggregate blocks, the EU and the ROW, as a consequence of the introduction of the HT technology. However, the structure of these functions implies that all 16 regions in the model are able to participate in the aggregate supply response to prices. While all regions certainly respond to a certain region-specific 'incentive price', in reality not all of them respond to (lower) world prices, owing to price interventions interfering in their domestic market.[2] This means that the technology-induced production surplus of those regions will not be exported on the world market, but will free up land allocated to sugar beets instead, so that their total production remains unchanged.[3]

$$\bar{\pi}_{i,j}(p,\rho) = A_{i,j} + \rho\alpha_{i,j} + \frac{(1+\rho\beta_i)G_{i,j}}{1+\eta_i}p^{1+\eta_i} - \delta w_{i,j}(1+\rho\mu_{i,j}) \quad (1)$$

General parameters	Technology-specific parameters
p = sugar price	$\alpha = 16 \times 6$ matrix of coefficients of unit profit increase due to the HT technology
A, $G = 16 \times 6$ matrices of parameters subsuming all other input prices, presumed constant	$\beta = 16 \times 1$ vector of coefficients of yield change due to the HT technology
$\eta = 16 \times 1$ vector of elasticities of yield with respect to sugar price	$\rho \in [0,1]$ = adoption rate
$\delta w = 16 \times 6$ matrix of seed costs (δ = constant optimal density of seeds and w = seed price)	$\mu = 16 \times 6$ matrix of markups on HT seed price (reflecting technology fee)
$\theta = 16 \times 6$ matrix of elasticities of land supply with respect to sugar profit per hectare	

[2] We are grateful to Brent Borrell for pointing this out.

[3] In the short run, these surpluses will be added to the carry-over and 'precautionary' production to ensure quota fulfillment. In the medium and long run, farmers will adapt their land allocated to sugarbeet production.

Table 23.2. World sugar production and utilization (10^3 tonnes), average 1996–2000.

Raw sugar	Beet	Cane	Total	%	Export	%	Net export	Change stocks	Cons-umption
EU (15)	18,406	9	18,415	14	8,398	20	3,790	−10,475	14,261
Eastern Europe	7,984	0	7,984	6	1,611	4	−6,385	193	14,177
Other West. Eur.	3,137	0	3,137	2	425	1	−151	10,997	3,112
USA	4,056	3,126	7,181	6	226	1	−1,747	169	9,081
Japan	662	180	841	1	6	0	−1,588	−76	2,506
Canada	121	0	121	0	19	0	−1,108	−1	1,230
Australia	0	5,311	5,311	4	3,975	10	3,972	44	1,103
Brazil	0	17,517	17,517	13	8,356	20	8,356	189	9,293
India	0	16,959	16,959	13	457	1	−46	475	16,525
Indonesia	0	1,913	1,913	1	7	0	−1,507	110	3,243
Mexico	0	5,102	5,102	4	645	2	583	23	4,501
South Africa	0	2,628	2,628	2	1,249	3	1,152	78	4,070
Thailand	0	5,391	5,391	4	3,457	8	3,457	122	1,813
Other Centr. Am.	0	8,600	8,600	7	5,728	14	5,220	127	2,954
Other South Am.	483	6,270	6,753	5	1,687	4	592	152	6,012
Other Africa	623	5,901	6,524	5	2,487	6	−3,603	356	7,215
Other Asia	1,962	13,418	15,380	12	2,687	6	−8,955	192	24,151
Other Oceania	0	414	414	0	348	1	76	21	309
World	37,433	92,738	130,171	100	41,767	100	2,107	2,696	125,556
Share %	29	71	100						

Source: F.O. Licht (2001).

For those regions, we include this possibility by equalling their supply functions to their (constant) observed total production:

$$Q_{i,j}(p,\rho)=\overline{Q}_{i,j} \quad (4)$$

For regions i responding to world prices, we parameterize the introduction of HT sugarbeet using Equation 3. All quantities and prices are converted to their white sugar equivalent. The aggregate EU sugar supply function in year j can be modelled by imputing the country- and year-specific adoption rates $\rho_{i,j}$ in the variable ρ and adding up all country-specific supply functions. Note that this aggregate supply function contains a constant and a variable term, which is a function of world prices:

$$Q_{EU,j}(p,\rho_{EU,j})=\sum_{i=2}^{15} Q_{i,j}(p,\rho_{i,j})=\sum \overline{Q}_{i,j}+\sum Q_{i,j}(p,\rho_{i,j}) \quad (5)$$

In equation 5 $\rho_{EU,j}$ represents the 14×1 adoption vector of the new technology in the EU in year j, with elements $\rho_{i,j}$ ($i = 2, 3, \ldots, 15$). This aggregate sugar supply function is very detailed in that it contains 10 parameters per country, totalling 140 parameters, of which 56 are related to the new technology. In an analogous way, ROW aggregate supply in year j can be modelled as a function containing a constant and a variable term:

$$Q_{ROW,j}(p,\rho_{ROW,j})=\sum_{i=0}^{1} Q_{i,j}(p,\rho_{i,j})=\sum \overline{Q}_{i,j}+\sum Q_{i,j}(p,\rho_{i,j}) \quad (6)$$

In equation 6 $\rho_{ROW,j}$ represents the 2×1 adoption vector of the new technology in the ROW in year j with elements $\rho_{i,j}$ ($i = 0, 1$). The 16×1 adoption vector in the whole world in year j is denoted by $\rho_{W,j}$, containing elements $\rho_{i,j}$ ($i = 0, 1, \ldots, 15$).

Next, we model the innovation as occurring in a large, open economy with technology spillovers and shape the two-region framework of Alston *et al.* (1995) (p. 219) to the specific features of the EU's Common Market Organization (CMO) for sugar. In Fig. 23.1 the framework is represented. For each country, the four technology-specific parameters engender a pivotal, divergent shift of the supply curve. At the centre of the analysis is the calculation of a counterfactual world price p_j (after decline) in year j to isolate the effect of the

Table 23.3. Average global area and production of beets and beet sugar, 1996–2000.

Country	Area (10^3 ha)	%	Beet production (10^3 t)	%	Beet yield (t/ha)	Sugar yield (% white sugar)	Sugar production (10^3 t white sugar)	%
Austria	47	1	2,969	1	63	16	476	1
Belgium-Lux.	96	1	5,927	2	62	16	960	3
Denmark	63	1	3,369	1	54	16	532	2
Finland	33	1	1,108	0	33	14	153	0
France	440	7	31,259	12	71	14	4,410	13
Germany	480	7	26,480	10	55	16	4,211	12
Greece	45	1	2,663	1	59	11	286	1
Ireland	33	1	1,708	1	52	13	217	1
Italy	270	4	12,958	5	48	12	1,606	5
Netherlands	114	2	6,531	3	58	15	1,012	3
Portugal	6	0	361	0	58	15	53	0
Spain	137	2	8,110	3	59	14	1,136	3
Sweden	58	1	2,592	1	45	16	407	1
UK	184	3	9,786	4	53	15	1,475	4
EU-15	2,005	31	115,819	46	58 (av)	15 (av)	16,934	49
Hungary	69	1	2,973	1	43	15	432	1
Czech Rep.	74	1	3,244	1	44	15	488	1
Poland	368	6	13,951	5	38	15	2,044	6
Russia	761	12	13,697	5	18	11	1,500	4
Ukraine	879	14	15,188	6	17	13	1,975	6
Turkey	434	7	17,939	7	41	13	2,289	7
Other	443	7	12,580	5	29	12	1,503	4
Europe	5,034	78	195,391	77	39 (av)	14 (av)	27,165	79
US	568	9	27,959	11	49	13	3,731	11
China	406	6	11,410	4	28	10	1,103	3
Iran	178	3	4,784	2	27	12	564	2
Other	292	5	14,718	6	37	13	1,875	5
ROW	4,474	69	138,443	54	31 (av)	13 (av)	17,505	51
World	6,479	100	254,263	100	39 (av)	14 (av)	34,438	100

Source: FAO (2002), F.O. Licht (2001).

technology-induced supply shift from other exogenous changes in supply and demand. This price change differs from the observed change in world price if the technology is adopted as assumed. It rather represents what the world price would be if all supply and demand conditions are identical except for the introduction of the new technology (Falck-Zepeda *et al.*, 2000b). Hence, in our analysis we represent the world price as a function of the worldwide adoption vector: $p_j(\rho_{W,j})$. If we assume a constant elasticity EU demand function for sugar:

$$D_{EU,j}(p) = \kappa_{EU,j}\, p^{-\varepsilon_{EU,j}}, \tag{7}$$

the EU's export supply curve in year j can be modelled as

$$ES_j(p, \rho_{EU,j}) = Q_{EU,j}(p, \rho_{EU,j}) - D_{EU,j}(p) = Q_{EU,j}(p, \rho_{EU,j}) - C_j \tag{8}$$

with C_j the fixed consumption level in year j, due to fixed annual intervention prices. The world price reduction (from $p_j(0)$ to $p_j(\rho_{W,j})$ in Fig. 23.1) is a synergy of two forces. First, the EU's export supply expansion (from $ES_j(p,0)$ to $ES_j(p,\rho_{EU,j})$), due to a technology-induced pivotal shift of the EU's aggregate supply function (from $Q_{EU,j}(p,0)$ to $Q_{EU,j}(p,\rho_{EU,j})$), would cause the world price to decline from $p_j(0)$ to $p_j(\rho_{EU,j})$. This price decrease can be determined using a reduced form equation, extracted from the Food and Agricultural Policy Research Institute, University of Missouri (FAPRI's) world

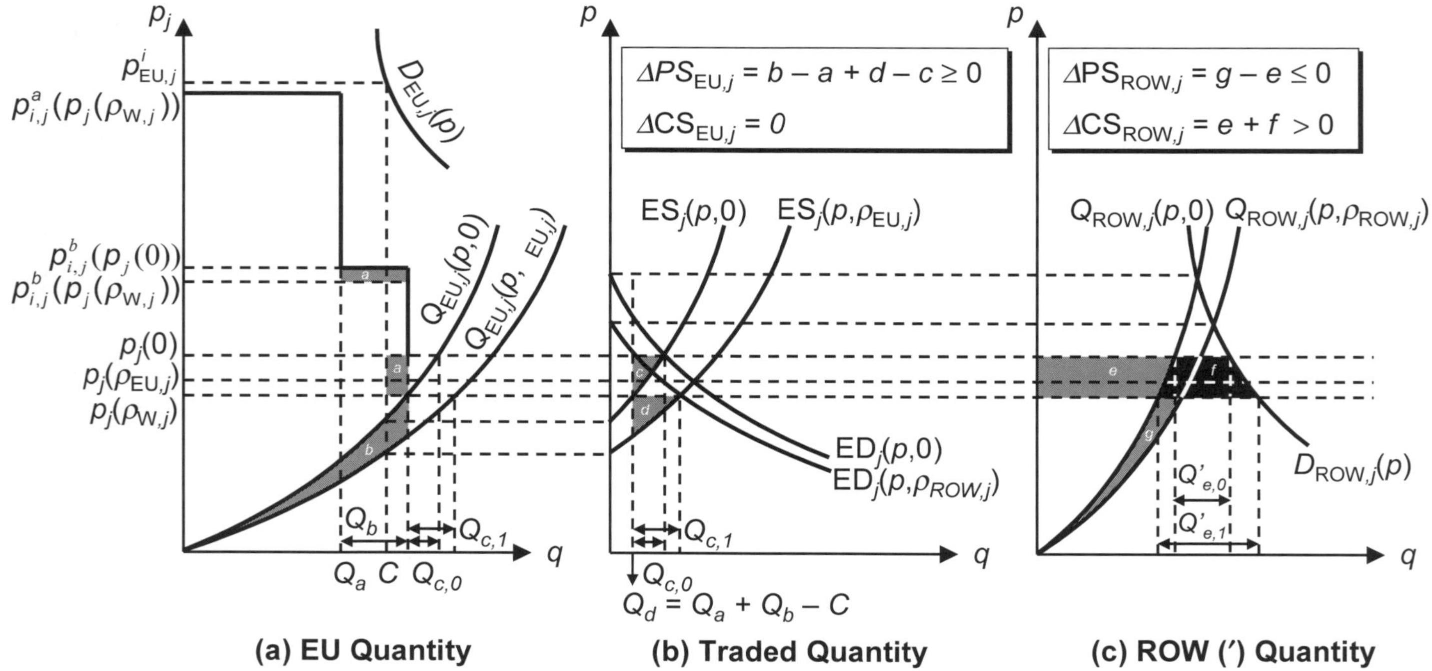

Fig. 23.1. Distribution of R&D benefits in the EU's sugar sector, with technology spillovers to the ROW.

sugar model by Poonyth *et al.* (2000), which calculates the world sugar price as a function of actual and lagged EU net sugar exports. By taking the first differential, and if we assume that imports are not affected by the innovation, due to fixed ACP (African, Caribbean, Pacific) import arrangements, we can calculate the world price as a function of the EU's technology--induced export supply expansion. For each year j the model transforms the observed world price into the price that would result from the EU's technology-induced export expansion in year j and $j-1$ (Equation 9, see bottom of page).

The short-run flexibility σ_1 is –1 and the long-run flexibility is approximately half that of the short-run ($\sigma_1 + \sigma_2 = -0.54$), reflecting sugar export demand elasticities that are twice as large in the long run as in the short run (Poonyth *et al.*, 2000). The positive value for the coefficient of the lagged export supply expansion term reflects the output contraction of the ROW as a reaction on the world price decline from $p_j(0)$ to $p_j(\rho_{EU,j})$. Inclusion of this reaction transforms our model into a dynamic equilibrium displacement model.

Secondly, the ROW technology-induced output expansion, which equals the export demand contraction, would further reduce the world price from $p_j(\rho_{EU,j})$ to the counterfactual world price $p_j(\rho_{W,j})$. We assume a constant elasticity ROW demand function for sugar:

$$D_{ROW,j}(p) = \kappa_{ROW,j}\, p^{-\varepsilon_{ROW,j}} \qquad (10)$$

The positive ROW supply shift (from $Q_{ROW,j}(p,0)$ to $Q_{ROW,j}(p,\rho_{ROW,j})$ in Fig. 23.1) translates into a negative export demand shift (from $ED_j(p,0)$ to $ED(p,\rho_{ROW,j})$):

$$ED_j(p,0) = D_{ROW,j}(p) - Q_{ROW,j}(p,0) \qquad (11)$$

$$ED_j(p,\rho_{ROW,j}) = D_{ROW,j}(p) - Q_{ROW,j}(p,\rho_{ROW,j}) \qquad (12)$$

Market clearing at equilibrium in the world market implies:

$$MC_j(p,\rho_{W,j}) = ES_j(p,\rho_{EU,j}) - ED_j(p,\rho_{ROW,j}) = 0 \qquad (13)$$

Root calculation of the market clearing constraint in equation 13 finally yields an estimate of the counterfactual world price $p_j(\rho_{W,j})$, which is essentially a function of the global adoption vector $\rho_{W,j}$:

$$p_j(\rho_{W,j}) = \mathrm{root}[MC_j(p,\rho_{W,j}),p] \qquad (14)$$

The overall world price change (from $p_j(0)$ to $p_j(\rho_{W,j})$) can now be transmitted to EU domestic prices using the principles of the EU's Common Market Organization (CMO) for sugar, which came into full effect in 1968. Each year j, the Council fixes an *intervention price* ($p^i_{EU,j}$) for sugar and *minimum prices* for beet. Anticipating an increase in consumption, the quotas ($\overline{Q}_{a,j}$ and $\overline{Q}_{b,j}$) are set at a higher level than internal consumption C_j, the internal demand ($D_{EU,j}$) at the intervention price $p^i_{EU,j}$ (Fig. 23.1). This overproduction $\overline{Q}_{d,j}$ ($= \overline{Q}_{a,j} + \overline{Q}_{b,j} - C_j$), although receiving a guaranteed B sugar price, is exported on the world market and hence subsidized. This export subsidy system is completely auto-financed by levies on A and B quota production. Consumers, who pay a high internal intervention price, subsidize the internal within-quota production. A levy τ^a_j of maximum 2% of the intervention price applies on the entire quota. Moreover, B quota production receives an additional, more variable, levy τ^b_j of maximum 37.5% of the intervention price. Both levies serve to satisfy the auto-financing constraint AFC_j, which is a function of the world price, while the latter is a function of worldwide adoption (Combette *et al.*, 1997) (Equation 15, see bottom of page).

$$p_j(\rho_{EU,j}) = p_j(0)\left[1 + \sigma_1 \frac{ES_j(p_j(0),\rho_{EU,j}) - ES_j(p_j(0),0)}{ES_j(p_j(0),0)} + \sigma_2 \frac{ES_{j-1}(p_{j-1}(0),\rho_{EU,j-1}) - ES_{j-1}(p_{j-1}(0),0)}{ES_{j-1}(p_{j-1}(0),0)}\right]$$

with $\sigma_1 = -1.0$ and $\sigma_2 = 0.46$ (9)

$$AFC_j(p_j(\rho_{W,j})) = p^i_{EU,j}\tau^a_j(p_j(\rho_{W,j}))(\overline{Q}_{a,j} + \overline{Q}_{b,j}) + p^i_{EU,j}\tau^b_j(p_j(\rho_{W,j}))\overline{Q}_{b,j} - (\overline{Q}_{a,j} + \overline{Q}_{b,j} - C_j)(p^i_{EU,j} - p_j(\rho_{W,j})) = 0 \qquad (15)$$

The levies have to fill the gap between the world price and the high internal price for quota production in excess of consumption and exported on the world market. If the auto-financing constraint does not solve by combining Equations 15 and 16, the system of Equations 15 and 17 is solved. Finally, when the latter neither yields a solution, a multiplicator α is defined solving the system of Equations 15 and 18 (Equations 16, 17 and 18, see bottom of page).

By imputing the technology-induced world price $p_j(\rho_{W,j})$ into Equation 15, the system of Equations 15 to 18 yields an estimate of the levies that have to be imposed on quota-production to satisfy the auto-financing constraint. This specification clearly visualizes how the levies are a function of the world price, while the world price in its turn is a function of worldwide adoption. For each Member State, A and B quota prices can be deducted from the State's intervention price $p^i_{i,j}$ and the levies:

$$p^a_{i,j}(p_j(\rho_{W,j})) = p^i_{i,j}[1-\tau^a_j(p_j(\rho_{W,j}))], \text{ and} \quad (19)$$

$$p^b_{i,j}(p_j(\rho_{W,j})) = p^i_{i,j}[1-\tau^a_j(p_j(\rho_{W,j}))- \tau^b_j(p_j(\rho_{W,j}))]. \quad (20)$$

By imputing $p_j(\rho_{W,j})$ into Equations 19 and 20, the model allows us to transform technology-induced changes in world price into domestic quota price changes. Thus, the producer price is endogenous since it depends on sugar production, internal demand and the gap between the intervention and the world price. All out of quota production is called 'C sugar' and can either be: (i) stocked to be carried over to the following marketing year, enabling to smooth out annual production variations; or (ii) exported on the world market at the world price, i.e. without[4] export subsidies.

Finally, the EU's CMO for sugar contains some additional features, such as the ACP import arrangements, conferring free access to the EU market for ACP countries, up to a certain maximum limit. These arrangements are essentially aid flows accruing to ACP countries and are omitted from our welfare framework, since they do not affect the flow of research benefits.[5] The same argument holds for the EU's stocking and carrying-over policy, at least in the medium and long run.[6]

The opposite effects of technology-induced cost-reduction and depression of world and domestic prices are transmitted to average land rents through Equation 1 by imputing the corresponding prices and adoption rates. Note that the land rents are a function of Equation 1 the region-specific and Equation 2 the world-wide adoption rates, the latter through the world price: $\bar{\pi}_{i,j}[p^a_{i,j}(p_j(\rho_{W,j})), \rho_{i,j}]$ for A quota, $\bar{\pi}_{i,j}[p^b_{i,j}(p_j(\rho_{W,j})), \rho_{i,j}]$ for B quota, and $\bar{\pi}_{i,j}[p_j(\rho_{W,j}), \rho_{i,j}]$ for C sugar beets. If $L_{i,j}$ $(\bar{\pi})$ denotes the optimal allocation of land to sugarbeet in country i in year j, the variation in producer surplus (relative to the benchmark without adoption) due to the innovation can be measured in the land market (Moschini *et al.*, 2000), graphically represented in Fig. 23.2. For the detailed formulas, we refer to Demont and Tollens (2002). The producer surplus change strongly depends on the country's competitiveness in sugar production. The change in producer surplus of a high-cost country i that only produces A sugar, without fulfilling its A quota (S_0 in Fig. 23.2), can be computed as area a. Portugal and Greece are the only examples. Note that the benefit resulting from the technology not only depends on the adoption within the region, but also on worldwide adoption rates through the technology-induced world price depreciation. The innovation rents of medium-cost countries, fulfilling their A quota but not

$$(16)\quad \left|\begin{matrix}\tau^a_j(p_j(\rho_{W,j})) \in [0, 0.02] \\ \tau^b_j(p_j(\rho_{W,j})) = 0\end{matrix}\right| \quad (17)\quad \begin{matrix}\tau^a_j(p_j(\rho_{W,j})) = 0.02 \\ \tau^b_j(p_j(\rho_{W,j})) \in [0, 0.375]\end{matrix} \quad (18)\quad \left|\begin{matrix}\tau^a_j(p_j(\rho_{W,j})) = (1+\alpha)\ 0.375 \\ \tau^b_j(p_j(\rho_{W,j})) = (1+\alpha)\ 0.375\end{matrix}\right|$$

[4] It can be argued that even C sugar is implicitly subsidized since fixed costs of exporting producers are already covered by the high within-quota prices (Harris and Tangermann, 1993).

[5] Ivan Roberts correctly points out that this is so as long as the aid is maintained. But if it were to be discontinued, it would raise world prices, influencing C-sugar and B-sugar returns.

[6] In the short run, producers could stock surpluses generated by the innovation, but this 'hold-up' of R&D benefits is temporary, since the stocks are limited to 20% of the A quota (European Commission, 1996).

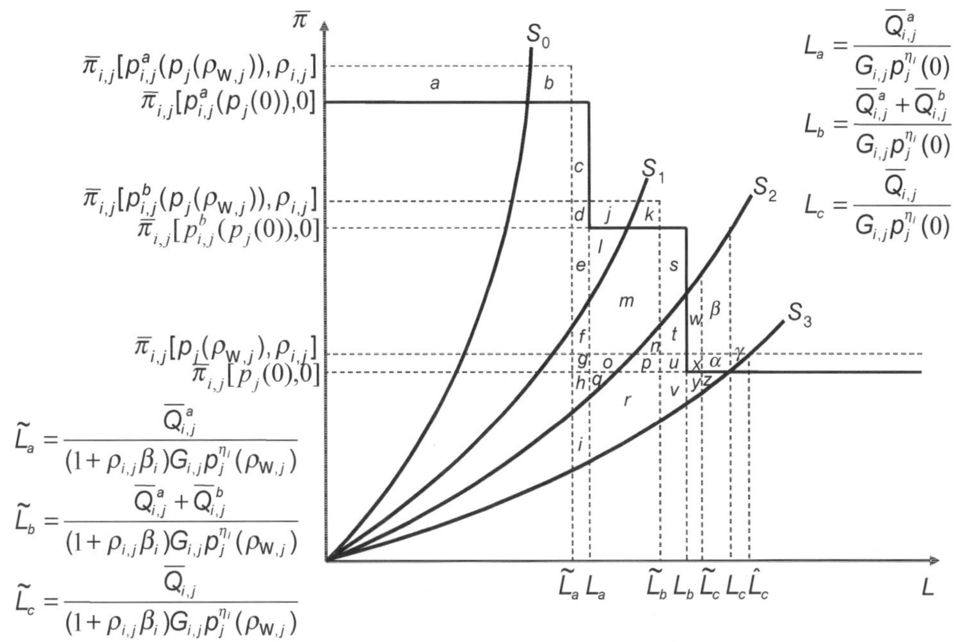

Fig. 23.2. Innovation rents measured in the land market.

their B quota (S_1 in Fig. 23.2), can be calculated as area $a + b - c + j$. For exporting low-cost EU countries responding to world prices (S_3 in Fig. 23.2), the change in producers' surplus is area $a + b - c + j + k - s - t + x + \alpha + \gamma$. According to Frandsen *et al.* (2001), only four EU countries fulfil this criterion: Austria, France, Germany and the UK. Figure 23.1 illustrates graphically how the benefits are split up in (i) a within-quota (area $b - a$), and (ii) an out-of-quota part (area $d - c$), earned on the world market.

For regions in the ROW responding to world prices, innovation rents can be calculated as:

$$\Delta \mathrm{PS}_{i,j}(p_j(\rho_{W,j}),\rho_{i,j}) = \int_{\bar{\pi}_{i,j}(p_j(0),0)}^{\bar{\pi}_{i,j}(p_j(\rho_{W,j}),\rho_{i,j})} L_{i,j}(\bar{\pi})d\bar{\pi} \quad (21)$$

The ROW cane industry is assumed to respond to world prices, but no technology-induced surplus is generated by the model. In specifying Equation 4, we assume that exporting low-cost EU or ROW regions not responding to world prices have no possibilities for output expansion. Instead, they will respond to new technologies by freeing up land allocated to sugarbeet, so that their total production remains unchanged. In Fig. 23.2, the post-innovation case is indicated with the symbol $\sim$. For those regions, we include this possibility by equalling their land supply function to their constant total production, divided by the (optimal) yield function:

$$L_{i,j}(p,\rho_{i,j}) = \frac{\overline{Q}_{i,j}}{(1+\rho_{i,j}\beta_i)G_{i,j}p^{\eta_i}} \quad (22)$$

The change in producers' surplus for exporting low-cost EU regions not responding to world prices can be graphically visualized in Fig. 23.2. The land supply function S_2 shows that these regions would normally not supply C-sugar, since the rents of the latter are not sufficient to cover production costs. However, in order to ensure that their quotas are fulfilled, even in low-yield years, farmers choose to accept a minimal precautionary overproduction. This overproduction leads to a financial loss (areas $w + x + \alpha + \beta$ in the pre-innovation case and areas $t + w$ in the post-innovation case), which can be considered as a risk premium, paid by the farmer to ensure his quota-fulfilment. Graphically, innovation rents would be calculated as area $a + b - c + j + k - s - t + x + \alpha + \beta$. Note that these innovation rents equal the innovation rents of price-responsive regions minus the area γ, plus the area β. The area γ can be interpreted as the rents that would be captured by having the possibility to expand land (from L_c to $\tilde{L}_c$ in Fig. 23.2), purely in response to the profit increase, disregarding any yield-effect. The area β is a part of the risk premium that is eliminated by the land-contracting effect of the new technology. Area β is difficult to measure and depends strongly on farmers' and processors' risk aversion. While we observe full quota fulfilment and C-sugar production for these countries, we know that their land supply function is more closely related to S_2 than to S_3. Since data is lacking for precise vertical positioning of the latter, we will assume the same land supply function for price-responsive and price-irresponsive regions. The measurement error that results from this simplification equals area $\beta - \gamma$ and is assumed to be small for price-irresponsive regions. For regions in the ROW not responding to world prices finally, innovation rents are computed with Equation 21 for the same reason. Non-EU European countries are expected to be part of this group. In addition, the US sugar sector is highly protected by a tariff quota system, eliminating any link between domestic prices and supply and world prices (Roberts and Wish-Wilson, 1991). Therefore, we assume that the highly protected ROW beet region will not export its technology-induced surplus on the world market, but instead will free up land allocated to sugar beets. The EU's aggregate welfare increase is simply the sum of all production blocks' producer surplus changes:

$$\Delta PS_{EU,j}(p_j(\rho_{W,j}),\rho_{EU,j}) = \sum_{i=2}^{15} \Delta PS_{i,j}(p_j(\rho_{W,j}),\rho_{i,j}) \tag{23}$$

In Fig. 23.1, the aggregate benefit for the EU can be assessed by a pivotal shift of the aggregate EU supply function. The exported surplus Q_d is subsidized, since it receives the guaranteed B quota price, while it is exported at the world price. Decline of the world price from $p_j(0)$ to $p_j(\rho_{W,j})$ raises subsidy costs up to Q_d $(p_j(0) - p_j(\rho_{W,j}))$, represented by the lower area a. These extra costs have to be borne by the producers via increased levies on their quota production (Equations 15 to 20). In most cases, adapting only the B quota levy is sufficient, visualized in Fig. 23.1 through a decline of the B quota price. Hence, the cost for the producers equals $Q_b[p^b_{i,j}(p_j(0)) - p^b_{i,j}(p_j(\rho_{W,j}))]$, represented by the upper area a, which is essentially the same as the lower area a. Thus, the total within-quota benefits equal area $b - a$. To these rents, out-of-quota benefits have to be added, represented by the difference between areas d and c. The EU's change in consumer surplus can be modelled as:

$$\Delta CS_{EU,j}(p_j(\rho_{W,j}),\rho_{i,j}) = \int_{p^i_{EU,j}(p_j(\rho_{W,j}),\rho_{EU,j})}^{p^i_{EU,j}(p_j(0),0)} D_{EU,j}(p)dp = 0 \tag{24}$$

In our model however, the EU's intervention price is fixed, so it is neither a function of the world price, nor of the adoption rate within the EU:

$$p^i_{EU,j}(p_j(\rho_{W,j}),\rho_{EU,j}) = \bar{p}^i_{EU,j} \tag{25}$$

This means that technology-induced welfare effects for consumers would only be possible within the CMO for sugar if the EU endogenized world prices and/or technology adoption rates in their intervention price. In contrast, world price changes are endogenous to producer prices through the auto-financing constraint. Analogous to Equation 23, the ROW aggregate innovation rents (area $g - e$ in Fig. 23.1) are simply the sum of cane ($i = 0$) and beet ($i = 1$) producers' surplus changes:

$$\Delta PS_{ROW,j}(p_j(\rho_{W,j}),\rho_{ROW,j}) = \sum_{i=0}^{1} \Delta PS_{i,j}(p_j(\rho_{W,j}),\rho_{i,j}) \tag{26}$$

The ROW consumers' surplus change (area e + area f in Fig. 23.1) equals:

$$\Delta CS_{ROW,j}(p_j(\rho_{W,j}),\rho_{ROW,j}) = \int_{p_j(\rho_{W,j})}^{p_j(0)} D_{ROW,j}(p)dp \tag{27}$$

Finally, to calculate the profit of the input suppliers, we need an estimate for all regions i of the supply of land to the sugar beet industry in equilibrium: $L_{i,j}[p_j(\rho_{W,j})\rho_{i,j}]$. Note again the double dependence of land supply on local as well as global adoption rates, the latter through the technology-induced world price depreciation. Again, we include the possibility for some regions not responding to world prices, to respond to the new technology by freeing up land allocated to sugarbeet instead. In that case the yield-increasing effect of the new technology negatively affects its demand, due to the quota system. The profit of the input suppliers can now be computed as:

$$\Pi_j(p_j(\rho_{W,j}),\rho_{W,j}) = \sum_{i=0}^{15} \rho_{i,j} L_{i,j}(p_j(\rho_{W,j}),\rho_{i,j})\mu_{i,j}\delta w_{i,j} \qquad (28)$$

Total welfare increase is simply the sum of all welfare increases. Finally, by using a risk adjusted rate of return of 10.5%, derived from the capital asset pricing model (CAPM), we can aggregate all year-specific welfare changes and actualize them to the year 2001/02.

Data and Model Calibration

In our simulation model we assume hypothetically that the EU's sugar industry, as a competitive player in the world market, and the ROW beet region embraced the new technology since the marketing year 1996/97, and progressively adopted it up to 2000/01. Our model is calibrated on the observed production data from this period. First of all we need a rough[7] estimate of the observed land rent $\bar{\pi}_{i,j}$ in all regions. Then, observed yields ($y_{i,j}$), 'incentive prices' ($\hat{p}_{i,j}$ which can be $p^a_{i,j}(p_j(0))$, $p^b_{i,j}(p_j(0))$ or $p_j(0)$), quantities ($\bar{Q}_{i,j}$) and quota ($\bar{Q}^a_{i,j}$ and $\bar{Q}^b_{i,j}$) are taken from various sources (European Commission, 1999; F.O. Licht, 2001; FAO, 2002). Using stochastic sensitivity analysis via @RISK, subjective prior distributions of non-deterministic parameters (elasticities, yield increases, cost reductions, technology price mark-up, etc.) are included to generate posterior distributions of the outcomes (counterfactual world price and research benefits) of the model. In this chapter however, we only report the means of the obtained distributions.

The estimate of the cost reduction induced by the introduction of the new technology is crucial to the economic surplus calculation. Due to the absence of farm-level adoption in the EU, we combine information from field trials with production cost data from national farm surveys and Eurostat to calibrate the technology-specific parameters $\alpha_{i,j}$ and β_i. Field trials suggest that yield boosts (β_i) vary from 0% to 8% (Wevers, 1998; Bückmann *et al.*, 2000; Dewar, 2000; Jassem, 2000). Hence, for this parameter we define a conservative triangular distribution with a minimum of zero, a most likely value of 2% and a maximum of 5%. Average herbicide costs and application costs for all EU countries are reported by Hermann (1996, 1997). The change in weeding costs ($\alpha_{i,j}$) is calculated by taking the difference between the conventional reported average herbicide and application costs and the costs that would be generated in a comparable system in which the combination of glyphosate[8] and HT sugarbeet seed is used. For the northern countries (Belgium + Luxembourg, Denmark, Germany, France, Ireland, Italy, The Netherlands, Austria, Finland, Sweden, UK), characterized by the use of at least 2.5 herbicide applications, the HT system is based on a glyphosate dose of 6 l/ha, sprayed through an average of 2.5 applications (2 times 3 l/ha of 3 times 2 l/ha). Southern countries (Greece, Spain and Portugal), use at most 1.5 applications on average. In these cases, the counterfactual HT system is assumed to be a one-pass application of 3 l/ha glyphosate. We further assume an exogenously fixed price decline of 20% in the market of conventional herbicides, due to the competition effect between the conventional and the new technology. The fixed per hectare profitability parameter $\alpha_{i,j}$ is a result of the before-mentioned factors. As a first step, we do not include any distribution for this parameter, but shift all uncertainty to the potential price markup of HT sugarbeet. Due to the very close connection between $\alpha_{i,j}$, the adoption pattern, the conventional herbicide price decline, and the potential price mark-up, a wide distribution for the latter is used to incorporate all uncertainty regarding the potential average per hectare profitability of the new technology.

Since nowhere in the world has any market developed yet for HT sugar beet seeds, no information is available on price premiums in this non-competitive market. We assume that the observed price mark-up of 40% for US Roundup Ready® soybeans represents an

[7] After an extensive sensitivity analysis it appears that this is just an inconsequential scaling parameter, which is in line with the observations of Moschini *et al.* (2000).

[8] We consider this case and assume that the profitability of both brands converges after their introduction.

upper limit. Using a static *ex ante* framework for France, Lemarié *et al.* (2001) also found an optimal price mark-up of 40% for the commercialization of HT sugar beet. We expect price premiums to be lower in the EU, due to the negative public opinion, than in the US. We also assume that the input industry would sufficiently lower its prices in the early adoption stage (even to zero) to penetrate the market. Due to the large uncertainty regarding price premiums ($\mu_{i,j}$), we incorporate a wide triangular distribution of potential price premiums with a minimum of zero, a most likely value of 20% and a maximum of 40%. Inspired by the Roundup Ready® soybeans case in the USA, we assume that the input supplier will not apply any regional price differentiation in the EU, but only between the EU and the ROW.

Supply and demand elasticities and their respective standard errors are taken from literature. The work of Poonyth *et al.* (2000) is particularly interesting since it reports estimates for each EU member country's elasticity of land supply with respect to sugar beet prices, defined as $\psi = (\partial L/\partial p)(p/L)$. Given these estimates, the parameter θ is calibrated as:

$$\theta_{i,j} = \frac{\psi_i \hat{\pi}_{i,j}}{\hat{p}_{i,j} y_{i,j}} \tag{29}$$

Devadoss and Kropf (1996) report supply elasticities for all major sugar producers in the world. For the ROW cane and ROW beet regions in our model, a production-weighted average is calculated of the reported supply elasticities. Since these elasticities already incorporate yield response to prices, we set $\eta_i = 0$ in these regions. For EU regions we set $\eta_i = 0.05$, inspired by Moschini *et al.* (2000). Given the assumed, estimated and retrieved parameters, structural parameters, such as $A_{i,j}$, $G_{i,j}$, and $\lambda_{i,j}$ will be calibrated so as to retrieve acreage, quantity, yield and price data for each year j:

$$A_{i,j} = \hat{\pi}_{i,j} + \delta w_{i,j} = \frac{\hat{p}_{i,j} y_{i,j}}{1+\eta_i} \tag{30}$$

$$y_{i,j} = \frac{y_{i,j}}{\hat{p}_{i,j}^{\hat{\pi}_i}} \tag{31}$$

$$\lambda_{i,j} = \frac{\overline{Q}_{i,j}}{\hat{\pi}_{i,j}^{\theta_{i,j}} y_{i,j}} \tag{32}$$

The sugar demand elasticity of the ROW $\varepsilon_{ROW,j}$ is calibrated on the export demand elasticities in FAPRI's world sugar model (Equation 9), reported by Poonyth *et al.* (2000). For this calibration step we force the market to clear (Equation 13) after the EU technology-induced world price decline, without adoption in the ROW:

$$MC_j(p_j(\rho_{EU,j}), \rho_{EU,j}) = ES_j(p_j(\rho_{EU,j}), \rho_{EU,j}) - ED(p_j(\rho_{EU,j}), 0) = 0 \tag{33}$$

The scale parameter κ_j is calibrated on the observed sugar demand in the ROW $\hat{D}_{ROW,j}$ and the 'incentive price' $\hat{p}_{i,j}$, which is the world price:

$$\kappa_i = \frac{\hat{D}_{ROW,j}}{\hat{p}_{i,j}^{-\varepsilon_{ROW,j}}} \tag{34}$$

The sugar demand elasticity of the ROW $\varepsilon_{ROW,j}$ can now be endogenously calibrated:

$$\varepsilon_{ROW,j} = \frac{-\ln\left(\dfrac{ES_j(p_j(\rho_{EU,j}), \rho_{EU,j}) + \overline{Q}_{ROW,j}}{\hat{D}_{ROW,j}}\right)}{\ln\left(\dfrac{p_j(\rho_{EU,j})}{p_j(0)}\right)} \tag{35}$$

We finally introduce technological change into the model by assuming an exogenous logistic adoption curve (Griliches, 1957):

$$\rho_{i,j} = K_i / (1 + e^{a_i + b_i j}) \tag{36}$$

To have a comparing point, we first estimate the parameters of the adoption curve of a comparable biotechnology innovation in the USA. We believe that the US case of HT Roundup Ready® soybeans is comparable to the EU's case of HT sugarbeet, because of: (i) the common herbicide tolerance technology; (ii) the importance of the crop in total production and in most Member States; and (iii) the importance of the export of the refined products of both crops. Assuming an adoption ceiling of 75% we find estimates of 2.76 for a and –0.85 for the adoption speed b. Since we do not have any information on the potential adoption curve of HT sugar beets in the EU, we assume that the observed adoption pattern of Roundup Ready® soybeans in the US is an upper limit. Since no significant adoption is expected to occur before 2005 (Krick, 2000), we assume a hypothetical adoption pattern with half the speed, i.e. b = –0.43, of the observed adoption pattern of Roundup Ready® soybeans in the USA. We allow technology spillovers to the ROW beet region, subject to the same hypothetical adoption pattern, but assume a *ceteris paribus* in the ROW cane region. Since we are only focusing on one technology in one sector, in our model the technology cannot 'spillover' to the ROW cane region. As a result, our estimated 'welfare effects forgone' have to be interpreted as functions, conditional on the assumed exogenous adoption pattern.

Results

Table 23.4 summarizes the results. Surprisingly, the largest share (53%) of the benefits is accruing to the ROW if we assume that beet producers in these (mostly industrial) countries: (i) are able to achieve the same efficiency-enhancing effects through the use of the new technology; and (ii) are not able to export the technology-induced surplus on the world market and further significantly erode world market prices. Total producers' welfare increase is €949 million. Despite the fact that the EU and the ROW produce roughly the same quantity of sugar, the technology rents are not equally shared, respectively €345 million and €605 million or 36% and 64%. The innovation engenders an important fixed per hectare benefit $\alpha_{i,j}$, such that the benefit sharing reflects the land sharing, respectively

Table 23.4. Average price and welfare effects (in million €) due to the introduction of herbicide tolerant sugarbeet in the European Union and the rest of the world.

Year	1996/97 Benchmark	1996/97	1997/98	1998/99	1999/2000	2000/01	2001/02 Aggregate
World price (%)	100	99.88	99.88	99.81	99.76	99.71	–
A beet price (%)	100	100.00	100.00	100.00	100.00	100.00	–
B beet price (%)	100	99.95	99.96	99.95	99.96	99.92	–
ROW cane	0	–38.6	–36.8	–39.7	–43.8	–85.8	–316.1
ROW beet	0	50.9	70.8	86.9	104.5	160.9	604.8
Belgium–Lux.	0	1.1	1.6	2.5	2.7	3.9	15.0
Denmark	0	0.8	1.2	1.5	2.0	2.8	10.3
Germany	0	6.1	8.8	11.2	15.1	21.6	79.5
Greece	0	0.8	1.1	1.4	2.1	2.7	10.2
Spain	0	2.6	3.6	4.6	5.8	8.6	32.2
France	0	4.8	7.6	8.6	11.3	17.6	63.2
Ireland	0	0.3	0.4	0.5	0.6	1.0	3.5
Italy	0	6.4	7.1	7.4	10.0	25.8	71.0
Netherlands	0	0.9	1.3	2.1	2.1	3.2	12.2
Austria	0	0.9	1.3	1.6	2.1	2.8	11.0
Portugal	0	0.0	0.3	0.4	0.6	0.7	2.4
Finland	0	0.6	0.8	1.2	1.5	2.0	7.8
Sweden	0	0.6	0.9	1.1	1.5	2.2	8.0
United Kingdom	0	1.4	2.2	2.5	3.3	5.0	18.4
EU producers	0	27.2	38.1	46.6	60.7	99.7	344.7
EU consumers	0	0.0	0.0	0.0	0.0	0.0	0.0
ROW producers	0	12.3	34.0	47.2	60.7	75.1	288.7
ROW consumers	0	39.3	37.3	41.6	47.3	84.3	323.2
Input suppliers	0	15.7	21.1	28.9	39.1	47.6	194.4
Total	0	94.6	130.5	164.3	207.8	306.7	1150.9
EU Prod. (%)	–	29	29	28	29	33	30
EU Cons. (%)	–	0	0	0	0	0	0
Net ROW (%)	–	55	55	54	52	52	53
Input Suppl. (%)	–	17	16	18	19	16	17
Total	–	100	100	100	100	100	100

ROW, rest of world.

31% and 69% (Table 23.3). The depressing effect on world prices, engendered by innovating world price responsive regions, causes ROW consumers to gain €323 million, but is completely offset by ROW cane growers' loss of €316 million. Since we assume no technology-induced export expansion in the world price irresponsive ROW beet area, spillovers do not depress world prices nor affect the EU. Instead, the world price responsive EU is able to erode its own profitability (immiserizing growth) through technology adoption, but this effect is small. The model suggests that a minor world price decline of 0.29% is expected to occur after 5 years of adoption, given the assumed adoption pattern. Due to the auto-financing system, this price decline is only partially (28%) transmitted to domestic B sugar prices, declining by only 0.08% during the same period. As a comparison, Combette *et al.* (1997) report price transmission coefficients between 0 and 11% for A sugar and between 11% and 62% for B sugar while Devadoss and Kropf (1996) report an overall price transmission coefficient of 48%. The EU sugar beet industry captures the next largest share of the benefits (€345 million or 30%). Since minimum beet prices are fixed, no important price declines are possible. As a result, the benefits essentially flow to farmers without affecting processors. The smallest share (€194 million or 17%), accrues to the seed suppliers and gene developers. The limited ability of the input industry to extract a large part of the benefits can be explained by the fact that in a quota system, producers irresponsive to world prices will decrease their land supply to the sugar industry, rather than increase it. This negatively affects demand for the new technology. Since EU intervention prices are exogenously fixed each year, no domestic price declines are engendered by the introduction of the technology. As a result, EU consumers do not gain from the innovation. The global welfare effects finally accumulate to about €1 billion after 5 years of adoption.

Conclusion

We develop a welfare framework shaped to the European sugar sector to assess the size and distribution of the benefits of transgenic sugar beet adoption in the EU and the ROW. Our model results suggest that the ROW captures the largest share of the benefits (53%). The EU beet growers absorb the next largest share (30%), while the smallest share (17%) accrues to seed suppliers and gene developers. Cane growers in the ROW lose due to the depressing effect of the technology on world sugar prices. Remarkably, consumers outside the EU gain while EU citizens continue to subsidize the sector through high sugar prices under the Common Market Organization for sugar, despite the innovation. Therefore, trade policies should at least endogenize the effects of new technologies, such as agricultural biotechnology, that have an important impact on societal welfare.

References

Achard, F.C. (1799) *Ausfuhrliche Beschreibung der Methode, Nach Welcher bie der Kultur de Runkelrube Verfahren Werden Muss.* C.S. Spener, Berlin.

Alston, J.M., Norton, G.W. and Pardey, P.G. (1995) *Science Under Scarcity: Principles and Practice of Agricultural Research Evaluation and Priority Setting.* Cornell University Press, Ithaca, New York.

Bückmann, H., Petersen, J., Schlinker, G. and Märländer, B. (2000) Weed control in genetically modified sugar beet – two years experiences of a field trial series in Germany. *Zeitschrift für PflanzenKrankheit und PflanzenSchutzung* 17, 353–362.

Combette, P., Giraud-Heraud, E. and Réquillart, V. (1997) La politique sucrière européenne après les accords du GATT: Une analyse de quelques scénarios d'évolution. *Economie et Prévision* 127, 1–13.

Demont, M. and Tollens, E. (2002) *Impact of Agricultural Biotechnology in the European Union's Sugar Industry.* Working Paper, No 61, Katholieke Universiteit Leuven, Leuven.

Devadoss, S. and Kropf, J. (1996) Impacts of trade liberalizations under the Uruguay Round on the world sugar market. *Agricultural Economics* 15, 83–96.

Dewar, A.M. (2000) Control of pests by manipulation of weeds in GM herbicide-tolerant sugar beet. *IIRB 63rd Congress Proceedings.* IIRB, Brussels, pp. 187–198.

Dewar, A.M., May, M.J. and Pidgeon, J. (2000) GM Sugar beet – the present situation. *British Sugar Beet Review* 68, 22–27.

Duff, A. (1999) Economic aspects of GM beet and cane. Paper presented at *Agricultural Outlook Forum 1999*, Arlington, Virginia, 22–23 February.

European Commission (1996) *Sugar. CAP Working Notes 1996/97* .Office for Official Publications of the European Communities, Luxembourg.

European Commission (1999) *The Agricultural Situation in the European Union: 1998 Report*. Office for Official Publications of the European Communities, Luxembourg.

F.O. Licht (2001) *World Sugar Statistics 2002*. F.O. Licht, Kent.

Falck-Zepeda, J.B., Traxler, G. and Nelson, R.G. (2000a) Rent creation and distribution from biotechnology innovations: the case of Bt cotton and herbicide-tolerant soybeans in 1997. *Agribusiness* 16, 21–32.

Falck-Zepeda, J.B., Traxler, G. and Nelson, R.G. (2000b) Surplus distribution from the introduction of a biotechnology innovation. *American Journal of Agricultural Economics* 82, 360–369.

FAO (2002) *FAOSTAT Agriculture Data*.

Frandsen, S.E., Jensen, H.G., Yu, W. and Walter-Jørgensen, A. (2001) *Modelling the EU Sugar Policy: a Preliminary Study of Policy Reform Scenarios*. Working Paper, No. 13, SFJI (Danish Institute of Agricultural and Fisheries Economics).

Griliches, Z. (1957) Hybrid corn: an exploration in the economics of technological change. *Econometrica* 25, 501–522.

Harris, S.A. and Tangermann, S. (1993) A review of the EC sugar regime. In: Marks, S.V. and Maskus, K.E. (ed.) *The Economics and Politics of World Sugar Policies*. University of Michigan Press, Ann Arbor, pp. 109–133.

Hermann, O. (1996) Enquête sur les parasites et sur le coût des traitements phytosanitaires dans la culture de la betterave sucrière en Europe. *CIBE XXXVIth Congress Proceedings*. CIBE, Paris, pp. 1–8.

Hermann, O. (1997) Enquête sur le coût de la lutte phytosanitaire dans la culture de la betterave sucrière en Europe. *IIRB 60th Congress Proceedings*. IIRB, Brussels, pp. 469–475.

International Policy Council (1996) *Sugar Policy in the Post-Uruguay Round Era*. Position Paper, No. 1. International Policy Council on Agriculture Food and Trade, Washington, DC.

Jassem, M. (2000) Genetically modified sugar beets – survey of benefits, prospects and risks. *Electronic Journal of Polish Agricultural Universities* 3, 1–10.

Krick, A. (2000) Table of situation and research by country. Paper presented at *Challenges for Sugar Farmers in the New Millennium, 7th World Sugar Farmers' Conference*, Durban, South Africa, 10–15 September .

Lemarié, S., Desquilbet, M., Diemer, A., Marette, S., Levert, F., Carrère, M. and Bullock, D.S. (2001) *Les Répartitions Possibles entre les Acteurs de la Filière Agro-alimentaire des Gains Éventuels Tirés des Plantes Transgéniques en France*. Rapport au commissariat Général du Plan, INRA, Rennes.

Märländer, B. and Bückmann, H. (1999) Genetically modified varieties in Germany – status and prospects with special respect of a sustainable sugar beet cultivation. *Zuckerindustrie* 124, 943–946.

Moschini, G. and Lapan, H. (1997) Intellectual property rights and the welfare effects of agricultural R&D. *American Journal of Agricultural Economics* 79, 1229–1242.

Moschini, G., Lapan, H. and Sobolevsky, A. (2000) Roundup ready soybeans and welfare effects in the soybean complex. *Agribusiness* 16, 33–55.

Poonyth, D., Westhoff, P., Womack, A. and Adams, G. (2000) Impacts of WTO restrictions on subsidized EU sugar exports. *Agricultural Economics* 22, 233–245.

Roberts, I. and Wish-Wilson, P. (1991) Domestic and world market effects of EC sugar policies. *ABARE Discussion Paper, No. 91.1*, ABARE, Canberra.

Schäufele, W.R. (2000) Chemische Unkrautbekämpfung in Zuckerrüben im Wandel – Ergebnisse einer Befragung in der IIRB-Arbeitsgruppe 'Unkrautregulierung'. *IIRB 63rd Congress Proceedings*. IIRB, Brussels, pp. 93–109.

Thirtle, C. (1999) Productivity and the returns to levy-funded R&D for sugar production in the eastern counties of England. *Journal of Agricultural Economics* 50, 450–467.

Wevers, J.D.A. (1998) Agronomic and environmental aspects of herbicide-resistant sugar beet in the Netherlands. *Aspects of Applied Biology* 52, 393–399.

24 The Economic Impacts of Agricultural Biotechnology on International Trade, Consumers, and Producers: the Case of Maize and Soybeans in the USA

Andrew P. Barkley
Department of Agricultural Economics, Kansas State University, Manhattan, Kansas, USA

Abstract

The development and commercial adoption of agricultural biotechnology has resulted in a large increase in agricultural productivity. In many nations, numerous output-enhancing genetically modified organisms (GMOs) have been developed, are rapidly passing through governmental regulatory approval procedures, and are nearly ready for release and adoption by commodity producers. This continuing process of development, release, and commercial adoption will result in significant changes in commodity prices, production, consumption, and trade levels. For traded commodities, these impacts will be large, both in nations that adopt biotechnology, and nations that do not. The goal of this research is to identify and quantify the economic impact of biotechnology adoption in US maize and soybean production on prices, production, consumption, and international trade levels.

Following previous literature on the impacts of research on agricultural commodities, a partial equilibrium, aggregate-level, international trade model of three nations is developed to assess and measure the economic consequences of biotechnology. The model specifies supply and demand relationships for the traded agricultural commodities of maize and soybeans. Commercial adoption of genetically modified grain seed shifts the supply curve of the adopting nation by the amount of research-induced cost reduction attributable to the technological change. These supply shifts are the foundation of an economic simulation model to measure the economic impacts of biotechnology adoption. Three scenarios are considered:

- Location-specific biotechnology: a single nation (USA) adopts a cost-reducing GMO that is not transferable to other nations.
- Location-general biotechnology: a nation (USA) adopts a cost-reducing GMO, and other nations also adopt the GMO.
- Consumer opposition to biotechnology: a single nation (USA) adopts a cost-reducing GMO, but due to a lack of consumer acceptance, the demand for the GMO commodity declines in other nations.

The model will be applied to maize and soybean markets to calculate the economic effects on producers and consumers in the USA, the European Union (EU), and the rest of the world (ROW). Simulation model results demonstrate that the adoption of biotechnology in USA maize and soybean production has led to relatively small changes in production levels, prices, and trade flows, with the largest changes for commodities that have the largest price elasticity of demand. Import reductions in the EU, due to either trade restrictions or reduced consumer purchases of GM food, are economically insignificant for maize, but would result in large price decreases if extended to soybeans or other nations. The simulation model results suggest that if USA exports to the EU were to be eliminated, soybean prices would decrease 16% and maize prices would decrease 12%.

Technological transfer of biotechnology to other nations is not likely to greatly alter world prices or international trade: price decreases of 0.8% would result if current USA adoption rates were to transferred to other nations. In the extreme case of universal adoption of biotechnology at 100% adoption rates by all nations, maize and soybean prices are projected to decrease by 1–2%.

Introduction

The development and commercial adoption of agricultural biotechnology has resulted in significant increases in agricultural productivity (Shoemaker *et al.*, 2001). In many nations, numerous output-enhancing genetically modified organisms (GMOs) have been developed, are rapidly passing through governmental regulatory approval procedures, and are nearing release and adoption by commodity producers (Roberts *et al.*, 2001). This continuing process of development, release, and commercial adoption will result in significant changes in commodity prices, quantities supplied, and quantities demanded. For traded commodities, these impacts will be large, both in nations that adopt biotechnology, and nations that do not. The goal of this research is to identify and quantify the economic consequences of biotechnology development and adoption on prices and quantities supplied, demanded, and traded of GMOs in both adopting and non-adopting nations, using maize and soybean production in the USA as a case study.

The adoption of biotechnology in agriculture has increased output per unit of input, and reduced input costs for a large number of agricultural producers (Shoemaker *et al.*, 2001). Burton *et al.* (2002) state that the transgenic (genetically modified, or GM) traits of the 'first generation' GM crops were herbicide tolerance, insect and viral resistance, and hybrid technology. Fulton and Keyowski (1999) explain that these input traits lower many costs of production through a combination of: (i) reduced costs of control of weed, pest and disease infestations; (ii) reduced crop losses from these infestations; and (iii) increased yields.

Globally, GMO adoption has increased dramatically: in 2001, 5.5 million farmers in 13 countries grew an estimated 52.6 million acres (21.3 million ha) of GM crops (James, 2002). The number of acres planted to GM crops increased 19% between 2000 and 2001, compared with an increase of 11% between 1999 and 2000. To date, the adoption of biotechnology has been geographically concentrated, with 99% of all GM crops produced by only four nations: USA, Argentina, Canada, and China. Globally, GM crops accounted for 46% of all soybean acres planted, and 7% of all maize acres (James 2002).

In the 6-year period, GM maize increased from 4.4% of all planted acres in the USA in 1996 to 32% in 2002; GM soybean acres in the USA increased from 7.4% in 1996 to 74% in 2002 (James, 1999; USDA/NASS, 2002, *Prospective Plantings*). In the past year, growth in planted acres of GM soybeans has slowed from 68% planted acres in 2001 to 74% planted acres in 2002. Genetically modified maize acres, however, continue to grow rapidly, from 26% in 2001 to 32% in 2002 (USDA/NASS, 2002, *Prospective Plantings*).

The adoption of biotechnology in agricultural production has been widespread, particularly in the USA, and is expected to spread to other nations in the next few years. Consumer acceptance of food from GM sources, however, has not kept pace with producer adoption rates. As a result, consumer concerns have led to lower consumption levels in some regions, particularly the European Union (EU), (Gaskel *et al.*, 1998; Kalaitzandonakes, 2000; Evenson *et al.*, 2002). Estimates of the economic consequences of this significant technological advancement are thus informative and interesting to policy makers, agricultural researchers, and commodity market stakeholders, particularly those involved in agricultural research system design and conduct.

Following previous literature on the impacts of research on agricultural commodities, a partial equilibrium, aggregate-level, international trade model of three regions (USA, EU, and the rest of the world (ROW)) is developed to assess and measure the economic impacts of biotechnology in both adopting and non-adopting nations. The model specifies supply and demand relationships for the traded agricultural commodities, maize and soybeans. The export demand relationships are econometrically estimated with a cross-section of nations for the period

1989–2000 for the EU and the ROW. The domestic (USA) supply and demand are estimated for the period 1981–2000.

Commercial adoption of genetically modified grain seed shifts the supply curve of the adopting nation by the amount of research-induced cost reduction attributable to the technological change. Supply shifts for maize and soybeans are the basis for a simulation of the impact of biotechnology development and adoption on global prices, production, consumption, and international trade levels. Three scenarios are quantified with the economic simulation model:

- Location-specific biotechnology: a single nation (USA) adopts a cost-reducing GMO, that is not transferable to other nations because of physiological, legal or other limitations.
- Location-general biotechnology: a nation (USA) adopts a cost-reducing GMO, and other nations also adopt the GMO.
- Consumer opposition to biotechnology: a single nation (USA) adopts a cost-reducing GMO, but owing to a lack of consumer acceptance, the demand for the GMO commodity declines in other nations.

The results suggest that the optimal research investment in biotechnology depends on: (i) if the GMO is location-specific, or transferable; (ii) the elasticity of demand for the GM crop; and (iii) consumer acceptance of biotechnology, and the potential for trade barriers and/or decreased demand as a result of consumer concerns. The details of the economic model of maize and soybean production, consumption, and trade is specified in the next section.

Simulation Model

To quantify the impact of the adoption of biotechnology in the USA on global markets for maize and soybeans, we employ a simulation model similar to those used by Shahid and Gempesaw (2002) and Kinnucan *et al.* (1995). The model focuses on the production and consumption of US maize and soybeans: production occurs in the USA, and consumption is derived from three sources: (i) domestic (USA) demand; (ii) export demand from the EU; and (iii) export demand from the ROW.

Maize and soybeans are interesting and important commodities for study for several reasons (Ballenger *et al.*, 2000). First, the two crops are economically important, contributing US\$19.8 billion (maize) and US\$12.5 billion (soybeans) to the US agricultural economy in 2001 (Dittrick). Second, these crops are widely traded: 22% of maize production and 33% of soybean production are exported (UNFAO). Both *Bt* maize, which resists the maize rootworm, and Roundup Ready® soybeans have experienced high rates of adoption by farmers in the USA. Interestingly, maize and soybeans are substitutes in production, often grown in rotation, and are related in consumption, as they are both major ingredients in livestock feed. In the partial equilibrium, aggregate model presented below, the subscript (i = c, s) identifies the commodity, where c = corn (maize) and s = soybeans. The superscripts are: d = domestic demand, x = export demand, and s = supply. The first five equations below represent: USA domestic maize demand (Equation 1), export demand (Equations 2 and 3), USA maize supply (Equation 4), and market-clearing equilibrium (Equation 5). Equations (6 to 10) are the (identical) soybean market component of the model.

$$Q^{d}_{cUSA} = f_c(P_c, P_s) \tag{1}$$

$$Q^{x}_{cEU} = g_{cEU}(P_c, P_s, T_{cEU}) \tag{2}$$

$$Q^{x}_{cROW} = g_{cROW}(P_c, P_s) \tag{3}$$

$$Q^{s}_{cUSA} = h_c[(P_c - BIO_{cUSA}), P_s] \tag{4}$$

$$Q_c = Q^{s}_{cUSA} = Q^{d}_{cUSA} + Q^{x}_{cEU} + Q^{x}_{cROW} \tag{5}$$

$$Q^{d}_{sUSA} = f_s(P_s, P_c) \tag{6}$$

$$Q^{x}_{sEU} = g_{sEU}(P_s, P_c, T_{sEU}) \tag{7}$$

$$Q^{x}_{sROW} = g_{sROW}(Ps, P_c) \tag{8}$$

$$Q^{s}_{sUSA} = h_s[(P_s - BIO_{sUSA}), P_c] \tag{9}$$

$$Q_s = Q^{s}_{sUSA} = Q^{d}_{sUSA} + Q^{x}_{sEU} + Q^{x}_{sROW} \tag{10}$$

Where Q^{d}_{iUSA} is the domestic demand for commodity maize (Equation 1) or soybeans (Equation 6) in the USA, P_c is the world price of maize, and P_s is the world price of soybeans. Export demand for US maize is the sum of two components: Q^{x}_{cEU} is the EU export demand for

US maize (Equation 2), and Q^{x}_{cROW} is the ROW export demand for US maize (Equation 3), and similarly for soybeans (Equations 7 and 8). The term T_{cEU} is an exogenous reduction in maize imports (Equation 2). The reduction could be due to either a quantitative trade restriction on maize (percentage quota) imposed by the EU, or a reduction in EU maize imports due to reduced consumer purchases of GM crops due to concern for the impact of GM food on human health and/or the environment. Similarly, the soybean import trade reduction in the EU is T_{sEU} (Equation 7).

The production of maize and soybeans in the USA (Q^{s}_{iEU} in Equations 4 and 9) is assumed to be a function of both maize and soybean prices. The production of maize and soybeans in the US is enhanced by biotechnology, captured by the exogenous variable BIO_{iUSA}, (i = c, s), defined as the per-unit cost reduction in crop production due to the adoption of biotechnology. Totally differentiating Equations 1 to 10 yields:

$$\mathrm{dln}(Q^{d}_{cUSA}) = N_{ccUSA}\mathrm{dln}(P_c) + N_{csUSA}\mathrm{dln}(P_s) \quad (11)$$

$$\mathrm{dln}(Q^{x}_{cEU}) = N_{ccEU}\mathrm{dln}(P_c) + N_{csEU}\mathrm{dln}(P_s) + N_{cTEU}\mathrm{dln}(T_{cEU}) \quad (12)$$

$$\mathrm{dln}(Q^{x}_{cROW}) = N_{ccROW}\mathrm{dln}(P_c) + N_{csROW}\mathrm{dln}(P_s) \quad (13)$$

$$\mathrm{dln}(Q^{s}_{cUSA}) = E_{ccUSA}[\mathrm{dln}(P_c) - b_c] + E_{csUSA}\mathrm{dln}(P_s) \quad (14)$$

$$\mathrm{dln}(Q^{s}_{cUSA}) = k_{cUSA}\mathrm{dln}(Q^{d}_{cUSA}) + k_{cEU}\mathrm{dln}(Q^{x}_{cEU}) + k_{cROW}\mathrm{dln}(Q^{x}_{cROW}) \quad (15)$$

$$\mathrm{dln}(Q^{d}_{sUSA}) = N_{ssUSA}\mathrm{dln}(P_s) + N_{scUSA}\mathrm{dln}(P_c) \quad (16)$$

$$\mathrm{dln}(Q^{x}_{sEU}) = N_{ssEU}\mathrm{dln}(P_s) + N_{scEU}\mathrm{dln}(P_c) + N_{sTEU}\mathrm{dln}(T_{sEU}) \quad (17)$$

$$\mathrm{dln}(Q^{x}_{sROW}) = N_{ssROW}\mathrm{dln}(P_s) + N_{scROW}\mathrm{dln}(P_c) \quad (18)$$

$$\mathrm{dln}(Q^{s}_{sUSA}) = E_{ssUSA}[\mathrm{dln}(P_s) - b_s] + E_{scUSA}\mathrm{dln}(P_c) \quad (19)$$

$$\mathrm{dln}(Q^{s}_{sUSA}) = k_{sUSA}\mathrm{dln}(Q^{d}_{sUSA}) + k_{sEU}\mathrm{dln}(Q^{x}_{sEU}) + k_{sROW}\mathrm{dln}(Q^{x}_{sROW}) \quad (20)$$

where N_{ijn} is the price elasticity of demand for good i with respect to the price of j in nation n (n = USA, EU, ROW). N_{iTEU} is the elasticity of export demand of good i to the EU with respect to the EU quantitative restriction, T_{iEU}. If the restriction is fully enforced and transmitted to the market, then $N_{iTEU} = 1$. The term E_{ijUSA} is the price elasticity of supply of good i with respect to the price of j. The share of total supply of good i that is consumed domestically is k_{iUSA}; k_{iEU} and k_{iROW} are the export shares of good i to the EU and ROW. The impact of biotechnology on maize production is $b_c = \mathrm{dBIO}_{cUSA}/P_c$ (Equation 14), or the change in the per-unit cost reduction due to biotechnology as a percentage of the initial price of maize. Similarly, biotechnology's impact on soybean supply is given by $b_s = \mathrm{dBIO}_{sUSA}/P_s$ (Equation 19).

The model represented by the system of simultaneous linear Equations 11 to 20 has 10 equations and 10 unknowns. The model can be solved using linear algebra for a change in an exogenous variable, resulting in estimates of percentage changes of the endogenous variables. The exogenous variables studied here are the estimated supply shifts ($b_i > 0$) and potential trade reductions in the EU ($T_{iEU} > 0$). The demand elasticity coefficients N_{ijn} and supply elasticity coefficients E_{ijUSA} are statistically estimated using data from US maize and soybean markets, as explained in detail below.

Elasticity Estimation Procedures

The econometric supply models for US maize and soybeans are characterized by a simple structure following previous literature, and reported in Equation 21.

$$Q^{s}_{iUSAt} = h_i(P_{ct-1}, P_{st-1}, GP_{ct}, P_{int}, T) \qquad i = c, s \quad (21)$$

The production of maize (i = c) and soybeans (i = s) are considered to be a function of lagged market prices (P_{ct-1}, P_{st-1}), government programme prices (GP_{ct}, GP_{st}), and input costs (P_{int}). Programme prices are commodity loan rates, and input costs are represented by an index of cost of production items, interest, taxes, and wage rates (PITW). A trend variable is included to capture any time-specific supply determinants not accounted for by the included variables. All price variables are deflated by the Gross Domestic Product (GDP) deflator

(IMF, 2000b, *World Economic Outlook*). Data on maize and soybean prices and quantities for the period 1980–2000 are from USDA/ERS *Agricultural Outlook*, and the input cost index is from USDA/NASS *Agricultural Statistics*.

The domestic demand of US feed grains is specified in Equation 22, where the domestic use of maize and soybeans is a function of prices, and the price of beef (P_b).

$$Q^{d}_{iUSAt} = f_i(P_{ct}, P_{st}, P_{bt}, T) \quad i = c, s \qquad (22)$$

The inclusion of livestock price reflects that the major component of maize and soybean demand is the derived demand (value marginal product) of the demand for livestock (beef). The included price is the price received index of all livestock products (USDA/NASS *Agricultural Statistics*). A trend variable is included. The source for quantity and price data is USDA/ERS *Agricultural Outlook*. All price variables were deflated with the GDP deflator (IMF, 2000a, *Yearbook*). Given the highly correlated nature of the time series data employed in the estimation of the supply and demand models, multicollinearity is anticipated (Belsley *et al*., 1980).

The foreign demand for feed grains is given by Equation 23, where i = c,s; N = EU, ROW; j = nation j; and t = year t.

$$Q^{x}_{ijNt} = g_{iN}(P_{ct}, P_{st}, P_{bjt}, EX_{jt}, NATION_j) \qquad (23)$$

The import demand of maize and soybeans is postulated to be a function of real (deflated) feed grain prices (P_{ct}, P_{st}), livestock prices (P_{bjt}), the exchange rate (EX_{jt}), and qualitative variables ($NATION_j$) for each nation included in the regression. The relationship specified in Equation 23 is used to estimate four import demand regressions: EU maize and soybeans, and ROW imports of both feed grains. Import quantity data are from USDA/ERS/FATUS for the 12-year period 1989–2000. Price data are intended to reflect world prices, and are thus US prices of maize and soybeans (USDA/ERS *Agricultural Outlook*). The livestock price is the unit value of livestock exports to nation j in time t, reflecting the nation-specific value of livestock exports. The unit value is the dollar value of livestock exports divided by the quantity exported, also reported by USDA/ERS/FATUS. The exchange rate is from the IMF *World Economic Outlook*, and GDP deflator is from the IMF *Yearbook*.

Elasticity Estimation Results

Regression results for the supply and demand models are presented in Tables 24.1 to 24.6. The soybean supply model (equation 21) was subject to degrading multicollinearity (Belsley *et al.*, 1980). Therefore, two alternative models were estimated, as shown in Table 24.1. Model 1 excludes the maize loan rate, and Model 2 omits the lagged maize price in the attempt to reduce multicollinearity, and thereby increase the level of statistical significance of the price coefficients. The regression results indicate that soybean production in the USA during the time period 1981 to 2000 was significantly associated with both market and programme prices for maize and soybeans. The simple models performed well, with 79% (Model 1) and 77% (Model 2) of the variation in soybean production explained by the included independent variables. The input cost index was not statistically significant in the four estimated supply models. The trend variable, however, was statistically significant in all supply regressions, indicating that soybean and maize production was increasing over the 19-year period. Soybean supply price elasticities (E_{ss} and E_{sc}) were calculated at the mean values, and are reported in Table 24.1. The calculated elasticities for soybeans in Model 1 demonstrate a relatively inelastic response of soybean production to changes in market prices. The own-price elasticity of soybean supply (E_{ss}) was estimated to be 0.193, and the cross-price elasticity of soybean supply with respect to the lagged maize price equalled 0.215. These estimates are utilized in the economic simulation model of biotechnology adoption, since the relevant maize prices in Model 2 were statistically insignificant. Maize production elasticities were also calculated at the mean values, reported in Table 24.1: the maize own-price elasticity of supply (E_{cc}) equalled 0.551, and the maize cross-price elasticity of supply (E_{cs}) was not significantly different from zero.

Demand regression results are presented in Table 24.2, together with summary statistics of the dependent and included independent variables. As in the supply models, the domestic demand regressions were subject to multicollinearity, leading to insignificant parameter

Table 24.1. US maize and soybean supply regression results.[a]

Summary statistics

Variable	Mean	SD	Minimum	Maximum
Soybean production (million t)	58.544	9.837	42.153	75.055
Maize production (million t)	204.620	40.183	106.030	255.293
Soybean loan rate ($/bu)	3.352	0.557	2.804	4.593
Soybean price ($/bu, $t-1$)	4.311	1.095	2.490	6.751
Maize loan rate ($/bu)	1.370	0.460	1.007	2.196
Maize price ($/bu, $t-1$)	1.732	0.505	1.027	2.704
Input Cost index (1914 = 100)	869.985	49.368	822.485	989.115
Trend (1990 = 1)	10.500	5.916	1.000	20.000

US soybean supply

	Model 1			Model 2		
Variable	Estimate	*t*-test[b,c]	Elasticity	Estimate	*t*-test[b,c]	Elasticity
Intercept	–18.676	–0.884	–	–21.982	–1.033	–
Soy loan rate	13.308	4.939***	0.762	13.113	1.910**	–
Soy price ($t-1$)	2.627	2.160**	0.193	5.027	3.263***	–0.977
Maize loan rate	–	–	–	3.672	0.568	10.374***
Maize price ($t-1$)	7.278	2.108**	0.215	–	–	–
Input Cost Index	–0.028	–0.977	–	–0.028	0.751	–
Trend	3.123	15.235***	–	3.240	0.370	–
Dependent variable mean		58.544			58.544	
Root MSE		4.505			4.721	
R^2		0.846			0.830	
Adjusted R^2		0.790			0.770	
F-value		15.318***			13.700***	
Observations		20			20	

US maize supply

	Model 1			Model 2		
Variable	Estimate	*t*-test[b,c]	Elasticity	Estimate	*t*-test[b,c]	Elasticity
Intercept	238.539	1.574	–	147.263	0.929	–
Soy loan rate	–	–	–	34.186	0.682	–
Soy price ($t-1$)	12.103	1.031	–	–	–	–
Maize loan rate	6.108	0.184	–	–25.326	–0.423	–
Maize price ($t-1$)	65.107	1.931**	0.551	81.228	2.697***	0.688
Input Cost Index	–0.337	–1.761*	–1.443	–0.291	–1.538	–
Trend	8.196	6.222***	–	8.539	4.479***	–
Dependent variable mean		204.620			204.620	
Root MSE		31.138			31.239	
R^2		0.558			0.555	
Adjusted R^2		0.400			0.396	
F-value		3.528			3.488	
Observations		20			20	

[a]Significance levels: *** refers to $P < 0.01$; ** refers to $P < 0.05$; * refers to $P < 0.10$.
[b]Reported *t*-tests are heteroscedasticity consistent (White, 1980).
[c]Degrading collinearity exists between all of the variables (Belsley *et al.*, 1980).
MSE, mean squared error.

Table 24.2. US maize and soybean domestic demand regression results.[a]

Summary statistics

Variable	Mean	SD	Minimum	Maximum
Maize domestic use (billion bu)	5.625	1.023	4.329	7.468
Soy domestic use (billion bu)	131.434	32.184	98.450	198.247
Soy/beef price ratio	0.064	0.010	0.048	0.078
Maize/beef price ratio	0.026	0.005	0.017	0.036
Soybean price ($/bu)	4.100	1.003	2.490	6.335
Maize price ($/bu)	1.646	0.476	0.991	2.673
Livestock price index (1990–92 = 100)	63.552	8.702	51.775	81.435
Trend (1990 = 1)	10.500	5.916	1.000	20.000

US soybean domestic demand

	Model 1			Model 2		
Variable	Estimate	*t*-test[b,c]	Elasticity	Estimate	*t*-test[b,c]	Elasticity
Intercept	0.210	0.164	–	8.804	6.293***	–
Soy/beef price ratio	–	–	–	–49.709	–2.265**	–0.566
Soybean price	–0.131	–0.895	–	–	–	–
Maize price	0.177	0.901	–	–	–	–
Beef price index	0.051	2.544**	0.571	–	–	–
Trend	0.233	10.295***	–	–	–	–
Dependent variable mean		5.625			5.625	
Root MSE		0.191			0.930	
R^2		0.973			0.217	
Adjusted R^2		0.965			0.173	
F-value		132.968***			4.981***	
Observations		20			20	

US maize domestic demand

	Model 1			Model 2		
Variable	Estimate	*t*-test[b,c]	Elasticity	Estimate	*t*-test[b,c]	Elasticity
Intercept	–62.590	–0.515	–	200.743	5.654***	–
Maize/beef price ratio	–	–	–	–2702.158	–2.149**	–0.535
Maize price	–3.785	–0.198	–	–	–	–
Soybean price	–2.510	–0.196	–	–	–	–
Beef price index	2.160	1.175	–	–	–	–
Trend	6.979	3.216***	–	–	–	–
Dependent variable mean		131.434			131.434	
Root MSE		19.355			29.788	
R^2		0.715			0.189	
Adjusted R^2		0.638			0.143	
F-value		9.384***			4.180***	
Observations		20			20	

[a]Significance levels: *** refers to $P < 0.01$; ** refers to $P < 0.05$; * refers to $P < 0.10$.
[b]Reported *t*-tests are heteroscedasticity consistent (White, 1980).
[c]Degrading collinearity exists between all of the variables (Belsley *et al.*, 1980).
MSE, mean squared error.

estimates for maize and soybean prices. Since these estimates are required for elasticity calculations, an alternative specification was employed (Model 2), including the ratio of feed grain prices to beef price, in contrast to individual prices included as separate variables (Model 1). The ratio of soybean and maize prices to the beef price was found to be statistically significant at the 5% level (Model 2) for both maize and soybeans for the period 1981–2000, when all other variables were omitted to reduce multicollinearity.

The explanatory power of the regression models fell significantly when the other price variables were omitted (Model 2). However, reasonable domestic demand elasticity estimates were forthcoming from Model 2. Calculated at the means, the own price elasticity of domestic demand for soybeans (N_{ssUSA}) equalled –0.566, and the own-price elasticity for maize (N_{ccUSA}) was estimated to be –0.535. The cross-price elasticities were assumed to equal zero: $N_{scUSA} = 0$, and $N_{csUSA} = 0$. The trend variable was statistically significant in Model 1 for both maize and soybeans, but was not included in Model 2 in the attempt to reduce multicollinearity.

The export demand equations represented in Equation 23 differ from the domestic supply and demand equations discussed above, owing to data availability. Data on annual exports of maize and soybeans to individual nations are available from USDA/ERS/FATUS for the time period 1989 to 2000. These data allow for the estimation of export demand elasticities, using a cross-sectional time series model, with 12 years of data for each nation that imports maize and/or soybeans from the USA. The regressions do not represent complete coverage of all US feed grain exports, because of missing and incomplete observations. The included nations had complete data available for the entire 12-year period for all variables, and represent the major importers of US maize and soybeans.

The regression model for the EU import demand for US soybeans is presented in Table 24.3. The model explained over 92% of the variation in EU soybean imports from the USA. The livestock unit value and exchange rate variables were not statistically significant, possible due to degrading multicollinearity. The soybean price to maize price ratio was statistically significant. The own-price elasticity of export demand calculated at the means equalled –0.857, providing estimates for N_{ssEU} and N_{scEU}. The Netherlands imported the largest quantity of soybeans, and the estimated coefficients for the NATION qualitative variables measure the statistically significant differences in estimated intercept values for each importing nation.

The EU maize import demand regression (Table 24.4) had similar results to the EU soybean model, but the price ratio coefficient was not significantly different from 0. This is intuitively reassuring, since US maize exports to the EU have decreased to nearly zero since 1997, reflecting EU trade policies and consumer concerns over GM food. Therefore, EU maize imports from the USA are currently insignificant, both statistically and economically. The unresponsiveness to changes in relative prices is reflected in the regression results. The simulation model was calibrated by setting the EU import demand elasticities equal to a small positive number, reflecting substitution possibilities: $N_{ccEU} = -0.1$, $N_{csEU} = 0.1$. Although the estimated import demand elasticities were not statistically different from zero, the system of simultaneous equations was subject to linear dependence when the values used to calibrate the simulation model were set equal to zero. Therefore, small values were included to eliminate the linear dependence and make model results possible.

The ROW is defined as all nations that import maize and/or soybeans from the USA, excluding all EU nations. The ROW export demand for US soybeans model estimates are in Table 24.5. The explanatory ability of the model is high, with an adjusted R^2 equal to 0.935. The livestock unit value and exchange rate variables are statistically significant. The exchange rate coefficient, however, is positive, opposite to the anticipated sign. The calculated elasticity, however, is quite small (0.10). The estimated coefficient on the price ratio is statistically significant at the 0.15 level, and of the expected sign. The elasticity calculated at the means for ROW soybean exports equals –0.517, providing plausible elasticity estimates for both own-price (N_{ssROW}) and cross-price (N_{scROW}) elasticities.

ROW maize import regression results are found in Table 24.6. In this model, the livestock unit value and exchange rate variables

Table 24.3. EU import demand for US soybeans regression results.[a]

Summary statistics

Variable	Mean	SD	Minimum	Maximum
EU imports of US soy (million t)	820,956.080	879,649.040	86,771	3,901,986
USA price of soybeans ($/bu)	4.739	1.625	2.468	8.283
USA price of maize ($/bu)	2.915	1.004	1.536	4.811
Soybean/maize price ratio	1.636	0.131	1.319	1.794
Livestock unit value ($/cct)	4261.730	2207.290	0	9712
Exchange rate (currency/$)	172.054	364.187	0.433	1363.860
Belgium	0.111	0.316	0	1
France	0.111	0.316	0	1
Germany	0.111	0.316	0	1
Greece	0.111	0.316	0	1
Italy	0.111	0.316	0	1
Netherlands	0.111	0.316	0	1
Portugal	0.111	0.316	0	1
Spain	0.111	0.316	0	1
UK	0.111	0.316	0	1

EU import demand for US soybeans

Variable	Estimate	*t*-test[b,c]	Elasticity
Soybean/maize price ratio	–429883**	–2.151	–0.857
Livestock unit value	–2.934	–0.357	0
Exchange rate	298.400	0.552	0
Belgium	–2,335,879***	–14.413	–
France	–2,726,253***	–17.593	–
Germany	–1,996,610***	–11.837	–
Greece	–2,842,985***	–16.687	–
Italy	–2,954,905***	–4.453	–
Netherlands	3,684,817***	9.443	–
Portugal	–2,685,704***	–16.022	–
Spain	–1,623,836***	–9.498	–
UK	–2,629,216***	–17.056	–
Root MSE	235606.075		
Dependent variable mean	820956.083		
R^2	0.936		
Adjusted R^2	0.928		
F-value	126.866***		
Observations	108		

[a]Significance levels: *** refers to $P < 0.01$; ** refers to $P < 0.05$; * refers to $P < 0.10$.
[b]Reported *t*-tests are heteroscedasticity consistent (White, 1980).
[c]Degrading collinearity exists between all of the variables (Belsley *et al.*, 1980).
MSE, mean squared error; cct, hundred weight.

were statistically insignificant, probably due to multicollinearity. The price ratio, however, was highly statistically significant, yielding own-price and cross-price elasticity estimates: $E_{ccROW} = -10.592$ and $E_{csROW} = 10.592$. These highly elastic estimates of US maize exports in response to price changes reflect that USA maize exports are a small fraction relative to global trade in maize. As a result, ROW importers have other supply opportunities, and are responsive to price changes from an individual nation such as the USA.

Table 24.4. EU import demand for US maize regression results.[a]

Summary statistics

Variable	Mean	SD	Minimum	Maximum
EU demand for US maize (million t)	248,887.960	456,518.740	0	2,411,748
USA price of soybeans ($/bu)	4.739	1.628	2.468	8.283
USA price of maize ($/bu)	2.915	1.005	1.536	4.811
Soybean/maize price ratio	1.636	0.131	1.319	1.794
Livestock unit value ($/cct)	4487.340	2233.090	0.000	9712.000
Exchange rate (currency/$)	203.527	406.811	0.433	1363.860
Belgium	0.143	0.352	0	1
Germany	0.143	0.352	0	1
Italy	0.143	0.352	0	1
Netherlands	0.143	0.352	0	1
Portugal	0.143	0.352	0	1
Spain	0.143	0.352	0	1
UK	0.143	0.352	0	1

EU import demand for US maize

Variable	Estimate	*t*-test[b,c]	Elasticity
Soybean/maize price ratio	–309870	–1.487	0
Livestock unit value	40.123***	3.221	0.723
Exchange rate	24.387	0.085	0
Belgium	–920,508***	–4.949	–
Germany	–958,755***	–5.058	–
Italy	–1,049,776***	–2.743	–
Netherlands	–1,008,638***	–5.490	–
Portugal	–721,629***	–3.559	–
Spain	1,380,215**	2.986	–
UK	–1,006,803***	–5.366	–
Root MSE	318,724.792		
Dependent variable mean	248,887.964		
R^2	0.565		
Adjusted R^2	0.513		
F-value	10.698***		
Observations	84		

[a]Significance levels: *** refers to $P < 0.01$; ** refers to $P < 0.05$; * refers to $P < 0.10$.
[b]Reported *t*-tests are heteroscedasticity consistent (White, 1980).
[c]Degrading collinearity exists between all of the variables (Belsley *et al.*, 1980).
MSE, mean squared error; cct, hundred weight.

The estimated elasticities from the regression results reported in Tables 24.1 to 24.6 are summarized in Table 24.7, together with the demand shares of US maize and soybean production for the three purchasers of US feed grains: USA, EU, and ROW. The point elasticity estimates reported in Table 24.7 comprise the components of the economic simulation model of biotechnology adoption in USA feed grains presented in the next section.

Simulation Model Procedures

The simulation model (Equations 11 to 20) was used to calculate the impact of a change in an exogenous variable on the ten endogenous variables included in the system of equations. The exogenous variables are of two types. First, supply shifters that capture the impact of biotechnology adoption (b_{iUSA}) for maize and soybeans (i = c,s). Second, export demand shifters

Table 24.5. ROW import demand for US soybeans regression results.[a]

Summary statistics

Variable	Mean	SD	Minimum	Maximum
ROW imports of US soy (million t)	486,964.220	971,171.470	0	4,050,565
USA price of soybeans ($/bu)	4.739	1.622	2.468	8.283
USA price of maize ($/bu)	2.915	1.001	1.536	4.811
Soybean/maize price ratio	1.636	0.131	1.319	1.794
Livestock unit value ($/cct)	2868.150	2455.630	187.548	13,085.460
Exchange rate (currency/$)	229.957	601.683	1.426	3310.720
Nation qualitative variables	0.056	0.230	0	1

ROW import demand for US soybeans

Variable	Estimate	*t*-test[b,c]	Elasticity
Soybean/maize price ratio	−154018#	−1.518	−0.517
Livestock unit value	−39.159***	−3.445	−0.231
Exchange rate	227.772*	1.882	0.108
Australia	−3,614,068***	−43.413	—
Canada	−3,494,910***	−43.140	–
Colombia	−3,802,354***	43.043	–
Costa Rica	−3,641,476***	−45.816	–
Dominican Republic	−3,745,071***	−45.199	–
Egypt	−3,642,130***	−45.899	–
Guatemala	−3,748,856***	−45.821	–
Indonesia	−3,781,719***	−12.555	–
Jamaica	−3,742,538***	−43.918	–
Japan	4,052,244***	22.336	–
Malaysia	−3,341,780***	−37.277	–
Mexico	−1,548,008***	−6.200	–
New Zealand	−3,691,136***	−48.521	–
Philippines	−3,551,628***	−46.276	–
South Korea	−2,751,756***	−20.933	–
Switzerland	−3,418,254***	−33.582	–
Thailand	−3,407,715***	−32.421	–
Trinidad and Tobago	−3,637,813***	−46.147	–
Root MSE	248113.685		
Dependent variable mean	486964.218		
R^2	0.941		
Adjusted R^2	0.935		
F-value	154.952***		
Observations	216		

[a]Significance levels: *** refers to $P < 0.01$; ** refers to $P < 0.05$; * refers to $P < 0.10$.
[b]Reported *t*-tests are heteroscedasticity consistent (White, 1980).
[c]Degrading collinearity exists between all of the variables (Belsley *et al.*, 1980).
MSE, mean squared error; cct, hundred weight.

(T_{iEU}) that reflect possible trade reductions brought about by consumer concerns for GM food. The cost savings associated with biotechnology were incorporated into the model by using the estimated cost reductions of US$9.5/acre for maize and US$6/acre for soybeans (Marra *et al.*, 1998). Reductions in production costs due to biotechnology arise from savings in pesticides, herbicides, and average yield increases, which vary across location and time, and dependent on insect infestations and weather. The dollar value per acre

Table 24.6. ROW import demand for US maize regression results.[a]

Summary statistics

Variable	Mean	SD	Minimum	Maximum
ROW imports of US maize (million t)	1,120,509	2,919,669	0	15,968,039
USA price of soybeans ($/bu)	4.739	1.621	2.468	8.283
USA price of maize ($/bu)	2.915	1.000	1.536	4.811
Soybean/maize price ratio	1.636	0.130	1.319	1.797
Livestock unit value ($/cct)	2712.360	1848.920	27.172	12,114.260
Exchange rate (currency/$)	176.454	514.019	0.097	3310.720
Nation qualitative variables	0.038	0.193	0	1

ROW import demand for US maize

Variable	Estimate	t-test[b,c]	Elasticity
Soybean/maize price ratio	7,255,364***	18.032	10.592
Livestock unit value	81.904	1.538	–
Exchange rate	–756.444	–0.927	–
Australia	–12,147,691***	–16.484	–
Bahrain	–12,118,984***	–16.791	–
Barbados	–12,182,157***	–16.606	–
Canada	–11,267,871***	–15.542	–
Colombia	–10,809,227***	–13.796	–
Costa Rica	–11,591,747***	–16.312	–
Dominican Republic	–11,321,805***	–15.751	–
Egypt	–9,976,587***	–12.683	–
Ghana	–11,620,037***	–15.268	–
Grenada	–12,162,607***	–16.713	–
Guatemala	–11,757,756***	–16.350	–
Hong Kong	–12,100,002***	–16.793	–
Indonesia	–10,053,591***	–4.566	–
Jamaica	–11,729,613***	–16.306	–
Jordan	–12,074,709***	–15.898	–
Malaysia	–12,001,365***	–14.982	–
Mexico	–8,605,098***	–8.916	–
New Zealand	–12,082,354***	–16.793	–
Panama	–11,923,131***	–16.561	–
Philippines	–12,026,591***	–16.341	–
Poland	–11,735,112***	–16.080	–
Saudi Arabia	–11,290,179***	–15.536	–
South Korea	–7,243,043***	–4.759	–
Trinidad and Tobago	–11,933,232***	–16.571	–
United Arab Emirates	–12,221,657***	–16.415	–
Root MSE	1,415,675.620		
Dependent variable mean	1,120,509.359		
R^2	0.813		
Adjusted R^2	0.795		
F-value	44.081***		
Observations	312		

[a]Significance levels: *** refers to $P < 0.01$; ** refers to $P < 0.05$; * refers to $P < 0.10$.
[b]Reported t-tests are heteroscedasticity consistent (White, 1980).
[c]Degrading collinearity exists between all of the variables (Belsley *et al.*, 1980).
MSE, mean squared error; cct, hundred weight.

Table 24.7. Elasticity estimates used in maize and soybean export simulation model.

	Maize		Soybean	
USA supply elasticities[a]	E_{ccUSA}	0.551	E_{ssUSA}	0.193
	E_{csUSA}	0	E_{scUSA}	0.215
USA domestic demand elasticities[b]	N_{ccUSA}	−0.535	N_{ssUSA}	−0.566
	N_{csUSA}	0	N_{scUSA}	0
EU import demand elasticities[b]	N_{ccEU}	−0.1	N_{ssEU}	−0.857
	N_{csEU}	0.1	N_{scEU}	0.857
ROW import demand elasticities[b]	N_{ccROW}	−10.592	N_{ssROW}	−0.517
	N_{csROW}	10.592	N_{scROW}	0.517
Demand shares of US production[c]	k_{cUSA}	0.780	k_{sUSA}	0.669
	k_{cEU}	0.009	k_{sEU}	0.121
	k_{cROW}	0.212	k_{sROW}	0.210

[a]E_{ijUSA} = price elasticity of supply of good *i* with respect to good *j*.
[b]N_{ijn} = price elasticity of demand of good *i* with respect to good *j*.
[c]Averages for the 12-year period 1989–2000 (USDA/ERS/FATUS).

estimates were converted into percentage cost reduction terms by using average price and yield data for each crop. The percentage cost savings are 3.4% for maize and 3.6% for soybeans, assuming P_c = US$2/bu, maize yield = 138 bu/acre, P_s = $4.40/bu, and soybean yield = 38.1 bu/acre (2002 values, USDA/ERS *Agricultural Outlook*). These figures are aggregate and approximate: the impact of biotechnology may vary widely across location, climatic zone, and weather conditions.

In 2002, the adoption rate of maize biotechnology in the USA was estimated to equal 32% (James, 2002). Therefore, the decrease in maize prices equals the cost savings (3.4%) multiplied by the adoption rate (32%), or $\mathrm{dln}P_c = b_c = -0.034*0.32 = -0.011$. The exogenous shift variable is equal to $b_cE_{cc} = -0.011*0.551 = -0.006$, using the own-price maize supply elasticity (Table 24.7). Simulation scenarios were also considered for the (future) possibility of complete (100%) adoption of biotechnology. In this case, $\mathrm{dln}P_c = b_c = -0.034$, and $b_cE_{cc} = -0.019$. The term b_cE_{cc} is used in the simulation model scenarios presented in the next section.

The exogenous shift in soybean supply associated with biotechnology adoption was also calculated. With an adoption rate of 74% (James, 2002), the cost savings in soybean production of 3.6% result in an overall soybean price decrease equal to $\mathrm{dln}P_s = b_s = -0.027$, and $b_sE_{ss} = -0.005$. If all soybean producers were to adopt biotechnology, then $\mathrm{dln}P_s = b_s = -0.036$, and $b_sE_{ss} = -0.007$.

Interestingly, the exogenous increase in maize production due to the adoption of GM seeds and practices (= 0.006) is slightly greater than that of soybeans (= 0.005). The reason is the greater responsiveness of maize consumers (primarily livestock producers) to changes in price: maize has more substitutes (sorghum) than soybeans. Also, maize producers are more responsive to price changes than soybean producers (E_{cc} = 0.551; E_{ss} = 0.193), reflecting greater opportunities for shifting out of maize and into alternative crops than shifting out of soybeans into other crops.

The exogenous changes in export demand (T_{iEU}) reflect demand reductions by foreign buyers, and are limited to the EU to reflect real-world concern over GM food in Europe. Only the extreme case of outright import ban is considered, to reflect the elimination of US maize exports to the EU since 1997, and the possibility of a future soybean import ban. This form of trade reduction can originate from either a trade policy of import ban, or a market-based demand decrease based on reduced consumer purchases of non-GM food. In either case, we set $T_{iEU} = -1$, reflecting a 100% decrease in import demand ($\mathrm{dln}Q_i^x = -1$). The elasticity of export demand with respect to the trade reduction (N_{iTEU}) is set equal to 1 to reflect complete enforcement of the trade reduction: the full demand decrease of *T* is transmitted completely to the market. To the extent that trade reductions are not completely enforced, this elasticity could be equal to less than 1, but this case is not considered here.

Three scenarios are considered in the simulation model exercised in the next section: no bans ($T_{i\text{EU}} = 0,\ i = \text{c,s}$), a maize ban by the EU ($T_{c\text{EU}} = -1,\ i = \text{c}$), and a ban of all GM food in the European Union ($T_{i\text{EU}} = -1,\ i = \text{c,s}$). The model is limited by the usual assumptions of market models: aggregate level data, linear supply and demand curves, assumptions about consumer and producer elasticities, and parallel shifts in the supply function.

Simulation Model Results

The simulation model results of the impact of biotechnology adoption on feed grain markets are reported in Tables 24.8 to 24.11. Nine scenarios were considered, given a variety of assumptions about the real-world extent of GM technological change, and potential consumer reaction to the appearance of GM food in the marketplace. The results are in terms of per cent changes in the ten endogenous variables relative to the baseline case of no biotechnology adoption. The first three simulation models quantify the impact of biotechnology adoption for: (i) maize only; (ii) soybeans only; and (iii) both maize; and soybeans. These three scenarios assume the current (2002) adoption rates and levels of cost savings cited by Marra *et al.* (1998). The intuitively appealing results indicate small price decreases for crops characterized by biotechnology adoption, reflecting increased production, or cost savings. Since maize and soybeans are substitutes, when maize supply shifts (Simulation 1, Table 24.8), maize exports increase to both the EU and ROW, while soybean exports decrease slightly, soybean production falls, and soybean price increase slightly. When soybean prices fall because of stimulated supply (Simulation 2, Table 24.8), soybean exports increase at the expense of maize exports. In Simulation 3, both maize and soybean producers adopt GM technology, trade flows change only slightly, and

Table 24.8. US maize and soybean biotechnology simulation results.

	Simulation 1: US biotech adoption maize	Simulation 2: US biotech adoption soybean	Simulation 3: US biotech adoption maize and soybean
Assumptions:			
Biotechnology cost savings ($/acre)[a]	9.5	6	(1+2)
Biotechnology cost savings (%)[b]	3.4	3.6	(1+2)
Biotechnology adoption rate (%)[c]	32	74	(1+2)
Simulation results (% changes):			
Maize			
P_c = world maize price	–0.186	–0.445	–0.631
Q^d_{cUSA} = US maize domestic demand	0.100	0.238	0.337
Q^x_{cEU} = EU maize export demand	0.019	–0.019	0.000
Q^x_{cROW} = ROW maize export demand	1.979	–2.030	–0.051
Q^s_{cUSA} = US maize supply	0.497	–0.245	0.252
Soybeans			
P_s = world soybean price	0.001	–0.636	–0.636
Q^d_{sUSA} = US soy domestic demand	0.000	0.360	0.360
Q^x_{sEU} = EU soybean export demand	–0.160	0.164	0.004
Q^x_{sROW} = ROW soy export demand	–0.097	0.099	0.002
Q^s_{sUSA} = US soybean supply	–0.040	0.282	0.242

[a]Marra *et al.*, 1998.
[b]Assuming: maize yield = 138 bu/acre, maize price = $2/bu, soybean yield = 38.1 bu/acre, and soybean price = $4.40/bu (USDA/ERS *Agricultural Outlook*).
[c]Estimates of 2002 acres planted to GM crops (USDA/NASS, 2002, *Prospective Plantings*).

domestic consumption increases by 0.34% (maize) and 0.36% (soybeans). Changes in maize and soybean prices are relatively minor: P_c decreases 0.63% and P_s falls 0.64% (Simulation 3, Table 24.8).

The economic consequences of GM adoption together with trade reductions due to consumer concern over GM food on feed grain markets are reported in Table 24.9. Two scenarios are considered: (i) the realistic case of an EU maize ban (Simulation 4, Table 24.9), and a ban on all GM food sources in the EU (Simulation 5, Table 24.9). When the EU eliminates maize imports (Simulation 4, Table 24.9), world maize prices fall 0.91%, resulting in greater demand for maize in the USA (0.49%) and the ROW (2.92%). Maize supply increases approximately 1% as more maize is produced with the adopted GM technology. The soybean market is largely unaffected by a maize trade reduction, with a 0.64% reduction in soybean prices and minor decreases in soybean exports. Since the EU comprises a relatively small proportion of maize demand ($k_{cEU} = 0.009$, Table 24.7), the political implications of the ban in a dynamic trade policy environment are likely to outweigh the minor economic implications.

If the EU were to ban both maize and soybean imports originating from the USA, however, the model results suggest that the economic consequences would be large and significant. In this case, exports of maize and soybeans to the EU would be eliminated, resulting in a maize price decrease of 11.67%, and a soybean price decrease equal to16.03%, reflecting the significant proportion of soybean production that is currently sold to EU importers ($k_{sEU} = 0.121$, Table 24.7). Given these relatively large price decreases, the demand for soybeans in the USA and ROW increases. Maize exports to ROW are largely replaced by soybean exports, because of the large cross-price responsiveness of ROW

Table 24.9. US biotechnology and EU trade policy simulation results.

	Simulation 4: US biotech adoption maize and soybean EU maize ban	Simulation 5: US biotech adoption maize and soybean EU maize and soybean ban
Assumptions:		
Biotechnology cost savings ($/acre)[a]	Maize = 9.5	Soybeans = 6
Biotechnology cost savings (%)[b]	Maize = 3.4	Soybeans = 3.6
Biotechnology adoption rate (%)[c]	Maize = 32	Soybeans = 74
EU ban on US imports	Maize only	Maize and soybeans
Simulation results (% changes):		
Maize		
P_c = world maize price	–0.910	–11.670
Q^d_{cUSA} = US maize domestic demand	0.487	6.243
Q^x_{cEU} = EU maize export demand	–99.972	–100.436
Q^x_{cROW} = ROW maize export demand	2.918	–46.208
Q^s_{cUSA} = US maize supply	0.099	–5.830
Soybeans		
P_s = world soybean price	–0.635	–16.032
Q^d_{sUSA} = US soy domestic demand	0.359	9.074
Q^x_{sEU} = EU soybean export demand	–0.236	–96.261
Q^x_{sROW} = ROW soy export demand	–0.142	2.255
Q^s_{sUSA} = US soybean supply	0.182	–5.103

[a]Marra *et al.*, 1998.
[b]Assuming: maize yield = 138 bu/acre, maize price = $2/bu, soybean yield = 38.1 bu/acre, and soybean price = $4.40/bu (USDA/ERS *Agricultural Outlook*).
[c]Estimates of 2002 acres planted to GM crops (USDA/NASS, 2002, *Prospective Plantings*).

maize consumers to soybean prices (N_{csROW} = 10.592). Soybean exports to the ROW increase 2.26%. The supply of feed grains in the USA, however, falls by approximately 5.83% for maize, and 5.10% for soybeans.

To summarize the market impact of trade reductions brought about by either trade policies or market forces, the simulation model results suggest that if the reductions in feed grain exports are restricted to EU maize specifically, there are small changes in prices and quantities of maize and soybeans. If, on the other hand, both maize and soybean markets are subject to trade reductions by the EU, significant price decreases would occur in both maize and soybean global markets. Importing customers in the ROW would base import decisions on the resulting relative price changes, which would probably lead to a substitution out of maize and into soybeans.

The next two scenarios (Simulations 6 and 7, Table 24.10) consider the possibility that maize and soybean producers outside of the USA adopt biotechnology, and increase feed grain productivity as a result. These scenarios consider an exogenous decrease in export demand equal to the quantity increase that would result from the adoption of agricultural biotechnology in nations outside of the USA. The scenarios assume that the importing nations as a group are feed grain producers, and that the quantitative impact of GM technology is identical in all nations outside of the USA to the impact within the USA. Clearly, the two scenarios described here are not as closely calibrated to the real world as the previous scenarios. However, the scenarios are illustrative of the impacts of the possible outcomes of global technological transfer.

In Simulation 6 (Table 24.10), ROW producers adopt biotechnology at the same level of

Table 24.10. Global biotechnology adoption simulation results.

	Simulation 6: USA, ROW biotech maize and soybean adoption EU maize and soybean ban	Simulation 7: USA, ROW, EU biotech maize and soybean adoption No bans
Assumptions:		
Biotechnology cost savings (\$/acre)[a]	Maize = 9.5	Soybeans = 6
Biotechnology cost savings (%)[b]	Maize = 3.4	Soybeans = 3.6
Biotechnology adoption rate (%)[c]	Maize = 32	Soybeans = 74
EU ban on US imports	Maize and soybeans	None
Biotechnology adoption	USA, ROW	USA, ROW, EU
Simulation results (% changes):		
Maize		
P_c = world maize price	−11.803	−0.819
Q^d_{cUSA} = US maize domestic demand	6.314	0.438
Q^x_{cEU} = EU maize export demand	−100.436	−0.603
Q^x_{cROW} = ROW maize export demand	−46.814	−0.886
Q^s_{cUSA} = US maize supply	−5.903	0.149
Soybeans		
P_s = world soybean price	−16.166	−0.846
Q^d_{sUSA} = US soy domestic demand	9.150	0.479
Q^x_{sEU} = EU soybean export demand	−96.261	−0.477
Q^x_{sROW} = ROW soy export demand	1.756	−0.486
Q^s_{sUSA} = US soybean supply	−5.158	0.161

[a]Marra *et al.*, 1998.
[b]Assuming: maize yield = 138 bu/acre, maize price = \$2/bu, soybean yield = 38.1 bu/acre, and soybean price = \$4.40/bu (USDA/ERS *Agricultural Outlook*).
[c]Estimates of 2002 acres planted to GM crops (USDA/NASS, 2002, *Prospective Plantings*).

cost savings and adoption rates as in the USA, while the EU maintains import bans on both maize and soybeans. The model results are dominated by the EU trade reductions, with economic consequences very similar to those of Simulation 5 (Table 24.9). A more realistic future scenario is represented by Simulation 7, where all nations of the world adopt biotechnology: USA, EU, and ROW. Here, maize price decrease 0.82%, and soybean prices fall 0.85%. These price changes result in an increase in maize demand in the USA (0.44%) and a decrease in maize exports to the EU (0.60%) and the ROW (0.89%). The domestic demand of soybeans increases by 0.48%, but exports to the EU and ROW fall, resulting from productivity enhancement in the EU and ROW, and substitution into maize due to relative price changes. The supply of soybeans increases by 0.16%, and the supply of maize in the USA increases by 0.15%. The major implication of global adoption of biotechnology is that the economic impacts are roughly similar to the case of adoption in the USA alone. This is a result of the large proportion of the feed grain supply originating in the USA.

The final two scenarios (Simulations 8 and 9, Table 24.11) reflect global biotechnology adoption, but at complete adoption rates (100%), rather than the current rates of 32% for maize and 74% for soybeans. In the case of a maintained and enforced EU ban on maize and soybeans from the USA (Simulation 8, Table 24.11), the results are very similar to those of Simulations 5 and 6: the trade reduction dominates the economic consequences of supply shifts due to biotechnological use. The results of Simulation 9 (Table 24.11), however, demonstrate significant market changes if biotechnology were to be fully adopted throughout the world. In this case, maize prices fall 1.55%, and

Table 24.11. Complete biotechnology adoption simulation results.

	Simulation 8: USA, ROW biotech maize and soybean complete adoption EU maize and soybean ban	Simulation 9: USA, ROW, EU biotech maize and soybean complete adoption No bans
Assumptions:		
Biotechnology cost savings ($/acre)[a]	Maize = 9.5	Soybeans = 6
Biotechnology cost savings (%)[b]	Maize = 3.4	Soybeans = 3.6
Biotechnology adoption rate (%)[c]	Maize = 100	Soybeans = 100
EU ban on US imports	Maize and soybeans	None
Biotechnology adoption	USA, ROW	USA, ROW, EU
Simulation results (% changes):		
Maize		
P_c = world maize price	–12.251	–1.548
Q^d_{cUSA} = US maize domestic demand	6.554	0.828
Q^x_{cEU} = EU maize export demand	–100.403	–1.863
Q^x_{cROW} = ROW maize export demand	–42.732	1.969
Q^s_{cUSA} = US maize supply	–4.850	1.047
Soybeans		
P_s = world soybean price	–16.285	–1.183
Q^d_{sUSA} = US soy domestic demand	9.218	0.670
Q^x_{sEU} = EU soybean export demand	–96.543	–1.013
Q^x_{sROW} = ROW soy export demand	2.086	–0.889
Q^s_{sUSA} = US soybean supply	–5.077	0.139

[a]Marra *et al.*, 1998.
[b]Assuming: maize yield = 138 bu/acre, maize price = $2/bu, soybean yield = 38.1 bu/acre, and soybean price = $4.40/bu (USDA/ERS *Agricultural Outlook*).
[c]Estimates of 2002 acres planted to GM crops (USDA/NASS, 2002, *Prospective Plantings*).

soybean prices decrease 1.18%. Maize exports to the EU decrease by 1.86%, but ROW maize exports increase 1.97%. Maize production increases 1.05%. Soybean production, exports, and price do not change a great deal: soybean price falls 1.18%. The similarity of these results to those with current adoption rates demonstrates the relatively large percentage of US soybean producers who have already adopted biotechnology. Complete adoption would increase the adoption rate from 74% to 100%. At the assumed level of cost savings, this increase in adoption rates would alter soybean markets little, even if this form of technology were to be transferred to all soybean producers globally.

Sensitivity Analysis

Comprehensive sensitivity analyses were conducted to examine the impact of changes in each parameter value on the simulation model results. The model results were notably robust to changes in parameter values. Some 16 simulations were calculated, using the scenario of maize and soybean biotechnology adoption (Simulation 3) as the baseline scenario. Each of the 16 elasticities included in the model (Table 24.7) were altered and the simulation model results compared with the baseline results appearing in Simulation 3, Table 24.8. Parameter values were individually increased from the estimated values in Table 24.7 to equal 1, or unitary elasticity. The only exceptions were the highly responsive ROW elasticities ($N_{ccROW} = -10.592$ and $N_{csROW} = 10.592$), whose absolute values were increased to equal 20 in sensitivity trials.

Overall, very small changes occurred due to these parameter value changes. A few exceptions are worth noting. Maize exports to the ROW were sensitive to the assumed value of the own- and cross-price elasticities of demand (N_{ccROW} and N_{csROW}): the percentage change in maize exports to the ROW increased 1–2% when $E_{ccROW} = -1$, and when $N_{csROW} = -1$. Soybean exports to the ROW were also sensitive to the supply elasticities E_{ssUSA} and E_{scUSA}, changing 1–2% when these elasticities were increased to equal unity. These are the largest changes in model results due to changing parameter values. The sensitivity analyses provide evidence that the simulation model results are quite robust to the estimated parameter values underlying the reported model results.

Implications and Conclusions

This research investigated the economic consequences of the adoption of biotechnology in the US feed grain industry, a major exporter of maize and soybeans to the EU and the ROW. A three region aggregate model of the global supply and demand of maize and soybeans was developed to calculate the impact of GM crops on market prices, production, domestic demand, and exports of maize and soybeans produced in the USA. Econometric estimation of feed grain supply and demand provided elasticity estimates used to calibrate the global market model. The model was used to simulate several scenarios regarding biotechnology adoption and possible consumer reactions to GM food.

The simulation results demonstrated that producer adoption of biotechnology results in an increase in the supply of maize and soybeans, price reductions and increases in domestic demand and exports. The demand for maize produced in the USA is elastic; maize importers have alternative feed sources and other nations from which to import maize. Consequently, biotechnology results in decreased prices, which in turn promote large increases in exports. Since quantity increases are larger than price decreases, maize producers are better off. Soybean producers, on the other hand, face inelastic demand. Soybeans provide an important protein supplement to livestock feed rations, with few close substitutes. Also, the USA is a major exporter of soybeans, with few other nations to buy soybeans from in the global market. Therefore, biotechnology's impact on the soybean market results in lower prices, and smaller increases in soybean exports relative to maize.

A major finding of the simulation model results is the impact of trade reductions on international feed grain markets. Currently, there is considerable concern in the EU about the potential impacts of GM food on human health and the environment. As a result, maize shipments from the USA to the EU have fallen to nearly zero since 1997. Losses in export markets, whether they are due to trade policies such as import bans, or market-driven reductions in consumer purchases that reflect consumer preferences for non-GM food, can influence feed grain markets significantly. The simulation model results provide two major conclusions. First, elimination of maize imports by the EU has had very little, if any, impact on feed grain prices, production, and exports. This is because the EU comprised a small fraction of the international market for US maize. Second, if the trade reduction were extended to other crops and/or to other nations, significant losses in export markets and price reductions would be likely. Specific results demonstrate that a reduction of EU imports of maize and soybeans would result in a decrease in the price of maize by 11.67% and a decrease in the price of soybeans by 16.03%.

The simulation model was also used to investigate technological transfer of biotechnology to other nations. If the EU and ROW were to adopt biotechnology in the production of maize and soybeans at levels equal to those of the USA, global feed grain markets would not be greatly altered: prices, production levels, and international trade flows would remain similar to those when biotechnology is adopted only in the USA. Current adoption rates in the USA are 32% for maize and 74% in soybeans. Simulations were run for the case of 100% adoption by all maize and soybean producers worldwide. This scenario resulted in price decreases: 1.55% for maize and 1.18% in soybean markets. Also, maize exports to the EU fell 1.86%, and soybean exports to the EU fell by 1.01%. These results indicate that widespread adoption of biotechnology in maize and soybean production will reduce prices by less than 2%.

For the maize and soybean producers who can effectively use biotechnology products to reduce costs, the small decrease in global feed grain prices is likely to be more than offset by cost-reducing gains in efficiency. Thus, adoption rates in the USA are likely to continue increasing, leading to slightly lower global prices, and relatively larger gains in exports. The simulation model results and extensive sensitivity analyses confirm that the continuing adoption of biotechnology in the USA and the rest of the world will enhance production levels, placing continued downward pressure on feed grain prices. However, international trade is unlikely to undergo fundamental restructuring, but some substitution between livestock feed grains is likely to occur based on different rates of technological advance across crops, and concomitant relative price changes.

References

Ballenger, N., Bohman, M. and Gehlhar, M. (2000) Biotechnology: implications for US corn and soybean trade. United States Department of Agriculture, Economic Research Service (USDA/ERS). *Agricultural Outlook*, AGO-270.

Belsley, D., Kuh, E. and Welsch, R.E. (1980) *Regression Diagnostics*. John Wiley & Sons, New York.

Burton, M., James, S., Lindner, B. and Pluske, J. (2002) A way forward for Frankenstein foods. In: Santaniello, V., Evenson, R.E. and Zilberman, D. (eds) *Market Development for Genetically Modified Foods*. CAB International, Wallingford, UK, pp. 7–23.

Dittrick, J.M. *Key Indicators of the U.S. Farm Sector: Major Crops*. American Corn Growers Association. http://www.acga.org (last accessed 16 May 2002)

Evenson, R.E., Santaniello, V. and Zilberman, D. (2002) Introduction. In: Santaniello, V., Evenson, R.E. and Zilberman, D. (eds) *Market Development for Genetically Modified Foods*. CAB International, Wallingford, UK, pp. 1–5.

Fulton, M. and Keyowski, L. (1999) The producer benefits of herbicide-resistant canola. *AgBioForum* 2, 85–93. www.agbioforum.missouri.edu

Gaskel, G., Bauer, M. and Durant, J. (1998) Public perceptions of biotechnology: Eurobarometer 46.1. In: Durant, J., Bauer, M. and Gaskel, G. (eds) *Biotechnology in the Public Sphere*. Science Museum, London.

[IMF] International Monetary Fund (2002a) *International Financial Statistics Yearbook*. IMF, Washington, DC.

[IMF] International Monetary Fund (2002b) The World Economic Outlook (WEO) Databank. IMF, Washington, DC, www.imf.org

James, C. (1999) *Global Status of Transgenic Crops in 1999*. ISAAA Brief No. 12. International Service for the Acquisition of Agri-Biotech Applications, Ithaca, New York.

James, C. (2002) *Global Review of Commercialized Transgenic Crops: 2001*. ISAAA Brief No. 24. International Service for the Acquisition of Agri-Biotech Applications. Ithaca, New York.

Kalaitzandonakes, N. (2000) Agrobiotechnology and competitiveness. *American Journal of Agricultural Economics* 82, 1224–1233.
Kinnucan, W.H., Duffy, P.A. and Ackerman, K.Z. (1995) Effects of price versus nonprice promotion: the case of cotton. *Review of Agricultural Economics* 17, 91–100.
Marra, M., Carlson, G. and Hubbell, B. (1998) *Economic Impacts of the First Crop Biotechnologies*. Electronic publication, North Carolina State University, Department of Agricultural and Resource Economics. http://www.ag-econ.ncsu.edu/faculty/marra/FirstCrop/sld001.htm
Roberts, D., Unnevehr, L., Caswell, J., Sheldon, I., Wilson, J., Otsuki, T. and Orden, D. (2001) *Agriculture in the WTO: the Role of Product Attributes*. IATRC Commissioned Paper No. 17. IATRC, St Paul, Minnesota.
Shahid, A. and Gempesaw, C.M. (2002) The role of promotion programs for U.S. poultry exports. *Agricultural and Resource Economics Review* 31(April), 59–70.
Shoemaker, R., Harwood, J., Day-Rubenstein, K., Dunahay, T., Heisey, P., Hoffman, L., Klotz-Ingram, C., Lin, W., Mitchell, L., McBride, W. and Fernandez-Cornejo, J. (2001) *Economic Issues in Agricultural Biotechnology*. USDA/ERS Agriculture Information Bulletin No. 762, USDA, Washington, DC.
[UNFAO] United Nations Food and Agricultural Organisation *FAOSTAT Online Database*. http://www.fao.org (Last accessed 16 May 2002)
[USDA/ERS] United States Department of Agriculture. Economic Research Service *Agricultural Outlook*. Various issues. http://www.ers.usda.gov (Last accessed 16 May 2002)
[USDA/ERS/FATUS] United States Department of Agriculture. Economic Research Service Foreign Agricultural Trade of the United States. http://www.ers.usda.gov/db/FATUS (Last accessed 16 May 2002)
[USDA/NASS] United States Department of Agriculture. National Agricultural Statistical Service *Agricultural Statistics*. Various issues. http://www.usda.gov/nass (Last accessed 16 May 2002)
[USDA/NASS] United States Department of Agriculture. National Agricultural Statistics Service (2002) *Prospective Plantings*. 28 March. http://www.usda.gov/nass (Last accessed 16 May 2002)
White, H. (1980) A heteroskedasticity-consistent covariance matrix estimator and a direct test for heteroskedasticity. *Econometrica* 48, 817–838.

Index

Page numbers in *italics* indicate tables.